Tutorial Guide to X-ray and Gamma-ray Astronomy

Cosimo Bambi
Editor

Tutorial Guide to X-ray and Gamma-ray Astronomy

Data Reduction and Analysis

Springer

Editor
Cosimo Bambi
Department of Physics
Fudan University
Shanghai, China

ISBN 978-981-15-6336-2 ISBN 978-981-15-6337-9 (eBook)
https://doi.org/10.1007/978-981-15-6337-9

This Springer imprint is published by the registered company Springer Nature Singapore Pte Ltd.
The registered company address is: 152 Beach Road, #21-01/04 Gateway East, Singapore 189721, Singapore

Preface

X-ray and γ-ray astronomy, namely, the study of astrophysical objects in the X-ray and γ-ray bands, began in the early 1960s and opened a new window for the study of violent phenomena in the Universe. In the past 20 years, missions like *XMM-Newton*, *Chandra*, *NuSTAR*, *Swift*, and *Fermi*, just to cite some of them, have provided a large amount of data to study a number of astrophysical systems. For instance, X-ray and γ-ray radiation is emitted by material orbiting in the strong gravity region of black holes and can be used to study the physical properties of these objects as well as their astrophysical environment. The next generation of satellites, like *eXTP* and *ATHENA*, promises to provide unprecedented high-quality data to investigate a number of open questions about the physics and the astrophysics of the Universe.

Despite the importance of X-ray and γ-ray astronomy in modern physics and astrophysics, as well as the non-small communities working in this field, a manual for beginners, as well as a comprehensive reference for researchers, covering the main techniques of X-ray and γ-ray data reduction and analysis is missing in the literature. In most cases, one has to refer to online material spread over the web, and to rely on the help of advisors or colleagues.

The ambition of the present book is thus to try to provide a compact pedagogical manual on X-ray and γ-ray astronomy, where one can find all the necessary materials to quickly start to work in the field, and, in particular, to study black holes and the physical phenomena occurring in their strong gravity region. The book starts with a brief review on black holes and the emission mechanisms responsible for the generation of X-ray and γ-ray radiation. Then we discuss the observational facilities in X-ray and γ-ray astronomy, and how they work. The last part of the book is devoted to the discussion of X-ray and γ-ray data reduction and analysis. The book should provide the basic tools to be able to write a scientific paper with the material obtained after the analysis of a source.

Shanghai, China — Cosimo Bambi
January 2020

Contents

Contributors

William Alston Institute of Astronomy, Cambridge, UK

Cosimo Bambi Department of Physics, Fudan University, Shanghai, China

Peter Boorman Astronomical Institute, Academy of Sciences, Prague, Czech Republic;
Faculty of Physical Sciences and Engineering, Department of Physics & Astronomy, University of Southampton, Southampton, UK

Andrea Bulgarelli INAF OAS Bologna, Bologna, Italy

Valentina Fioretti INAF OAS Bologna, Bologna, Italy

Matteo Guainazzi European Space Agency, ESTEC, Noordwijk, The Netherlands

Jiachen Jiang Department of Astronomy, Tsinghua Univerisity, Beijing, China

Vinay L. Kashyap Center for Astrophysics, Harvard & Smithsonian, Cambridge, MA, USA

Sourabh Nampalliwar Theoretical Astrophysics, Eberhard-Karls Universität Tübingen, Tübingen, Germany

Michael Parker European Space Agency (ESA), European Space Astronomy Center (ESAC), Madrid, Spain

Dheeraj R. Pasham MIT Kavli Institute for Astrophysics and Space Research, MIT, Cambridge, MA, USA

Chapter 1
Fundamental Concepts

Cosimo Bambi and Sourabh Nampalliwar

1.1 Introduction

Beginning with the special theory of relativity in 1905, Albert Einstein soon realized that Newton's theory of gravity had to be superseded, to harmonize the equivalence principle and the special theory of relativity. After numerous insights, false alarms, and dead ends, the theory of general relativity was born in 1915 [14]. It took some years for it to take over Newton's theory as the leading framework for the description of gravitational effects in our Universe, and over the past century, it has become one of the bedrocks of modern physics.

Just a year after its proposition, Karl Schwarzschild was able to find an exact solution in general relativity, much to the surprise of Einstein himself, who only had approximate solutions by that time. The Schwarzschild solution [23] turned out to be much more astrophysically relevant than anyone could have imagined, and describes the simplest class of *black holes*[1] in Einstein's theory.

Roughly speaking, a black hole is a region in which gravity is so strong that nothing, not even light, can escape. A boundary, known as the *event horizon*, separates the interior of the black hole from the exterior region and acts as a one-way membrane:

[1]The origin of the term black hole is quite intriguing. While it is not clear who used the term first, it appeared for the first time in a publication in the January 18, 1964 issue of Science News Letter. It was on a report on a meeting of the American Association for the Advancement of Science by journalist Ann Ewing. The term became quickly very popular after it was used by John Wheeler at a lecture in New York in 1967.

C. Bambi (✉)
Department of Physics, Fudan University, 2005 Songhu Road,
Shanghai 200438, China
e-mail: bambi@fudan.edu.cn

S. Nampalliwar
Theoretical Astrophysics, Eberhard-Karls Universität Tübingen,
Auf der Morgenstelle 10, 72076 Tübingen, Germany
e-mail: sourabh.nampalliwar@uni-tuebingen.de

C. Bambi (ed.), *Tutorial Guide to X-ray and Gamma-ray Astronomy*,
https://doi.org/10.1007/978-981-15-6337-9_1

particle and radiation can enter the black hole but cannot exit from it. Remarkably, a primitive concept of black hole was already discussed at the end of the 18th century in the context of Newtonian mechanics by John Michell and Pierre-Simon Laplace. The starting point was the corpuscular theory of light developed in the 17th century. Here light is made of small particles traveling with a finite velocity, say c. Michell and Laplace noted that the escape velocity from the surface of a body of mass M and radius R exceeds c if $R < R_{\rm crit}$, where

$$R_{\rm crit} = \frac{2G_{\rm N}M}{c^2} \tag{1.1}$$

and $G_{\rm N}$ is Newton's gravitational constant. If such a compact object were to exist, it should not be able to emit radiation from its surface and should thus look black. This was the conclusion of Michell and Laplace and these objects were called dark stars.

The Schwarzschild type black holes are described by just one parameter, the *mass*, and it is the characteristic quantity setting the size of the system. The *gravitational radius* of an object of mass M is defined as

$$r_{\rm g} = \frac{G_{\rm N}M}{c^2} = 14.77\left(\frac{M}{10\,M_\odot}\right)\ \text{km} . \tag{1.2}$$

The associated characteristic time scale is

$$\tau = \frac{r_{\rm g}}{c} = 49.23\left(\frac{M}{10\,M_\odot}\right)\mu\text{s} . \tag{1.3}$$

For a 10 $M_\odot$ black hole, $r_{\rm g} \sim 15$ km and $\tau \sim 50\,\mu$s. We can thus expect that physical phenomena occurring around a similar object can have a variability timescale of the order of 0.1–1 ms. For a black hole with $M \sim 10^6\ M_\odot$, we find $r_{\rm g} \sim 10^6$ km and $\tau \sim 5$ s, so physical processes occurring near its gravitational radius can have a variability timescale of the order of 10–100 s. For the most supermassive black holes with $M \sim 10^9\ M_\odot$, we have $r_{\rm g} \sim 10^9$ km and $\tau \sim 1$ h.

The astrophysical implications of such black hole solutions were not taken very seriously for a long time. For example, influential scientists like Arthur Eddington argued that "some unknown mechanism" had to prevent the complete collapse of a massive body and the formation of a black hole in the Universe. The situation changed only in the 1960s with the advent of X-ray observations. Yakov Zel'dovich and, independently, Edwin Salpeter were the first, in 1964, to propose that quasars were powered by central supermassive black holes [22, 26]. In the early 1970s, Thomas Bolton and, independently, Louise Webster and Paul Murdin identified the X-ray source Cygnus X-1 as the first stellar-mass black hole candidate [10, 25]. The uncertainty of those times can be imagined by the scientific wager between Kip Thorne and Stephen Hawking, the latter claiming that Cygnus X-1 was in fact not a black hole. Hawking conceded the bet in 1990. In the past few decades, a large number of astronomical observations have pointed out the existence of stellar-mass black holes in some X-ray binaries [20] and of supermassive black holes at the center of many galaxies [17]. Thanks to X-ray and γ-ray missions like *XMM-Newton*, *Chandra*, *NuSTAR*, *Swift*, and *Fermi*, in the past 20 years there has been substantial progress in the study of these objects. In September 2015, the LIGO experiment detected, for the first time, the gravitational waves emitted from the coalescence of two black holes [1].

1.2 Black Holes in General Relativity

In 4-dimensional general relativity, black holes are relatively simple objects, in the sense that they are completely characterized by a small number of parameters: the mass M, the spin angular momentum J, and the electric charge Q. This is the result of the *no-hair theorem*, which holds under specific assumptions [12, 13, 16, 21]. The name "no-hair" refers to the fact black holes have only a small number of features (hairs). Violations of the no-hair theorem are possible in the presence of exotic fields, extra dimensions, or extensions of general relativity.

A *Schwarzschild black hole* is a spherically symmetric, non-rotating, and electrically uncharged black hole and is completely characterized by its mass. In the presence of a non-vanishing electric charge, we have a *Reissner-Nordström black hole*, which is completely specified by two parameters and describes a spherically symmetric and non-rotating black hole of mass M and electric charge Q. A *Kerr black hole* is an uncharged black hole of mass M and spin angular momentum J. The general case is represented by a *Kerr-Newman black hole*, which has a mass M, a spin angular momentum J, and an electric charge Q.

Astrophysically, black holes are expected to belong to the Kerr family. After the collapse of a massive body and the creation of an event horizon, the gravitational field of the remnant quickly reduces to that of a Kerr black hole by emitting gravitational waves [18, 19]. For astrophysical macroscopic objects, the electric charge is extremely small and can be ignored [2, 4]. The presence of an accretion disk around the black hole, as well as of stars orbiting the black hole, do not appreciably change the strong gravity region around the compact object [5–7]. Astrophysical black holes should thus be completely specified by their mass and spin angular momentum. It is often convenient to use the dimensionless spin parameter a_* instead of J. For a black hole of mass M and spin J, a_* is defined as

$$a_* = \frac{cJ}{G_N M^2} . \tag{1.4}$$

In general relativity, the choice of the coordinate system is arbitrary, and therefore the numerical values of the coordinates have no physical meaning. Nevertheless, they can often provide the correct length or time scale of the system. In Boyer–Lindquist coordinates, the typical coordinate system for Kerr black holes, the radius of the event horizon is

$$r_H = r_g \left(1 + \sqrt{1 - a_*^2}\right) , \tag{1.5}$$

and depends on M (via r_g and a_*) and J (via a_*). The radius of the event horizon thus ranges from $2\,r_g$ for a non-rotating black hole to r_g for a maximally rotating ($a_* = \pm 1$) black hole. Note that Eq. (1.5) requires that $|a_*| \leq 1$. Indeed for $|a_*| > 1$ there is no black hole and the Kerr solution describes the gravitational field of a *naked singularity*. In the context of astrophysical observations, the possibility of the existence of naked singularity is usually ignored, and this is also motivated by the considerations that (i) there is no known mechanism capable of creating a naked singularity, and (ii) even if created, the spacetime is likely unstable (for more details, see for instance Ref. [2]).

The properties of equatorial circular orbits around a black hole are important for astrophysical observations because they describe the orbits of the particles in a putative accretion disk around the compact object. In Newtonian mechanics, equatorial circular orbits (i.e. orbits in the plane perpendicular to the spin of the object) around a point-like object are always stable. However, this is not true for equatorial circular orbits around a Kerr black hole. Here we have the existence of an *innermost stable circular orbit*, often abbreviated to ISCO. In Boyer–Lindquist coordinates, the ISCO radius is [8]

$$\begin{aligned} r_{\rm ISCO} &= r_g \left[3 + Z_2 \mp \sqrt{(3 - Z_1)(3 + Z_1 + 2Z_2)}\right] , \\ Z_1 &= 1 + \left(1 - a_*^2\right)^{1/3} \left[(1 + a_*)^{1/3} + (1 - a_*)^{1/3}\right] , \\ Z_2 &= \sqrt{3a_*^2 + Z_1^2} \, . \end{aligned} \tag{1.6}$$

The ISCO radius turns out to be $6\,r_g$ for a Schwarzschild black hole and move to r_g ($9\,r_g$) for a maximally rotating black hole and a corotating (counterrotating) orbit, namely an orbit with angular momentum parallel (antiparallel) to the black hole spin. Figure 1.1 shows the radial values of the event horizon r_H and of the ISCO radius $r_{\rm ISCO}$ in Boyer–Lindquist coordinates as a function of the black hole spin parameter a_*.

1.3 Black Holes in Astrophysics

While we cannot observe any kind of radiation (neither electromagnetic, nor gravitational) from the region inside the event horizon, astrophysical black holes can be studied by detecting the electromagnetic and gravitational radiation produced in the vicinity of the event horizon. Gravitational radiation is generated by the interaction of matter/energy and the spacetime, and its frequency depends on the size of the system. In particular, the wavelength roughly scales as the linear size of the system

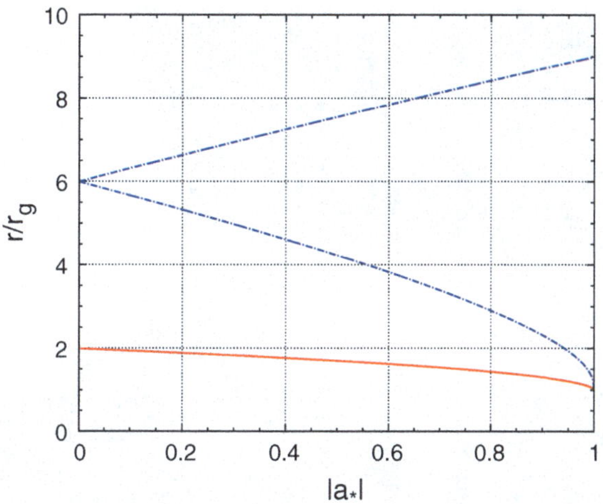

Fig. 1.1 Radius of the event horizon (red solid line) and of the ISCO (blue dash-dotted line) of a Kerr black hole in Boyer–Lindquist coordinates as a function of the spin parameter a_*. For the ISCO radius, the upper curve refers to counterrotating orbits and the lower curve to corotating orbits

emitting gravitational radiation. Gravitational radiation from black holes is expected to range from a few nHz, in the case of the merger of galaxies with supermassive black holes at their respective centers, to a few kHz, in the case of the merger and ringdown of stellar-mass black holes. Radiation of different wavelengths require different observational facilities to be detected.

Electromagnetic radiation can be emitted by the gas in the accretion disk, jet, and outflows, as well as by possible bodies (like stars) orbiting the black hole (see Fig. 1.2). The electromagnetic spectra of astrophysical black holes range from the radio to the γ-ray band (see Table 1.1 for the list of the bands of the electromagnetic spectrum). The photon energy is determined by the emission mechanism and the black hole environment. Photons with different wavelengths carry different information about the black hole and its environment, and require different observational facilities to be detected. Table 1.2 lists the possible components of the electromagnetic spectrum of a black hole system (more details on each component will be provided in the next chapter).

Among the various astrophysical processes, accretion onto a black hole can be an extremely efficient mechanism to convert mass into energy. If $\dot{M}$ is the mass accretion rate, the total power of the accretion process can be written as

$$P = \eta \dot{M} c^2 , \tag{1.7}$$

where η is the total efficiency. In general, the energy released in the accretion process will be converted into radiation and kinetic energy of jets/outflows, so we can write

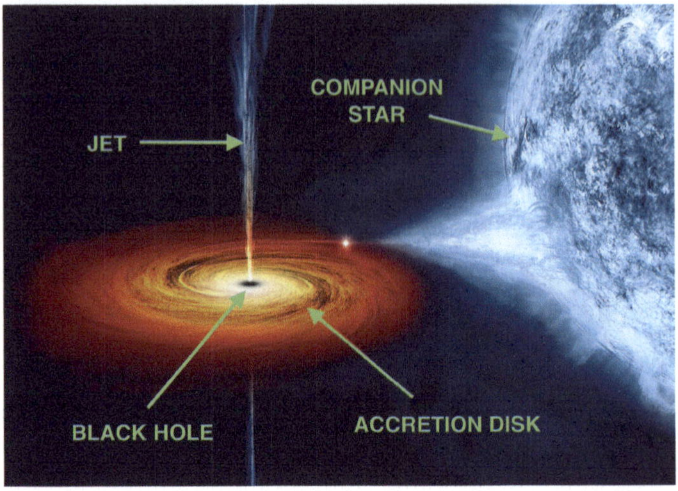

Fig. 1.2 An artist's illustration of Cygnus X-1. The stellar-mass black hole pulls material from a massive, blue companion star toward it. This material forms an accretion disk around the black hole. We also see a jet originating from the region close to the black hole. Credit: NASA

Table 1.1 Bands of the electromagnetic spectrum. Note that different authors may use slightly different definitions

Band	Wavelength	Frequency	Energy
Radio	> 0.1 m	< 3 GHz	$< 12.4\,\mu$eV
Microwave	1 mm–0.1 m	3–300 GHz	$12.4\,\mu$eV–1.24 meV
Infrared (IR)	700 nm–1 mm	300 GHz–430 THz	1.24 meV–1.7 eV
Visible	400–700 nm	430–790 THz	1.7–3.3 eV
Ultraviolet (UV)	10–400 nm	$7.9 \cdot 10^{14}$–$3 \cdot 10^{16}$ Hz	3.3–124 eV
X-Ray	0.01–10 nm	$3 \cdot 10^{16}$–$3 \cdot 10^{19}$ Hz	124 eV–124 keV
γ-Ray	< 0.01 nm	$> 3 \cdot 10^{19}$ Hz	> 124 keV

$$\eta = \eta_{\rm r} + \eta_{\rm k} \,, \tag{1.8}$$

where $\eta_{\rm r}$ is the radiative efficiency and can be measured from the bolometric luminosity $L_{\rm bol}$ from the equation $L_{\rm bol} = \eta_{\rm r} \dot{M} c^2$ if the mass accretion rate is known, and $\eta_{\rm k}$ is the fraction of gravitational energy converted into kinetic energy of jets/outflows. The actual efficiency depends on the morphology of the accretion flow. In the case of a Novikov–Thorne disk (see Sect. 2.3.1), the accretion disk is on the black hole equatorial plane, perpendicular to the spin of the compact object. The particles of the gas follow equatorial circular orbits, they lose energy and angular momentum, and they move to smaller and smaller radii. When the particles reach the ISCO radius, they quickly plunge onto the black hole, without significant emission of additional radiation. The efficiency of the process is thus given by

Table 1.2 Summary of the possible sources of electromagnetic radiation in black hole systems and typical energy bands for stellar-mass and supermassive black holes. For soft X-ray we mean the X-ray band below a few keV. Cold material orbiting the compact object and not belonging to the accretion disk is common in supermassive black holes: the emission lines can be narrow (broad) if the material is far (near) the compact object and moving with low (high) speed

Source	Emission	Stellar-mass black holes	Supermassive black holes
Accretion disk	Thermal	UV to soft X-ray	Visible to UV
Accretion disk	Reflection spectrum	X-ray	X-ray
Corona	Inverse compton	X-ray and γ-ray	X-ray and γ-ray
Jet	Synchrotron	Radio to soft X-ray	Radio to soft X-ray
Jet	Inverse Compton	X-ray and γ-ray	X-ray and γ-ray
Cold material	Emission lines	–	IR to X-ray
Companion star	Thermal	Visible and UV	–

$$\eta_{\rm NT} = 1 - E_{\rm ISCO} \, , \tag{1.9}$$

where $E_{\rm ISCO}$ is the specific energy of the gas at the ISCO radius, namely the energy per unit mass of the gas. For a Kerr black hole, the specific energy of a particle orbiting an equatorial circular orbit at the Boyer–Lindquist radial coordinate r is [2]

$$E = \frac{r^{3/2} - 2r_{\rm g}r^{1/2} \pm a_* r_{\rm g}^{3/2}}{r^{3/4}\sqrt{r^{3/2} - 3r_{\rm g}r^{1/2} \pm 2a_* r_{\rm g}^{3/2}}} \, . \tag{1.10}$$

If we plug the radial coordinate of the ISCO radius in Eq. (1.6), we find that the efficiency of the process is around 5.7% for a Schwarzschild black hole and monotonically increases (decreases) as the spin parameter increases (decreases) up to about 42.3% (3.8%) for $a_* = 1$ ($a_* = -1$):

$$\begin{aligned} \eta_{\rm NT}(a_* = 0) &= 1 - \frac{2\sqrt{2}}{3} \approx 0.057 \, , \\ \eta_{\rm NT}(a_* = 1) &= 1 - \frac{1}{\sqrt{3}} \approx 0.423 \quad (\text{corotating disk}) \, , \\ \eta_{\rm NT}(a_* = -1) &= 1 - \frac{5}{\sqrt{27}} \approx 0.038 \quad (\text{counterrotating disk}) \, . \end{aligned} \tag{1.11}$$

Figure 1.3 shows $\eta_{\rm NT}$ as a function of the spin parameter a_* for corotating (upper curve) and counterrotating (lower curve) disks. The efficiency of a Novikov–Thorne disk can be compared to other astrophysical processes. For instance, if we consider nuclear reactions inside the Sun, the main process is the fusion of protons to form helium-4 nuclei. The total mass of the final state is lower than the total mass of the initial state, and this difference is released into energy (electromagnetic radiation and

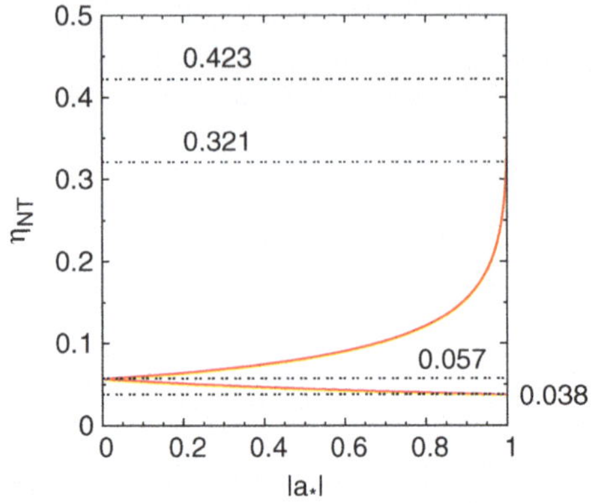

Fig. 1.3 Efficiency of a Novikov–Thorne disk $\eta_{\rm NT}$ as a function of the spin parameter a_* for Kerr black holes. The upper (lower) curve is for corotating (counterrotating) orbits. The dotted horizontal lines mark the radiative efficiencies for $a_* = 1$ ($\eta_{\rm NT} \approx 0.423$), $a_* = 0.998$ ($\eta_{\rm NT} \approx 0.321$), $a_* = 0$ ($\eta_{\rm NT} \approx 0.057$), and $a_* = -1$ ($\eta_{\rm NT} \approx 0.038$)

kinetic energy of the particles in the final state). The efficiency of the process is only around 0.7%, namely about 0.7% of the initial mass is converted into energy.

If the mass accretion rate is low and the accreting gas has a low angular momentum, the efficiency of the accretion process can be much smaller than 1, $\eta \ll 1$, because the particles of the gas simply fall onto the gravitational well of the black hole without releasing much electromagnetic radiation. Very low efficiencies are also possible in the case of very high mass accretion rate, and in this case it is because the particle density of the accretion flow is too high and the medium becomes optically thick to the radiation emitted by the gas, so everything is advected onto the black hole and lost after crossing the event horizon. An important concept in this regard is the *Eddington luminosity*. The concept is actually more general, and the Eddington luminosity refers to the maximum luminosity for an object, not necessarily a black hole. The Eddington luminosity $L_{\rm Edd}$ is reached when the pressure of the radiation luminosity on the emitting material balances the gravitational force towards the object. If a normal star has a luminosity $L > L_{\rm Edd}$, the pressure of the radiation luminosity drives an outflow. If the luminosity of the accretion flow of a black hole exceeds $L_{\rm Edd}$, the pressure of the radiation luminosity stops the accretion process, reducing the luminosity. Assuming that the emitting medium is a ionized gas of protons and electrons, the Eddington luminosity of an object of mass M is

$$L_{\rm Edd} = \frac{4\pi G_{\rm N} M m_p c}{\sigma_{\rm Th}} = 1.26 \cdot 10^{38} \left(\frac{M}{M_\odot}\right) \text{ erg/s}\,, \tag{1.12}$$

where m_p is the proton mass and $\sigma_{\rm Th}$ is the electron Thomson cross section. For an accreting black hole, we can define the Eddington mass accretion rate $\dot{M}_{\rm Edd}$ from

$$L_{\text{Edd}} = \eta_{\text{r}} \dot{M}_{\text{Edd}} c^2 , \tag{1.13}$$

where η_{r} is still the radiative efficiency.

1.4 X-Ray and γ-Ray Observatories

Our focus in this book is X-rays and γ-rays. There are a number of astrophysical sources emitting X-ray (0.1–100 keV) and γ-ray (>100 keV) radiation, such as galaxy clusters, compact objects, supernova remnants, and stars. X-ray radiation can be emitted by hot gas (10^6–10^9 K) or generated by bremsstrahlung, synchrotron processes, inverse Compton scattering, fluorescent emission, and nuclear decay. γ-ray radiation can be generated by the same processes at higher energies, as well as by electron-positron annihilation. As a back-of-the-envelope estimate, consider an electromagnetic particle falling onto a black hole, beginning from infinity at rest. In Newtonian mechanics, the energy of a particle is the sum of its kinetic and potential energy, and the sum is zero if the particle is at rest at infinity

$$E = \frac{1}{2} m v^2 - G_{\text{N}} \frac{M m}{r} \approx 0 , \tag{1.14}$$

where m and v are the mass and the velocity of the particle falling onto the black hole and M is the black hole mass. At the radial coordinate $r \sim 10\ r_{\text{g}}$, the kinetic energy of the particle is around 10% of its rest mass, namely around 100 MeV for protons and 50 keV for electrons. We can thus expect the emission of radiation with such an energy, which is indeed in the X-ray and γ-ray bands.

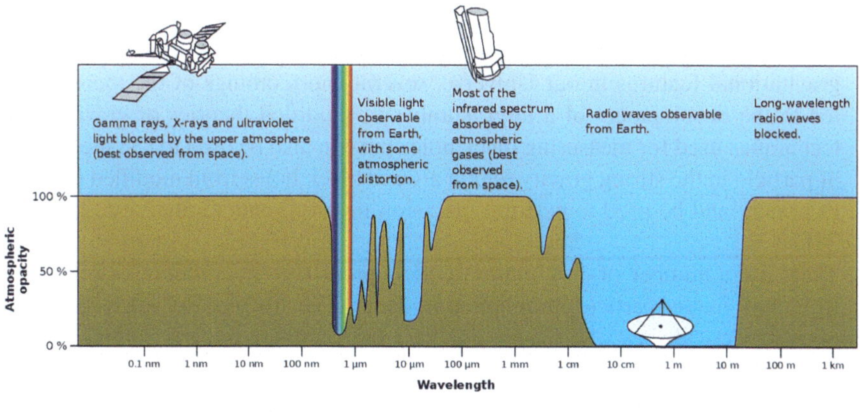

Fig. 1.4 Atmospheric opacity as a function of photon wavelength. Since the atmosphere is opaque at most wavelengths, only optical and radio telescopes can be at ground level on Earth. γ-ray, X-ray, UV, and IR observational facilities are required to be on board of rockets or satellites. Credit: NASA

A large part of the electromagnetic spectra is blocked by the Earth's atmosphere, see Fig. 1.4. If it were not so, life on Earth—at least as we know—would be impossible, because γ-rays, X-rays, and UV photons are harmful for any organism. To be able to observe X-rays and γ-rays, observatories must thus be on board rockets or satellites. The first X-ray observatory can be considered a V2 rocket launched in 1948, which was used to observe the Sun, the brightest X-ray source in the sky. The first extrasolar X-ray source was discovered in 1962 by a team led by Riccardo Giacconi with an X-ray detector on board of an Aerobee 150 sounding rocket [15]. The source, known as Scorpius X-1, is an X-ray binary with a neutron star of 1.4 $M_{\odot}$ and a companion of 0.42 $M_{\odot}$. Giacconi received the Nobel Prize in Physics in 2002 for pioneering the research field today called X-ray astronomy. Since the discovery of Scorpius X-1, a steady progress in technology, theory and analysis, has made X-ray astronomy a leading scientific field in astrophysics research. Tables 1.3 and 1.4 present some of the most important X-ray and γ-ray observatories from past, present, and future.

1.5 Open Problems and Future Directions

X-ray and γ-ray radiation have provided invaluable information about black holes and their astrophysical environments and breakthroughs in fundamental physics. In the case of accreting black holes, we can study of the accretion process in the strong gravity region, how the gas falls onto the compact object, and how jets and outflows are generated. In the past 10–15 years, a few X-ray techniques have been developed to measure black hole spins, and before the detection of gravitational waves these were the only techniques capable of measuring black hole spins.

While several puzzles have been answered, many new ones have appeared and remain unresolved. Some of them are as follows:

1. While Einstein's general relativity is the standard framework for describing the gravitational features in our Universe, several shortcomings of the theory have led to the development of a large number of modified theories of gravity. The techniques used for measuring black hole spin can also be used to test the motion of particle in the strong gravity region around black holes from modified theories of gravity and be used to perform precision tests of general relativity [2, 3, 11, 24].
2. There are a number of dark matter models predicting the production of γ-rays from dark matter particle annihilation or decay. The study of the γ-ray spectrum of astrophysical sites where there may be a large amount of dark matter particles is an indirect search for dark matter. If we detect an excess of γ-rays with respect to that expected from the pure astrophysical environment, as well as some specific feature in the γ-ray spectrum, this may be interpreted as an indirect evidence of dark matter particles.
3. What is the spin distribution among stellar-mass and supermassive black holes and how does the spin distribution change over cosmological times? In the case

Table 1.3 List of some of the most important X-ray missions from past, present, and future

Mission	Launch date	End of mission	Instruments
PAST			
Röntgensatellit (ROSAT)	1990	1999	XRT (0.1–2 keV)
Advanced Satellite for Cosmology and Astrophysics (ASCA)	1993	2000	GIS (0.7–10 keV) SIS (0.4–10 keV)
Rossi X-ray Timing Explorer (RXTE)	1995	2012	ASM (2–10 keV) PCA (2–60 keV) HEXTE (15–250 keV)
Suzaku	2005	2015	XRS (0.3–12 keV) XIS (0.2–12 keV) HXD (10–600 keV)
Hitomi	2016	2016	SXS (0.4–12 keV) SXI (0.3–12 keV) HXI (5–80 keV)
PRESENT			
Chandra X-ray Observatory (CXO)	1999	–	ACIS (0.2–10 keV) HRC (0.1–10 keV) LETG (0.08–2 keV) HETG (0.4–10 keV)
XMM-Newton	1999	–	EPIC-MOS (0.15–15 keV) EPIC-pn (0.15–15 keV) RGS (0.33–2.5 keV)
International Gamma-Ray Astrophysics Laboratory (INTEGRAL)	2002	–	IBIS (15 keV–10 MeV) SPI (18 keV–8 MeV) JEM-X (3–35 keV)

(continued)

of supermassive black holes, the spin distribution would also provide information about the evolution of their host galaxies [9].

4. What is the mechanism responsible for the production of jets in black holes?
5. What is the mechanism responsible of the observed quasi-periodic oscillations (QPOs) in the X-ray power density spectrum of black holes? Can we use QPOs for measuring black hole spins and test general relativity?
6. What is the exact origin of supermassive black holes and how do they grow so fast? In particular, we know supermassive black holes of billions of Solar masses at redshifts higher than 6 and we do not have a clear understanding of how such objects were created and were able to grow so fast in a relatively short time.

Table 1.3 (continued)

Mission	Launch date	End of mission	Instruments
Swift	2004	–	BAT (15–150 keV) XRT (0.2–10 keV)
Monitor of All-sky X-ray Image (MAXI)	2009	–	SSC (0.5–10 keV) GSC (2–30 keV)
Nuclear Spectroscopic Telescope Array (NuSTAR)	2012	–	FPMA (3–79 keV) FPMB (3–79 keV)
ASTROSAT	2015	–	SXT (0.3–80 keV) LAXPC (3–80 keV) CZTI (100–300 keV)
Neutron star Interior Composition Explorer (NICER)	2017	–	XTI (0.2–12 keV)
Hard X-ray Modulation Telescope (HXMT)	2017	–	HE (20–250 keV) ME (5–30 keV) LE (1–15 keV)
Spektrum-Roentgen-Gamma (Spektr-RG)	2019	–	eROSITA (0.3–10 keV) ART-XC (0.5–11 keV)
FUTURE			
X-Ray Imaging and Spectroscopy Mission (XRISM)	2022	–	Resolve (0.4–12 keV) Xtend (0.3–12 keV)
Enhanced X-ray Timing Polarization (eXTP)	2027	–	SFA (0.5–20 keV) LAD (1–30 keV)
Advanced Telescope for High Energy Astrophysics (ATHENA)	2031	–	X-IFU (0.2–12 keV) WFI (0.1–15 keV)

7. How does the host environment determine the properties of supermassive black holes? And how do supermassive black holes determine the properties of their host environment?
8. Do intermediate mass black holes exist? Do small primordial black holes created in the early Universe exist?

Table 1.4 List of some of the most important γ-ray missions from past and present

Mission	Launch date	End of mission	Instruments
PAST			
GRANAT	1989	1999	SIMGA (30–1300 keV) PHEBUS (0.1–100 MeV) KONUS-B (0.01–8 MeV) TOURNESOL (0.002–20 MeV)
Compton Gamma Ray Observatory (CGRO)	1991	2000	OSSE (0.06–10 MeV) COMPTEL (0.8–30 MeV)) EGRET (20–3000 MeV) BATSE (0.015–110 MeV)
PRESENT			
International Gamma-Ray Astrophysics Laboratory (INTEGRAL)	2002	–	SPI (0.02–8 MeV) IBIS (0.015–10 MeV)
Swift	2004	–	BAT (15–150 keV)
Astrorivelatore Gamma ad Immagini LEggero (AGILE)	2007	–	GRID (30 MeV–50 GeV) MC (0.25–200 MeV)
Fermi gamma ray space telescope	2008	–	LAT (20 Mev–300 GeV) GBM (8 keV–30 MeV)

References

1. B.P. Abbott et al., LIGO scientific and virgo collaborations. Phys. Rev. Lett. **116**, 061102 (2016). arXiv:1602.03837 [gr-qc]
2. C. Bambi, *Black Holes: A Laboratory for Testing Strong Gravity*, (Springer, Singapore, 2017). https://doi.org/10.1007/978-981-10-4524-0
3. C. Bambi, Rev. Mod. Phys. **89**, 025001 (2017). arXiv:1509.03884 [gr-qc]
4. C. Bambi, A.D. Dolgov, A.A. Petrov, JCAP **0909**, 013 (2009). arXiv:0806.3440 [astro-ph]
5. C. Bambi, D. Malafarina, N. Tsukamoto, Phys. Rev. D **89**, 127302 (2014). arXiv:1406.2181 [gr-qc]
6. C. Bambi et al., Universe **4**, 79 (2018). arXiv:1806.02141 [gr-qc]
7. E. Barausse, V. Cardoso, P. Pani, Phys. Rev. D **89**, 104059 (2014). arXiv:1404.7149 [gr-qc]
8. J.M. Bardeen, W.H. Press, S.A. Teukolsky, Astrophys. J. **178**, 347 (1972)
9. E. Berti, M. Volonteri, Astrophys. J. **684**, 822 (2008). arXiv:0802.0025 [astro-ph]
10. C.T. Bolton, Nature **235**, 271 (1972)
11. Z. Cao, S. Nampalliwar, C. Bambi, T. Dauser, J.A. Garcia, Phys. Rev. Lett. **120**, 051101 (2018). arXiv:1709.00219 [gr-qc]
12. B. Carter, Phys. Rev. Lett. **26**, 331 (1971)
13. P.T. Chrusciel, J.L. Costa, M. Heusler, Living Rev. Rel. **15**, 7 (2012). arXiv:1205.6112 [gr-qc]
14. A. Einstein, Annalen Phys. **49**, 769 (1916); Annalen Phys. **14**, 517 (2005)
15. R. Giacconi, H. Gursky, F.R. Paolini, B.B. Rossi, Phys. Rev. Lett. **9**, 439 (1962)

16. W. Israel, Phys. Rev. **164**, 1776 (1967)
17. J. Kormendy, D. Richstone, Ann. Rev. Astron. Astrophys. **33**, 581 (1995)
18. F. Pretorius, Phys. Rev. Lett. **95**, 121101 (2005). arXiv:gr-qc/0507014
19. R.H. Price, Phys. Rev. D **5**, 2419 (1972)
20. R.A. Remillard, J.E. McClintock, Ann. Rev. Astron. Astrophys. **44**, 49 (2006). arXiv:astro-ph/0606352
21. D.C. Robinson, Phys. Rev. Lett. **34**, 905 (1975)
22. E.E. Salpeter, Astrophys. J. **140**, 796 (1964)
23. K. Schwarzschild, Sitzungsber. Preuss. Akad. Wiss. Berlin (Math. Phys.) **1916**, 189 (1916) arXiv:physics/9905030
24. A. Tripathi, S. Nampalliwar, A.B. Abdikamalov, D. Ayzenberg, C. Bambi, T. Dauser, J.A. Garcia, A. Marinucci, Astrophys. J. **875**, 56 (2019). arXiv:1811.08148 [gr-qc]
25. B.L. Webster, P. Murdin, Nature **235**, 37 (1972)
26. Y.B. Zeldovich, Dokl. Akad. Nauk **155**, 67 (1964); Sov. Phys. Dokl. **9**, 195 (1964)

Chapter 2
Accreting Black Holes

Sourabh Nampalliwar and Cosimo Bambi

2.1 Theory of Black Holes: Formation and Masses

Black holes happen to be surprisingly simple objects. Only two parameters, the *mass* M and the *spin* J, are thought to be sufficient to characterize a black hole in our Universe [11]. The spin parameter cannot be arbitrary and must satisfy the constraint $J/M^2 \le 1$, which is the condition for the existence of the event horizon, as shown in Eq. (1.5). There are no theoretical constraints on the value of the mass of a black hole, which may thus be arbitrarily small as well as arbitrarily large.

From astronomical observations, we have strong evidence of two classes of astrophysical black holes:

1. *Stellar-mass black holes* [126], with masses $\sim 3-100 M_\odot$.
2. *Supermassive black holes* [72], with masses $> 10^5 M_\odot$.

One would expect, and there is some evidence, that black holes with masses in the intermediate range should exist [29]. These are termed *intermediate-mass black holes*. Each of these classes is theorized to have a different past, present and future. We will discuss them separately.

2.1.1 Stellar-Mass Black Holes

The most common formation channel for stellar-mass black holes is gravitational collapse. In lay terms, when a star runs out of fuel, the pressure inside is insufficient to hold the star against gravitational pull and the star collapses. For massive enough stars, the star collapses all the way to a singularity and a black hole is born.

S. Nampalliwar (✉)
Theoretical Astrophysics, Eberhard-Karls Universität Tübingen,
Auf der Morgenstelle 10, 72076 Tübingen, Germany
e-mail: sourabh.nampalliwar@uni-tuebingen.de

C. Bambi
Department of Physics, Fudan University, 2005 Songhu Road, Shanghai 200438, China
e-mail: bambi@fudan.edu.cn

C. Bambi (ed.), *Tutorial Guide to X-ray and Gamma-ray Astronomy*,
https://doi.org/10.1007/978-981-15-6337-9_2

The initial mass of a stellar-mass black hole depends on the properties of the progenitor: its mass, its evolution, and the supernova explosion mechanism [15]. Depending on these details, the supernova remnant could be a neutron star, where the quantum neutron pressure can hold against the gravitational collapse, or a black hole. In fact, the lower bound on the black hole initial mass may come from the maximum mass for a neutron star: the exact value is currently unknown, since it depends on the equation of state of matter at super-nuclear densities, but it should be in the range of 2–3 $M_\odot$. It is possible though, that a mass gap exists between the most massive neutron stars and the less massive black holes [36]. An upper bound on stellar-mass black holes may be derived from the progenitor's metallicity. The final mass of the remnant is determined by the mass loss rate by stellar winds, which increases with the metallicity because heavier elements have a larger cross section than lighter ones, and therefore they evaporate faster. For a low-metallicity progenitor [59, 60, 138], the mass of the black hole remnant may be $M \lesssim 50\ M_\odot$ or $M \gtrsim 150\ M_\odot$. As the metallicity increases, black holes with $M \gtrsim 150\ M_\odot$ disappear, because of the increased mass loss rate. Note, however, that some models do not find remnants with a mass above the gap, because stars with $M \gtrsim 150\ M_\odot$ may undergo a runaway thermonuclear explosion that completely destroys the system, without leaving any black hole remnant [59, 60]. Stellar-mass black holes may thus have a mass in the range of 3–100 $M_\odot$. Until now, all the known stellar-mass black holes in X-ray binaries have a mass $M \approx 3$–$20\ M_\odot$ [25]. Gravitational waves, on the other hand, have shown the existence of heavier stellar-mass black holes. In particular, the event called GW150914 was associated with the coalescence of two black holes with masses $M \approx 30\ M_\odot$ that merged to form a black hole with $M \approx 60\ M_\odot$ [1].

From stellar evolution studies, we expect that in our Galaxy there is a population of 10^8–10^9 black holes formed at the end of the evolution of heavy stars [155, 159], and the same number can be expected in similar galaxies. But with observations, we only know about 20 black holes with a dynamical measurement of the mass and about 50 without (it is thus possible that some of them are not black holes but neutron stars). This is because their detection is very challenging. The simplest scenario is when the black hole is in a binary system and has a companion star. The presence of a compact object can be discovered from the observation of a short timescale variability, the non-detection of a stellar spectrum, etc. The study of the orbital motion of the companion star can permit the measurement of the mass function [25]

$$f(M) = \frac{K_c^3 P_{\text{orb}}}{2\pi G_N} = \frac{M \sin^3 i}{(1+q)^2}, \tag{2.1}$$

where $K_c = v_c \sin i$, v_c is the velocity of the companion star, i is the angle between the normal of the orbital plane and our line of sight, P_{orb} is the orbital period of the system, $q = M_c/M$, M_c is the mass of the companion, and M is the mass of the dark object. If we can somehow estimate i and M_c, we can infer M, and in this case we talk about dynamical measurement of the mass. The dark object is a black hole if $M > 3\ M_\odot$ [65, 76, 128].

Note that, among astronomers, it is common to call "black hole" a compact object for which there is a dynamical measurement of its mass proving that $M > 3\,M_{\odot}$. The latter indeed guarantees that the object is too heavy for being a neutron star. "Black hole candidates" are instead compact objects that are supposed to be black holes, for instance because of the detection of spectral features typical of black holes, but for which there is no dynamical measurement of their mass.

Black holes in X-ray binaries (black hole binaries[1]) are grouped into two classes: *low-mass X-ray binaries* (LMXBs) and *high-mass X-ray binaries* (HMXBs). Here, "low" and "high" refers to the stellar companion, not to the black hole: in the case of LMXBs, the companion star normally has a mass $M_c < 3\,M_{\odot}$, while for HMXBs the companion star has $M_c > 10\,M_{\odot}$. Observationally, we can classify black hole binaries either as *transient X-ray sources* or *persistent X-ray sources*. LMXBs are usually transient sources, because the mass transfer is not continuous (for instance, at some point the surface of the companion star may expand and the black hole strips some gas): the system may be bright for a period ranging from some days to a few months and then be in a quiescent state for months or even decades. Every year we discover 1–2 new objects, when they pass from their quiescent state to an outburst (see Sect. 2.4.1). Overall, we expect 10^3–10^4 LMXBs in the Galaxy [66, 176]. HMXBs are persistent sources: the mass transfer from the companion star to the black hole is a relatively regular process (typically it is due to the stellar wind of the companion) and the binary is a bright source at any time without quiescent periods. Figure 2.1 shows 22 X-ray binaries with a stellar-mass black hole confirmed by dynamical measurements. To have an idea of the size of these systems, the figure also shows the Sun (whose radius is 0.7 million km) and the distance Sun-Mercury (about 50 million km). The black holes have a radius <100 km and cannot be seen, but we can clearly see their accretion disks formed from the transfer of material from the companion star. The latter may have a quite deformed shape (in particular, we can see some cusps) due to the the tidal force produced by the gravitational field of the black hole. Among the sources listed in the figure, Cygnus X-1 (Cyg X-1 in Fig. 2.1), LMC X-1, LMC X-3, and M33 X-7 are HMXBs, while all other systems are LMXBs. Among these HMXBs, only Cygnus X-1 is in our Galaxy. Among the LMXBs, there is GRS 1915+105, which is quite a peculiar source: since 1992, it is a bright X-ray source in the sky, so it can be considered a persistent source. This is probably because of its large accretion disk, which can provide enough material at any time.

Black holes in compact binary systems (black hole-black hole or black hole-neutron star) can be detected with gravitational waves when the signal is sufficiently strong. Figure 2.2 shows the first detections by the LIGO/Virgo collaboration. The name of the event is classified as GW (gravitational wave event) and then the date of detection: for example, GW150914 was detected on 14 September 2015. LVT151012

[1] Generally speaking, a *black hole binary* is a binary system in which at least one of the two bodies is a black hole, and a *binary black hole* is a binary system of two black holes.

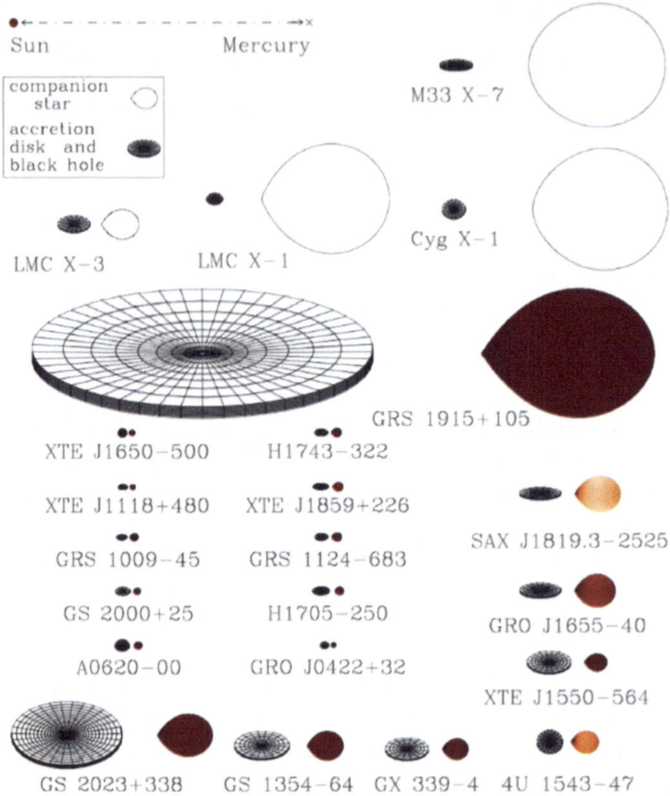

Fig. 2.1 Sketch of 22 X-ray binaries with a stellar-mass black hole confirmed by dynamical measurements. For every system, the black hole accretion disk is on the left and the companion star is on the right. The color of the companion star roughly indicates its surface temperature (from brown to white as the temperature increases). The orientation of the disks indicates the inclination angles of the binaries. For comparison, in the top left corner of the figure we see the system Sun-Mercury: the distance between the two bodies is about 50 million km and the radius of the Sun is about 0.7 million km. Figure courtesy of Jerome Orosz

is not classified as a gravitational wave event because the signal to noise ratio was not large enough to qualify as a detection.[2] For every event, the figure shows the two original black holes as well as the final one after merger.

Isolated black holes are much more elusive. In principle, they can be detected by observing the modulation of the light of background stars due to the gravitational lensing caused by the passage of a black hole along the line of sight of the observer [8].

[2]LVT stands for LIGO/Virgo transient.

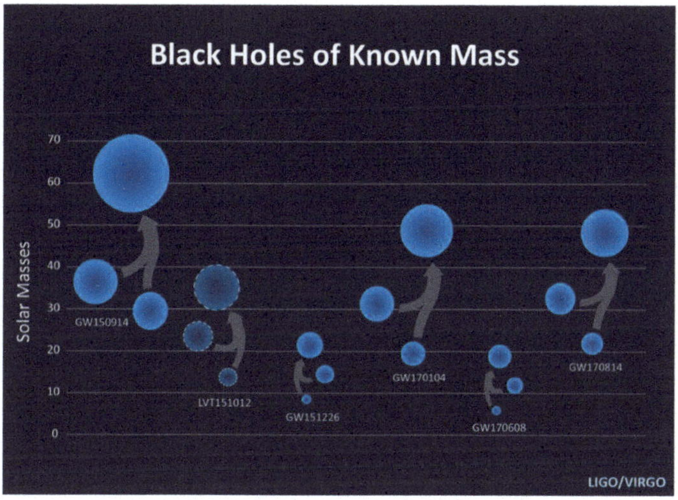

Fig. 2.2 Masses of the first black holes observed with gravitational waves, with the two initial objects merging into a larger one, as shown by the arrows. Image Credit: LIGO/NSF/Caltech/SSU Aurore Simmonet

2.1.2 *Supermassive Black Holes*

The formation channels of supermassive black holes are not well established. The gigantic masses of supermassive black holes are not thought to be natal, but acquired. Accretion has been shown to be an effective mechanism for growing the masses of black holes. In fact, some models suggest the possibility of super-Eddington accretion, and this may indeed be a possible path to the rapid growth of supermassive black holes [82]. Another possibility is merger of several black holes. But the question of the progenitor, or *seed*, remains open. See [162] for a review of the possible formation channels.

Astronomical observations show that at the center of many galaxies there is a large amount of mass in a relatively small volume. The standard interpretation is that these objects are supermassive black holes with $M \sim 10^5$–10^{10} $M_\odot$. Strong constraints come from the center of our Galaxy and NGC 4258 [83]. For our Galaxy, we can study the Newtonian motion of individual stars and infer that at the center there is an object with a mass of $4 \cdot 10^6$ $M_\odot$ (see Fig. 2.3). An upper bound on the size of this body can be obtained from the minimum distance approached by one of these stars, which is less than 45 AU and corresponds to $\sim$1,200 r_g for a $4 \cdot 10^6$ $M_\odot$ object. In the end, we can exclude the existence of a cluster of compact non-luminous bodies like neutron stars and therefore we can conclude that the most natural interpretation is that it is a supermassive black hole. In the case of NGC 4258, we can study the orbital motion of gas in the nucleus, and again we can conclude that the central object is too massive, compact, and old to be a cluster of neutron stars. In the case of other galaxies, it is not possible to put such constraints with the available data, but it is thought that

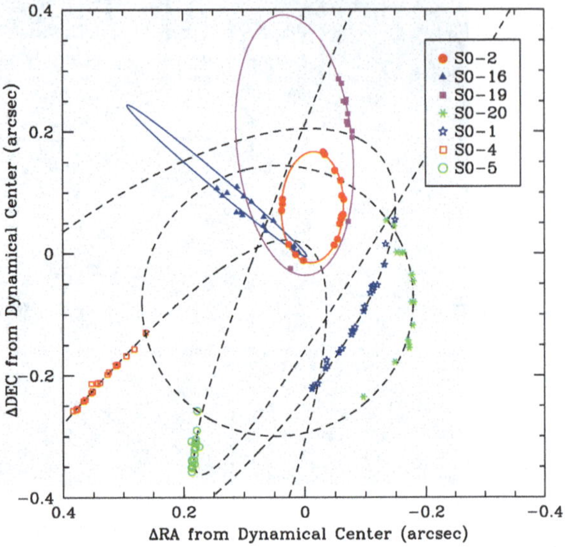

Fig. 2.3 Astrometric positions and orbital fits for seven stars orbiting the supermassive black hole at the center of the Milky Way. From [50]. ©AAS. Reproduced with permission

every mid-size (like the Milky Way) or large galaxy has a supermassive black hole at its center.[3] For smaller galaxies, the situation is more uncertain. Most models predict supermassive black holes at the center of lighter galaxies as well [162], but there exist predictions of faint low-mass galaxies with no supermassive black hole at their centers [163, 164]. Observations suggest that some small galaxies have a supermassive black hole and other small galaxies do not [39, 44].

2.1.3 Intermediate-Mass Black Holes

Intermediate-mass black holes are, by definition, black holes with a mass between the stellar-mass and the supermassive ones, say $M \sim 10^2$–$10^5\ M_\odot$. At the moment, there is no dynamical measurement of the mass of these objects, and their actual nature is still controversial. Among the possible formation channels, intermediate-mass black holes are expected to form at the center of dense stellar clusters, by mergers.

Observational evidence for intermediate-mass black holes is inconclusive. The presence of an intermediate-mass black hole at the center of stellar clusters should increase the velocity dispersion in the cluster. Some studies suggest that there are indeed intermediate-mass black holes at the center of certain globular clusters [48, 49]. Some intermediate-mass black hole candidates are associated with ultra luminous X-ray sources [28]. These objects have an X-ray luminosity $L_X > 10^{39}$ erg/s, which exceeds the Eddington luminosity of a stellar-mass object, and they may thus

[3]Exceptions may be possible: the galaxy A2261-BCG has a very large mass but it might not have any supermassive black hole at its center [118].

have a mass in the range 10^2–10^5 $M_\odot$. However, we cannot exclude the possibility that they are actually stellar-mass black holes (or neutron stars [10]) with non-isotropic emission and a moderate super-Eddington mass accretion rate [109]. The existence of intermediate-mass black holes is also suggested by the detection of some quasi-periodic oscillations (QPOs, see Sect. 2.5.3) in some ultra-luminous X-ray sources. QPOs are currently not well understood, but they are thought to be associated to the fundamental frequencies of the oscillation of a particle around a black hole. Since the size of the system scales as the black hole mass, QPOs should scale as $1/M$, and some observations indicate the existence of compact objects with masses in the range 10^2–10^5 $M_\odot$ [114].

2.2 Theory of Black Holes: Evolution and Spins

Apart from mass, a typical black hole is expected to have some spin. Generally speaking, the value of the spin parameter of a black hole can be expected to be determined by the competition of three physical processes: the event creating the object, mergers, and gas accretion.

2.2.1 Stellar-Mass Black Holes

In the case of black hole binaries, it is usually thought that the spin of a black hole is mainly natal and that the effect of the accretion process is negligible [67]. The argument is that a stellar-mass black hole has a mass around 10 $M_\odot$. If the stellar companion is a few Solar masses, the black hole cannot significantly change its mass and spin angular momentum even after swallowing the whole star. If the stellar companion is heavy, its lifetime is too short: even if the black hole accretes at the Eddington rate, there is not enough time to transfer the necessary amount of matter to significantly change the black hole spin parameter. One may expect that a black hole cannot swallow more than a few $M_\odot$ from the companion star, and for a 10 $M_\odot$ object this is not enough to significantly changes a_* [67]. If the black hole spin were mainly natal, its value should be explained by studying the gravitational collapse of massive stars. While there are still uncertainties in the angular momentum transport mechanisms of the progenitors of stellar-mass black holes, it is widely accepted that the gravitational collapse of a massive star with Solar metallicity cannot create fast-rotating remnants [172, 175]. The birth spin of these black holes is expected to be low (see e.g. [42] and references therein).

Observations of spins of stellar-mass black holes contradict the above hypothesis. For instance, in the case of LMXBs, the black hole in GRS 1915+105 has $a_* > 0.98$ [89] and $M = 12.4 \pm 2.0$ $M_\odot$ [120], while the stellar companion's mass is $M = 0.52 \pm 0.41$ $M_\odot$. In the case of HMXBs, the black hole in Cygnus X-1 has $a_* > 0.98$ [55, 56] and $M = 14.8 \pm 1.0$ $M_\odot$, while the stellar wind from the companion is not an efficient mechanism to transfer mass. Very high spin values are also measured for 4U 1630-472, GS 1354-645, MAXI J1535-571, and Swift J1658.2, see Table 2.2. While black holes in LMXBs and HMXBs should form in different environments, in both cases the origin of so high spin values is puzzling. In [42], the authors show that at least in the case of LMXBs, the accretion process immediately after the formation of a black hole binary may be very important and be responsible for the observed high spins. For HMXBs, possible channels for producing high spins are discussed in [119].

2.2.2 *Supermassive Black Holes*

The case of supermassive black holes in galactic nuclei is different. The initial value of their spin parameter is likely completely irrelevant: their mass has increased by several orders of magnitude from its original value, and the spin parameter has evolved accordingly.

There are two primary channels of mass acquisition for supermassive black holes, mergers and accretion. On average, the capture of small bodies (*minor merger*) in randomly oriented orbits should spin the black hole down, since the magnitude of the orbital angular momentum for corotating orbits is always smaller than the one for counterrotating orbits [62]. In the case of random merger of two black holes with comparable mass (*major merger*), the most probable final product is a black hole with $a_* \approx 0.70$, while fast-rotating objects with $a_* > 0.9$ should be rare [17]. On the other hand, accretion from a disk can potentially be a very efficient way to spin a compact object up[4] [17]. In this case, black holes in active galactic nuclei (AGNs) may have a spin parameter close to the Thorne limit (see next section). Such a possibility seems to be supported by some observations; see e.g. [170] and also the spin measurements from X-ray reflection spectroscopy in Table 2.3.

2.2.3 *Thorne Limit*

An accreting black hole changes its mass M and spin angular momentum J as it swallows more and more material from its disk. In the case of a Novikov–Thorne disk (see next section), it is relatively easy to calculate the evolution of these parameters. If we assume that the gas in the disk emits radiation until it reaches the radius of the innermost stable circular orbit (ISCO) and then quickly plunges onto the black hole, the evolution of the spin parameter a_* is governed by the following equation [154]

---

[4]Unless the accretion proceeds via short episodes (chaotic accretion) [69], in which case it is effectively like minor mergers.

$$\frac{da_*}{d\ln M} = \frac{c}{r_g}\frac{L_{\rm ISCO}}{E_{\rm ISCO}} - 2a_*\,, \tag{2.2}$$

where $E_{\rm ISCO}$ and $L_{\rm ISCO}$ are, respectively, the energy and the angular momentum per unit rest-mass of the gas at the ISCO radius. Assuming an initially non-rotating black hole of mass M_0, the solution of Eq. (2.2) is

$$a_* = \begin{cases} \sqrt{\frac{2}{3}}\frac{M_0}{M}\left[4 - \sqrt{18\frac{M_0^2}{M^2} - 2}\right] & \text{if } M \le \sqrt{6}\,M_0\,, \\ 1 & \text{if } M > \sqrt{6}\,M_0\,. \end{cases} \tag{2.3}$$

The black hole spin parameter a_* monotonically increases from 0 to 1 and then remains constant. $a_* = 1$ is the equilibrium spin parameter and is reached after the black hole has increased its mass by the factor $\sqrt{6} \approx 2.4$.

If we take into account the fact that the gas in the accretion disk emits radiation and that a fraction of this radiation is captured by the black hole, Eq. (2.2) becomes

$$\frac{da_*}{d\ln M} = \frac{c}{r_g}\frac{L_{\rm ISCO} + \zeta_{\rm L}}{E_{\rm ISCO} + \zeta_{\rm E}} - 2a_*\,, \tag{2.4}$$

where $\zeta_{\rm L}$ and $\zeta_{\rm E}$ are related to the amount of photons captured by the black holes and must be computed numerically. Now the equilibrium value of the spin parameter is not 1 but the so-called *Thorne limit* $a_*^{\rm Th} \approx 0.998$ (its exact numerical value depends on the emission properties of the gas in the disk) [154]. Intuitively, the Thorne limit is lower than 1 because retrograde photons (i.e., those photons with angular momentum antiparallel to the black hole spin) have larger capture cross section and therefore they contribute to reduce the black hole spin.

2.3 Accreting Black Holes in Nature: Modeling

Accretion is the process of material spiraling onto a black hole as a consequence of the gravitational pull of the black hole. It is quite commonplace in the Universe and is crucial for various techniques for studying black holes. Here we will discuss only the concepts relevant within the context of this book. For a review on the theory of black hole accretion models, see [4].

The morphology of the accretion flow is mainly determined by two factors: (*i*) the angular momentum of the accreting gas, and (*ii*) the mass accretion rate. Depending on the value of these two quantities, we have different accretion models and different electromagnetic spectra for the accretion flow. We will now discuss some of the accretion models.

In the case of spherically symmetric accretion (vanishing or negligible angular momentum of the accreting gas), we have the so-called *Bondi accretion* (Fig. 2.4 top sketch), which essentially describes particles in radial free fall. In the Bondi accretion

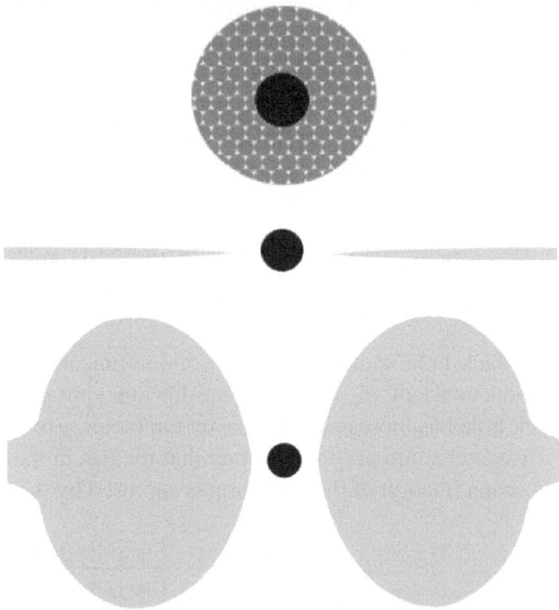

Fig. 2.4 Sketch of Bondi accretion (top), thin disk (middle), and thick disk (bottom). The black hole is indicated by the black filled circle and the accretion flow is represented by the gray shape

scenario, the radiative efficiency is very low [20, 134], i.e. the gravitational energy of the falling particles is mainly converted into their kinetic energy and lost into the black hole after crossing the event horizon.

In typical accretion flows around black holes the gas has a non-negligible angular momentum. In such cases, a disk is created. For stellar-mass black holes with a companion star, the disk is created by the mass transfer from the stellar companion to the black hole. In the case of supermassive black holes in galactic nuclei, the disk forms from the material in the interstellar medium [135] or as a result of galaxy merger [14, 87]. An accretion disk is *geometrically thin* (*thick*) if $h/r \ll 1$ ($h/r \sim 1$), where h is the semi-thickness of the disk at the radial coordinate r. The disk is *optically thin* (*thick*) if $h \ll \lambda$ ($h \gg \lambda$), where λ is the photon mean free path in the medium of the disk. If the disk is optically thick, we see the radiation emitted from the surface of the disk, like in the case of stars. The typical way to categorize disks is in terms of their Eddington ratio:

- *Thick disk* [2, 64]: If the mass accretion rate is super-Eddington ($\dot{M}/\dot{M}_{\rm Edd} > 1$), the gas pressure makes the disk inflate. The disk is thus geometrically thick (Fig. 2.4 bottom sketch). Since the particle density is high, the disk is optically thick and it cannot efficiently radiate away energy from its surface.
- *Slim disk* [3]: As the mass accretion rate decreases, the thickness of the disk decreases too. A slim disk describes the situation between a thick and a thin disk. For slim disks, roughly, $0.3 < \dot{M}/\dot{M}_{\rm Edd} < 1$.
- *Thin disk* [133]: For moderate accretion rates ($0.05 < \dot{M}/\dot{M}_{\rm Edd} < 0.3$), the gas pressure is negligible and we have a geometrically thin and optically thick disk

Table 2.1 Summary of the main scenarios of accretion processes around black holes and of their basic properties. The first column indicates the angular momentum of the accreting gas; the second column is for the mass accretion rate (in Eddington units); the third column indicates if the accretion disk is geometrically thick or thin; the fourth column shows if the accretion flow is optically thick or thin; the fifth column is for the radiative efficiency; the last column presents the name of the accretion model. See the text for more details

$Lc/G_{\rm N}M$	$\dot{M}/\dot{M}_{\rm Edd}$	h/r	h/λ	$\eta_{\rm r}$	Accretion model
$\ll$1	Any	–	Any	$\ll$0.1	Bondi accretion
>1	>1	~1	$\gg$1	$\ll$0.1	Thick disk
>1	0.3–1	<1	$\gg$1	<0.1	Slim disk
>1	0.05–0.3	$\ll$1	$\gg$1	~0.1	Thin disk
$\lesssim$1	$\ll$1	~1	$\ll$1	$\ll$0.1	ADAF

(Fig. 2.4 middle sketch). The radiative efficiency is high because the disk surface is large enough with respect to the mass accretion rate to radiate away the gravitational energy converted into heat. The standard model for thin disks around black holes is the Novikov–Thorne model [108, 110], which is described in Sect. 2.3.1.

- *Advection-dominated accretion flow (ADAF)* [104, 105]: If the mass accretion rate is very low ($\dot{M}/\dot{M}_{\rm Edd} < 0.05$), the disk evaporates and we have a low density accreting medium. Because of the low density, the medium is optically thin. The interaction rate between particles is low, so there is no efficient cooling mechanism: the gas temperature is very high, the gas is swallowed by the black hole without emitting much radiation, and the accretion luminosity is low.

Table 2.1 summarizes the accretion scenarios in terms of the specific angular momentum of the accreting gas L (in units of $r_{\rm g}c$), the mass accretion rate $\dot{M}$ (in Eddington units), the geometrical and optical thickness of the disk and the radiative efficiency.

2.3.1 Novikov–Thorne Disks

The Novikov–Thorne model is the standard framework for the description of geometrically thin and optically thick accretion disks around black holes. The main assumptions of the model are:

1. The accretion disk is geometrically thin ($h/r \ll 1$).
2. The accretion disk is perpendicular to the black hole spin.
3. The inner edge of the disk is at the ISCO.
4. The motion of the gas particles in the disk is determined by the gravitational field of the black hole, while the impact of the gas pressure is ignored.

For the full list of assumptions and a detailed discussion, see e.g., [11, 110] and references therein. Here we point out a few important considerations.

Assumption 1 just points out that the Novikov–Thorne model is strictly applicable only to thin disks, even though it is often used for all black hole X-ray sources. This is in part because it is the only simple analytic model in the market and in part because it is often difficult to estimate the Eddington-scaled mass accretion rate (in particular for AGNs, where mass and distance are usually poorly constrained).

Assumption 2's validity depends on the origin and the evolution of the system. The cases of stellar-mass black holes and of supermassive black holes are somewhat different. Let us begin with the former. If the stellar-mass black hole is the final product of the supernova explosion of a heavy star in a binary, its natal spin should be along the same direction as the progenitor star's spin. Assuming a symmetric explosion without strong shocks and kicks, this would mean a spin axis orthogonal to the orbital plane of the binary [41]. A misalignment may be introduced by a non-symmetric supernova explosion and/or shocks and kicks, as well as in those systems formed through multi-body interactions (binary capture or replacement), where the orientation of the spin of the black hole and that of the orbital angular momentum of the binary are initially uncorrelated.

Regardless of the natal spin and binary capture scenarios, at least the inner part of the disk—which plays the most important role in spin measurements—is expected to be equatorial, as a result of the Bardeen–Petterson effect [13, 74].[5] The alignment timescale of thin disks has been estimated to be in the range 10^6–10^8 yrs, and therefore the disk should be already adjusted in the black hole equatorial plane for not too young systems [139] (but see [85, 86] for more details). However, the actual timescale depends on parameters like the viscosity α which are usually not known [68, 79]. Moreover, some numerical simulations do not observe the adjustment of the alignment of the disk [40, 179]. Additionally, if the inner part of the disk is a hot, geometrically thick accretion flow, the inner disk precesses as a solid body about the black hole angular momentum axis [63]. Future X-ray spectropolarimetric measurements of the thermal spectrum of accretion disks will be able to check the validity of the assumption that the disk is in the equatorial plane (see, for instance, [11] and references therein).

In the case of supermassive black holes, the orientation of the accretion disk with respect to the black hole spin is expected to change during the evolution of the system, in particular because of galaxy mergers. However, in the absence of galactic mergers (or if enough time has passed since the previous merger), again the Bardeen–Petterson effect will make the inner part of the disk orthogonal to the black hole spin.

Assumption 3 is crucial for doing spin measurements. This is because, assuming the Kerr metric, there is a one-to-one correspondence between the spin parameter a_* and the ISCO radius (see Fig. 1.1). A measurement of the ISCO via the inner edge thus provides direct information about the spin of the black hole.

---

[5] *Bardeen–Petterson configuration* refers to a system in which the inner part of the disk is flat and perpendicular to the black hole spin, while the outer part is also flat but in the plane perpendicular to the angular momentum vector of the binary.

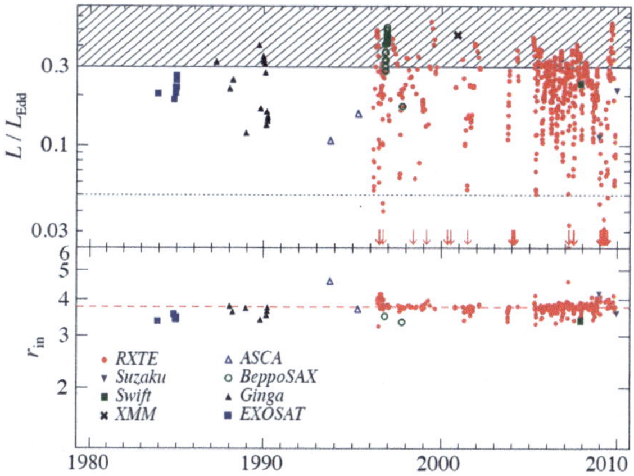

Fig. 2.5 Top panel: Accretion disk luminosity in Eddington units versus time for 766 spectra of LMC X-3. The shaded region does not satisfy the thin disk selection criterion $L/L_{\rm Edd} < 0.3$, as well as the data below the dotted line, which marks $L/L_{\rm Edd} = 0.05$. Bottom panel: fitted value of the inner disk radius of the 411 spectra in the top panel that can meet the thin disk selection criterion. See the text for more details. From [141]. ©AAS. Reproduced with permission

Observations show that the inner edge of the disk does not change appreciably over several years when the source is in the soft state (see Sect. 2.4.1 for the definition of soft state) with a luminosity between ~5% and ~30% of its Eddington limit. The most compelling evidence comes from LMC X-3. The analysis of many spectra collected during eight X-ray missions and spanning 26 years shows that the radius of the inner edge of the disk is nearly constant [141], see Fig. 2.5. The most natural interpretation is that the inner edge is associated to some intrinsic property of the geometry of the spacetime, namely the radius of the ISCO, and is not affected by variable phenomena like the accretion process.

Assumption 4 requires that the radial acceleration of the gas due to pressure gradients is negligible in comparison with the gravitational acceleration due to the black hole. This requires that, as the gas falls onto the black hole, its potential energy is transported away or radiated away, and only a negligible part is converted to internal energy of the gas [110]. This assumption holds in general for radiatively efficient accretion flows.

The accretion process in the Novikov–Thorne model can be summarized as follows. The particles of the accreting gas slowly fall onto the central black hole. When they reach the ISCO radius, they quickly plunge onto the black hole without emitting additional radiation. The total power of the accretion process is $L = \eta \dot{M} c^2$, where $\eta = \eta_{\rm r} + \eta_{\rm k}$ is the total efficiency, $\eta_{\rm r}$ is the radiative efficiency, and $\eta_{\rm k}$ is the fraction of gravitational energy converted to kinetic energy of jets/outflows. The Novikov–Thorne model assumes that $\eta_{\rm k}$ can be ignored, and therefore the radiative efficiency of a Novikov–Thorne accretion disk is

$$\eta_{\rm NT} = 1 - E_{\rm ISCO} , \tag{2.5}$$

where $E_{\rm ISCO}$ is the energy per unit rest-mass of the gas at the ISCO radius. The plot in Fig. 1.3 shows $\eta_{\rm NT}$ as a function of the spin parameter a_*.

In spin measurements, it is clearly very important to select the observations and the sources in which the disk is geometrically thin and its inner edge is at the ISCO radius. In the case of the continuum-fitting method, one usually selects sources in the soft state with a strong contribution from the thermal disk emission. The luminosity of the source should be between ~5% and ~30% of the Eddington limit [89]. At lower luminosities, the disk may be truncated. In such a case, the inner edge of the disk would be at a radius larger than the ISCO and between the inner edge of the disk and the black hole there is probably an accretion flow that can be described by ADAF. For higher accretion rates, the gas pressure becomes more important, and the disk is no longer thin. In such a case, the inner edge of the disk might be at a radius slightly smaller than the ISCO.

2.3.2 *Disk-Corona Model*

Observations have led us to theorize, in addition to a black hole with a disk, a corona around the black hole. An example of the disk-corona model is schematically illustrated in Fig. 2.6. The black hole accretes from a geometrically thin and optically thick disk. The disk emits as a blackbody locally and as a multi-color blackbody when integrated radially (*thermal component* indicated by the red arrows in Fig. 2.6) [99]. The *multi-color* feature comes from the fact that different parts of the disk have different temperatures. For a given radius of the disk, the temperature depends on the black hole mass and the mass accretion rate. Most of the radiation is emitted near the inner edge of the disk and is in the soft X-ray band (0.1–1 keV) for stellar-mass black holes and in the optical/UV band (1–10 eV) for supermassive black holes.

The *corona* is a hotter (~100 keV), usually compact and optically thin, cloud close to the black hole (the yellow region in Fig. 2.6), but its exact geometry is currently unknown [33, 57, 123]. Figure 2.7 shows some coronal geometries proposed in literature. The lamppost corona is a point-like source along the spin axis of the black hole [30]. Such a possibility may be realized in the case the corona is the base of the jet. In the sandwich geometry, the corona would be the atmosphere above the accretion disk [58]. In the cases of spherical or toroidal geometries, the corona would be the accretion flow between the inner edge of the disk and the black hole. In all cases, inverse Compton scattering of thermal photons from the accretion disk off the free electrons in the corona produces a *power-law component* (or *Comptonized component*, blue arrows in Fig. 2.6) with a cut-off energy that depends on the temperature of the corona ($E_{\rm cut} \sim$ 100–1000 keV) [148, 149].

The power-law component from the corona back-illuminates the accretion disk, producing a *reflection component* (green arrows in Fig. 2.6) with some fluorescent emission lines [45]. The strongest feature of the reflection component is usually the

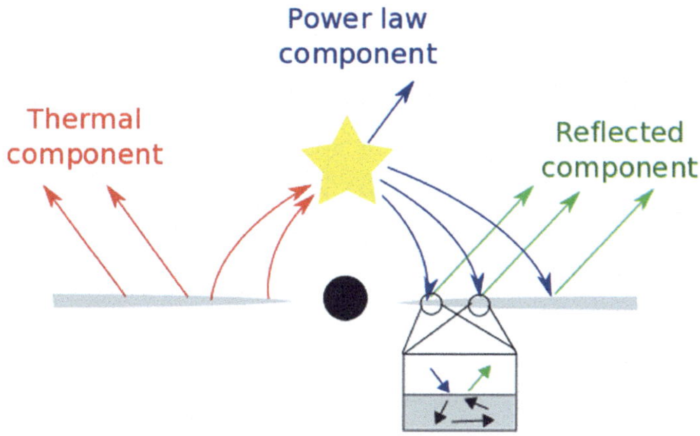

Fig. 2.6 The disk-corona model. The black hole is surrounded by a thin accretion disk with a multi-color blackbody spectrum (red arrows). Some thermal photons from the disk have inverse Compton scattering off free electrons in the corona, producing a power-law component (blue arrows). The latter also illuminates the disk, generating a reflection component (green arrows)

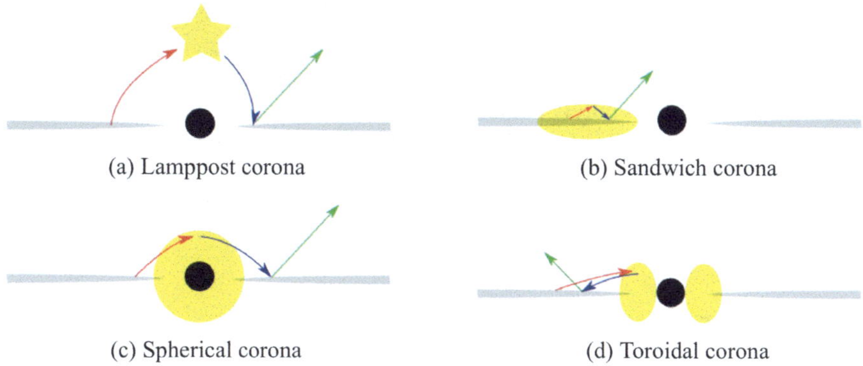

Fig. 2.7 Examples of possible coronal geometries: lamppost geometry (top left panel), sandwich geometry (top right panel), spherical geometry (bottom left panel), and toroidal geometry (bottom right panel)

iron Kα line, which is at 6.4 keV in the case of neutral or weakly ionized iron and shifts up to 6.97 keV for H-like iron ions, and the Compton hump at 10–30 keV. Figure 2.8 shows the resulting spectrum of the disk-corona geometry: we have the thermal component (in red) from the accretion disk, the power-law component (in blue) from the corona, and the reflection component (in green) from the illumination of the accretion disk by the power-law component. In the presence of jets, although the radiation from the jet is mostly in the radio/IR bands due to synchrotron radiation by accelerating particles, it may extend to the X-ray and γ-ray bands and then there is an additional jet component in the X-ray spectrum (see Sect. 2.6 for more details on jets).

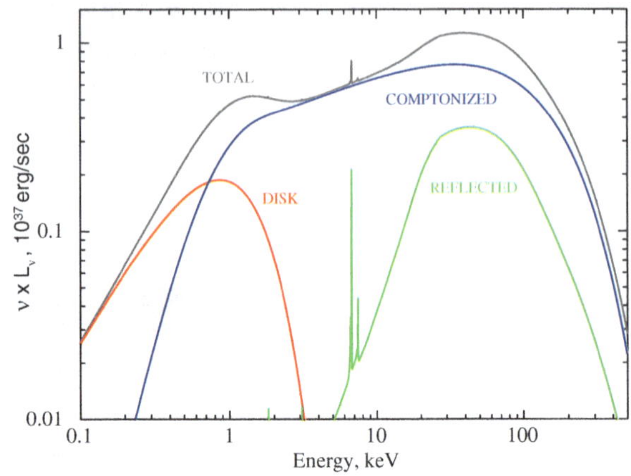

Fig. 2.8 Spectral components of an accreting black hole in the disk-corona model: disk's thermal spectrum (red), power-law component from the corona (blue), and reflection component from the illumination of the disk by the power-law component (green)

2.4 Accreting Black Holes in Nature: Classification

An accreting black hole can be found in different "spectral states", which are characterized by the luminosity of the source and by the relative contribution of its spectral components (thermal, power-law, reflection) [16, 61]. Although the spectral state classification is purely phenomenological, i.e. based on the observed X-ray spectrum, we expect a correlation (not completely understood as of now) to exist between spectral states and accretion flow configurations. The spectral classification of accreting black holes is different for stellar-mass and supermassive black holes.

2.4.1 Stellar-Mass Black Holes

Let us begin with the case of a stellar-mass black hole in an X-ray transient. The object typically spends most of the time in a *quiescent state* with a very low accretion luminosity ($L/L_{\rm Edd} < 10^{-6}$). At a certain point, the source has an *outburst* and becomes a bright X-ray source in the sky ($L/L_{\rm Edd} \sim 10^{-3}$–1). The quiescent state is determined by a very low mass accretion rate, namely a very low amount of material transfers from the companion star to the black hole. When there is a sudden increase of the mass accretion rate (for instance, the companion star inflates and the black hole strips material from the surface of the companion), the outburst happens. The object may be in a quiescent state for several months or even decades. An outburst typically lasts from some days to a few months (roughly the time that the black hole takes to swallow the material that produced the outburst). During an outburst, the spectrum of the source changes. The *hardness-intensity diagram* (HID) [16, 61] is a useful tool for describing the outburst. Figure 2.9 illustrates the typical life-cycle of an outburst. The x-axis is the source hardness, which is the ratio between its luminosity in the hard and soft X-ray bands (e.g., between the luminosity in the 6–10

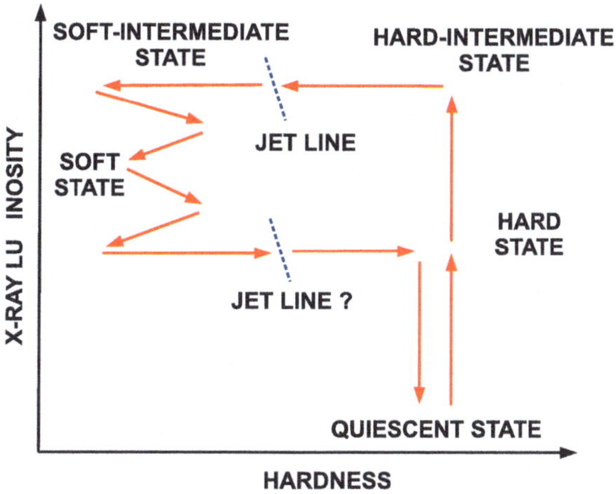

Fig. 2.9 Evolution of the prototype of an outburst in the hardness-intensity diagram. The source is initially in a quiescent state. At the beginning of the outburst, the source enters the hard state, then moves to some intermediate states, to the soft state, and eventually returns to a quiescent state. See the text for more details

and 2–6 keV bands). The y-axis can be any measure of the intensity of the source, e.g., the X-ray luminosity or the number of counts on the instrument. The specific hardness-intensity diagram depends on the source (e.g. the interstellar absorption) and on the instrument (e.g. its effective area at different energies), but, qualitatively it turns out to be very useful for studying transient sources.

The relation between spectral states and accretion flow can be understood by noting that the intensity of the thermal component is mainly determined by the mass accretion rate and the position of the inner edge of the accretion disk, while the contributions of the power-law and reflection components depend on the properties of the corona (its location, extension, geometry, etc.). In particular, the local flux of the disk's thermal component is approximately proportional to the mass accretion rate and the inverse of the cube of the disk's radius, $\mathscr{F}(r) \propto \dot{M}/r^3$ [75]. When the mass accretion rate is low and the inner edge of the disk is at large radii, the thermal component is weak, whereas, when the mass accretion rate is high and the the disk inner edge is close, the thermal component is strong. Similarly, the power-law and the reflection components are strong when the corona is large and close to the disk, whereas they are weak when the corona is small and far away from the disk. The relative contribution of all these components depends on the material around the black hole, and determines the spectral state.

Quiescent state—The source is initially in a quiescent state: the mass accretion rate and the luminosity are very low (the source may in fact be too faint to be detected) and the spectrum is hard. The inner edge of the accretion disk is truncated at a radius significantly larger than the ISCO and the accretion process around the black hole is described by ADAF. In this phase, the low-density accretion flow close to the black hole may act as the corona, which would thus be spherical and large.

Hard state (or corona-dominated state)—At the beginning of the outburst, the spectrum is hard and the source becomes brighter and brighter because the mass accretion rate increases ($L/L_{\rm Edd}$ starts from $\sim 10^{-3}$ and can reach values up to ~ 0.7 in some cases). The spectrum is dominated by the power-law and reflection components. The thermal component is subdominant, and the temperature of the inner part of the disk may be low (~0.1 keV or even lower), but it increases as the luminosity of the source rises. The inner edge of the disk is initially farther away than the ISCO, but it moves to the ISCO as the luminosity increases, and it may reach the ISCO at the end of the hard state. During the hard state, compact mildly relativistic steady jets are common, but the exact mechanism producing these jets is currently unknown. Observations point out a compact corona [33, 123], which may be the base of the jet in a lamppost geometry.

Intermediate states—The mass accretion rate rises, so the contribution of the thermal component increases. The power-law and the reflection components get weaker, probably because of a variation in the geometry/properties of the corona. As a consequence, the source moves to the left part of the HID. We first have the *hard-intermediate state* and then the *soft-intermediate state*. During the transition, transient highly relativistic jets are observed, which is denoted by the *jet line* in Fig. 2.9. If the hardness of the source oscillates near the jet line, we can observe several transient jets.

Soft state (or disk-dominated state)—The thermal spectrum of the disk is the dominant component in the spectrum and the inner part of the disk temperature is around 1 keV. If the luminosity of the source is between ~5% and ~30% of its Eddington luminosity, the disk inner edge is at the ISCO [89], and the accretion disk should be well described by the Novikov–Thorne model [73, 116]. In the soft state, we do not observe any kind of jet.[6] However, strong winds and outflows are common (while they are absent in the hard state). The luminosity of the source may somewhat decrease and the hardness may change, while remaining on the left side of the HID.

At a certain point, the transfer of material decreases, leading to the end of the outburst. The contribution of the thermal spectrum of the disk decreases and, as a consequence, the hardness of the source increases. The source re-enters the soft-intermediate state, the hard-intermediate state, then the hard state, and eventually, when the hardness is high, the luminosity drops down and the source returns to the quiescent state till the next outburst. Between the soft-intermediate and the hard-intermediate states, we may observe transient jets, but the existence of a jet line is not clear here. Every source follows the path shown in Fig. 2.9 counter-clockwise, but there are differences among different sources and even for the same source among different outbursts.

In the case of stellar-mass black holes in persistent X-ray sources, there is no outburst, but we can still use the HID. The most studied source is Cygnus X-1 (the other persistent sources are in nearby galaxies, so they are fainter and more difficult to study). This object spends most of the time in the hard state, but it occasionally moves to a softer state, which is usually interpreted as a soft state. LMC X-1 is always

[6]For instance, in the corona lamppost geometry, the corona may be the base of the jet. This could explain why, in the soft state, we do not see jets and the power-law and reflection components are weak.

in the soft state. LMC X-3 is usually observed in the soft state, rarely in the hard state, and there is no clear evidence that this source can be in an intermediate state.

2.4.2 *Supermassive Black Holes*

In the case of supermassive black holes, there are at least two important differences. First, because the size of the system scales as the mass (e.g., 1 day for a 10 $M_{\odot}$ black hole corresponds to 3,000 years for a 10^7 $M_{\odot}$ black hole) the study of the evolution of a specific system is rendered impossible on human timescales. Second, the temperature of the disk is in the optical/UV range for a supermassive black hole, as compared to stellar-mass black holes where it is the X-ray band. It is possible though to employ the same spectral state classification as above for supermassive black holes (see, for instance, [16] and references therein). Here, we will classify supermassive black holes according to their luminosity and spectral features.

Active galactic nuclei (AGNs) are very bright galactic nuclei, powered by the mass accretion onto their central supermassive black hole. The term AGN is usually used to indicate the same supermassive black hole as well. Figure 2.10 shows the AGN classification, groups and subgroups, and the corresponding fraction of members. While it is thought that most galaxies have a supermassive black hole at their center, only a small fraction of them host an AGN. In most galaxies, the central supermassive object is "dormant", like the supermassive black hole in our Galaxy, Sgr A*, which has a luminosity of the order of 10^{-7} in Eddington units.

Fig. 2.10 Sketch of the AGN family and of its subgroups. This classification has to be taken with caution, because different authors may use slightly different classifications. The diagram shows also the fraction of members in each subgroup. AGNs represent only 0.035% of the galactic nuclei. Most of the AGNs are radio-quiet and belong to the class of Seyfert galaxies

```
Galaxies
 |__Non-active galaxies [93%]
 |__Active galaxies [7%]
     |__LINERs [15%]
     |__Star-forming galaxies [85%]
     |__AGNs [0.5%]
         |__Radio-loud AGNs [10%]
         |   |__Radio galaxies
         |   |__Radio-loud quasars
         |   |__Blazars
         |       |__BL Lac objects
         |       |__OOV quasars
         |__Radio-quiet AGNs [90%]
             |__Seyfert-ELFs
             |__Radio-quiet quasars
             |__NELGs
             |__Seyfert galaxies [95%]
                 |__Seyfert 1s [30%]
                 |__Seyfert 1.5s, 1.8s, 1.9s [10%]
                 |__Seyfert 2s [60%]
```

About 93% of the galaxies are non-active. Among the 7% of the active galaxies, most of them are star-forming galaxies or low-ionization nuclear emission-line regions (LINERs). The latter are sometimes considered AGNs. Proper AGNs are relatively rare: they are in 0.5% of the active galaxies, which means only in 0.035% of all galaxies.

AGNs are mainly classified according to their luminosity and spectral features. It is thus useful to briefly review their possible spectral components:

1. *Radio emission* from jets with the typical spectrum from synchrotron radiation.
2. *IR emission* from the thermal spectrum of the accretion disk, which is reprocessed by gas and dust around the nucleus. This occurs when the accretion disk is obscured by gas and dust.
3. *Optical continuum* mainly from the thermal spectrum of the accretion disk, and in part from possible jets.
4. *Narrow optical lines* from cold material orbiting relatively far from the supermassive black hole. The orbital velocity of this material is 500–1,000 km/s.
5. *Broad optical lines* from cold material orbiting close to the supermassive black hole. The orbital velocity of this material is 1,000–5,000 km/s. The lines are broad due to Doppler boosting.
6. *X-ray continuum* from a hot corona and possible jets.
7. *X-ray lines* from fluorescence emission of the gas in the accretion disk illuminated by the X-ray continuum. The iron Kα line at 6.4 keV is usually one of the most prominent lines.

The AGN classification is sometimes confusing, some objects may not be easily associated to a specific group, and different authors may use different classifications. With reference to Fig. 2.10, we see that AGNs can be grouped into two categories, radio-quiet and radio-loud AGNs. In the radio-quiet AGN category, the jet component is absent or negligible, so the radio luminosity is low. Radio-quiet AGNs may be grouped into four classes: Seyfert extremely luminous far infrared galaxies (Seyfert-ELFs), Seyfert galaxies, narrow emission line galaxies (NELGs), and radio-quiet quasars. The classification is based on a number of properties. For instance, Seyfert galaxies have an optical continuum and emission lines. Seyfert 1s have both narrow and broad emission lines, while Seyfert 2s have only narrow emission lines. Seyfert galaxies of type 1.5, 1.8, and 1.9 are grouped according to their spectral appearance. The radio-loud AGN category has powerful jets, which may be powered by the black hole spin. Radio-loud AGNs can be grouped into three classes: radio galaxies, radio-loud quasars, and blazars. Blazars are characterized by rapid variability and by polarized optical, radio and X-ray emission. They are divided into BL Lacertae objects (BL Lac objects) and optically violent variable quasars (OVV quasars). OVV quasars have stronger broad emission lines than BL Lac objects.

According to the unified AGN model [157], all AGNs are essentially the same kind of objects. The difference appears because they are observed from different viewing angles. Figures 2.11 and 2.12 illustrate the idea of the unified AGN model.

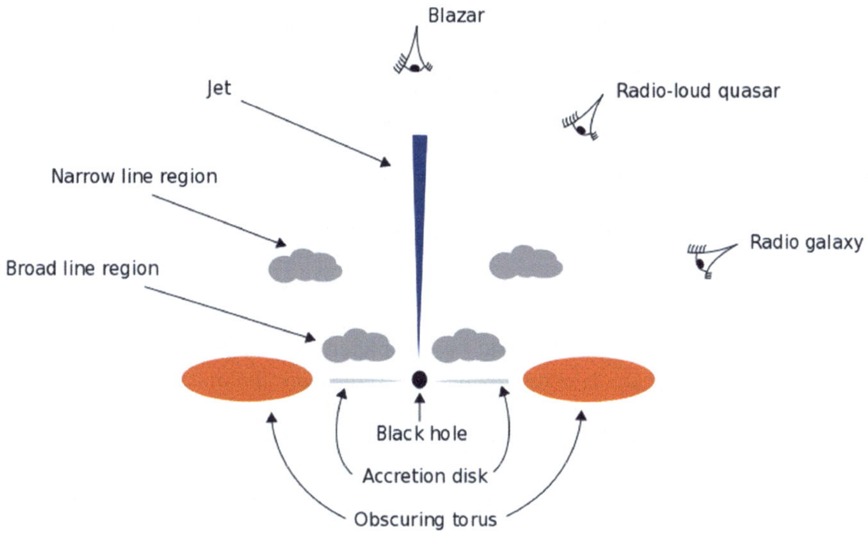

Fig. 2.11 Sketch of a radio-loud AGN according to the unified AGN model [157]. The black hole is surrounded by an accretion disk, which may be obscured by a dusty torus. The broad line region is close to the black hole and there are clouds orbiting with high velocity. The narrow line region is relatively far from the black hole and there are clouds moving at lower velocity. Depending on the angle between the jet and the line of sight of the observer, the AGN can appear as a blazar, as a radio-loud quasar, or as a radio galaxy, as shown in the picture

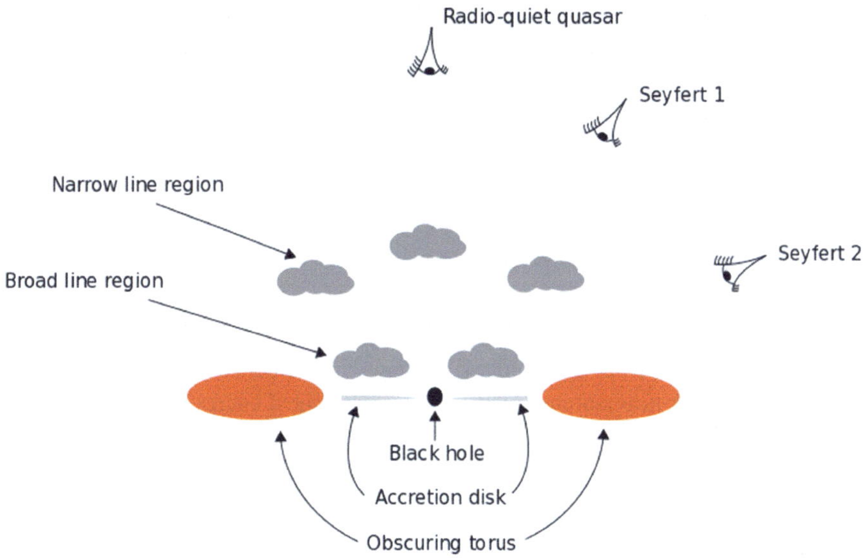

Fig. 2.12 As in Fig. 2.11, in the case of a radio-quiet AGN. Depending on the viewing angle of the observer, the AGN can appear as a radio-quiet quasar, as a Seyfert 1 galaxy, or as a Seyfert 2 galaxy

In the former figure, depending on the viewing angle, we observe a blazar, a radio-loud quasar and then a radio galaxy. In the latter figure, we have a similar situation. Depending on the viewing angle of the observer, the AGN can appear as a radio-quiet quasar, as a Seyfert 1 galaxy, or as a Seyfert 2 galaxy.

2.5 Accreting Black Holes in Nature: Observational Techniques

Any astrophysical black holes should be completely characterized by its mass M and its spin parameter a_*. It is relatively easy to measure the mass of a black hole, by studying the orbital motion of gas or of individual stars around the compact object. Spin measurements are much more challenging. The spin has no "gravitational effects" in Newtonian gravity. This is not the case in general relativity, and the spin alters the gravitational field around a massive body. However, any spin effect is strongly suppressed at larger radii, so black hole spin measurements require techniques that probe the strong gravity region close to the black hole event horizon. As of now, there are two leading techniques to measure black hole spins by studying the X-ray radiation emitted by the gas in the inner part of the accretion disk: the so-called *continuum-fitting method*, which requires analysis of the thermal spectrum of thin accretion disks and is usually applicable to stellar-mass black holes only, and *X-ray reflection spectroscopy* (or *iron line method*), which is based on the study of the disk's reflection spectrum, can be applied to both stellar-mass and supermassive black holes, and is currently the only available method to measure the spins of supermassive black holes. There are a few other proposed techniques for measuring black hole spins with electromagnetic radiation. Among these, the most promising one is probably the detection of *quasi-periodic oscillations* (QPOs).

2.5.1 Continuum-Fitting Method

Within the Novikov–Thorne model, we can derive the time-averaged radial structure of the accretion disk from the fundamental laws of the conservation of rest-mass, energy, and angular momentum [110]. The time-averaged energy flux emitted from the surface of the disk is

$$\mathscr{F}(r) = \frac{\dot{M} c^2}{4\pi r_{\rm g}^2} F(r) \,, \qquad (2.6)$$

where $\dot{M} = dM/dt$ is the time-averaged mass accretion rate, which is independent of the radial coordinate, and $F(r)$ is a dimensionless function of the radial coordinate that becomes roughly of order 1 at the disk inner edge (see [110] for more details). Assuming that the disk is in local thermal equilibrium, its emission is blackbody-like and at any radius we can define an effective temperature $T_{\rm eff}(r)$ from the time-averaged energy flux as $\mathscr{F} = \sigma T^4_{\rm eff}$, where σ is the Stefan-Boltzmann constant. Novikov–Thorne disks with the inner edge at the ISCO radius are realized when the accretion luminosity is between ∼5% and ∼30% of the Eddington limit of the object [89] and this is confirmed by theoretical [73, 116] and observational studies [141]. At lower luminosities, the disk is more likely truncated at a radius larger than the ISCO, and we have an ADAF between the inner edge of the disk and the black hole. At higher luminosities, the gas pressure becomes important, the inner part of the disk is not thin any longer, and the inner edge might be at a radius slightly smaller than the ISCO [7]. Requiring $\dot{M} \sim 0.1\, \dot{M}_{\rm Edd}$ as the condition for Novikov–Thorne disks, we can get a rough estimate of the effective temperature of the inner part of the accretion disk

$$T_{\rm eff} \sim \left(\frac{0.1\, \dot{M}_{\rm Edd} c^2}{4\pi\sigma r_{\rm g}^2} \right)^{1/4} \sim \left(\frac{10\, M_\odot}{M} \right)^{1/4} {\rm keV} \, , \tag{2.7}$$

and we can see that the disk's thermal spectrum is in the soft X-ray band for stellar-mass black holes and in the optical/UV band for the supermassive ones.

The continuum-fitting method is the analysis of the thermal spectrum of geometrically thin and optically thick accretion disks of black holes in order to measure the black hole spin parameter a_* [77, 88, 90, 178]. The technique is normally used for stellar-mass black holes only, because the spectrum of supermassive black holes is in the optical/UV band where dust absorption limits the capability of accurate measurements. The model describing the thermal spectrum of an accretion disk around a Kerr black hole depends on five parameters: the black hole mass M, the mass accretion rate $\dot{M}$, the inclination angle of the disk with respect to the line of sight of the observer i, the distance of the source from the observer D, and the black hole spin parameter a_*. The impact of these five parameters on the shape of the spectrum is illustrated in Fig. 2.13. Note that it is impossible to infer all the five model parameters from the data of the spectrum of a thin disk, because the spectrum's shape is too simple and there is a degeneracy. However, if we can get independent measurements of M, D, and i, usually from optical observations, it is possible to fit the data and measure a_* and $\dot{M}$. Presently, there are about ten stellar-mass black holes with a spin measurement obtained from the continuum-fitting method, see Table 2.2.

2.5.2 X-Ray Reflection Spectroscopy

X-ray reflection spectroscopy (or the iron line method) refers to the study of the reflection component. This technique can be applied to both stellar-mass and supermassive black holes and is currently the only available method to measure the spin of supermassive black holes [21, 127].

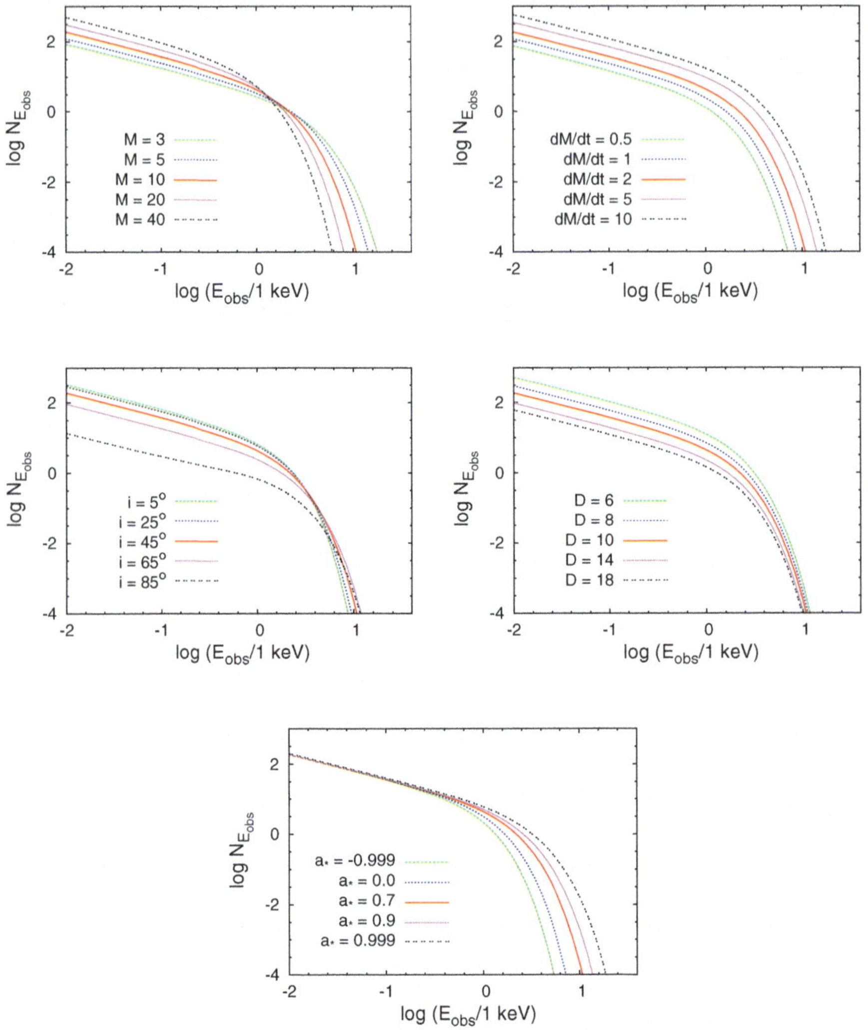

Fig. 2.13 Impact of the model parameters on the thermal spectrum of a thin disk: mass M (top left panel), mass accretion rate $\dot{M}$ (top right panel), viewing angle i (central left panel), distance D (central right panel), and spin parameter a_* (bottom panel). When not shown, the values of the parameters are: $M = 10\ M_\odot$, $\dot{M} = 2 \cdot 10^{18}$ g s^{-1}, $D = 10$ kiloparsec, $i = 45°$, and $a_* = 0.7$. M in units of $M_\odot$, $\dot{M}$ in units of 10^{18} g s^{-1}, D in kiloparsec, and flux density $N_{E_{\rm obs}}$ in photons keV^{-1} cm^{-2} s^{-1}

Table 2.2 Summary of the continuum-fitting and iron line measurements of the spin parameter of stellar-mass black holes. See the references in the last column for more details

BH binary	a_* (Continuum)	a_* (Iron)	Principal references
4U 1630-472	–	$0.985^{+0.005}_{-0.014}$	[70]
GRS 1915+105	>0.98	0.98 ± 0.01	[89, 94]
Cygnus X-1	>0.98	>0.95	[34, 55, 56, 112, 156, 168]
GS 1354-645	–	>0.98	[31]
MAXI J1535-571	–	>0.98	[96, 173]
Swift J1658.2	–	>0.96	[174]
LMC X-1	0.92 ± 0.06	$0.97^{+0.02}_{-0.25}$	[53, 145]
GX 339-4	<0.9	0.95 ± 0.03	[46, 71, 113, 122]
V404 Cyg	–	>0.92	[169]
GRS 1716-249	–	>0.92	[151]
XTE J1752-223	–	0.92 ± 0.06	[47, 125]
Swift J174540.2	–	>0.9	[100]
MAXI J1836-194	–	0.88 ± 0.03	[124]
XTE J1650-500	–	$0.84 \sim 0.98$	[166]
M33 X-7	0.84 ± 0.05	–	[78]
4U 1543-47	$0.80 \pm 0.10^{\star}$	–	[132]
GRS 1739-278	–	0.8 ± 0.2	[95]
IC10 X-1	$\gtrsim 0.7$	–	[143]
Swift J1753.5	–	$0.76^{+0.11}_{-0.15}$	[121]
GRO J1655-40	$0.70 \pm 0.10^{\star}$	>0.9	[121, 132]
GS 1124-683	$0.63^{+0.16}_{-0.19}$	–	[26]
XTE J1652-453	–	<0.5	[27]
XTE J1550-564	0.34 ± 0.28	$0.55^{+0.15}_{-0.22}$	[144]
LMC X-3	0.25 ± 0.15	–	[140]
H1743-322	0.2 ± 0.3	–	[142]
A0620-00	0.12 ± 0.19	–	[54]
XMMU J004243.6	< -0.2	–	[93]

Note *These sources were studied in an early work of the continuum-fitting method, within a more simple model, and therefore the published 1-σ error estimates are doubled following [88]

The most prominent feature of the reflection spectrum is usually the iron Kα line.[7] This is because iron is more abundant than other heavy elements (the iron-56 nucleus is more tightly bound than lighter and heavier elements, so it is the final product of

[7] A Kα line results from the transition of an electron from a p orbit of the L shell (quantum number 2) to the K shell (quantum number 1). The line is actually a doublet with slightly different energies, Kα_1 and Kα_2, respectively for the transitions $2p_{1/2} \rightarrow 1s$ and $2p_{3/2} \rightarrow 1s$ using the atomic notation. A Kβ line results from the transition of an electron from a p orbit of the M shell (quantum number 3) to the K shell. An Lα line is emitted from transition of an electron from a d orbit of the M shell to a p orbit of the L shell.

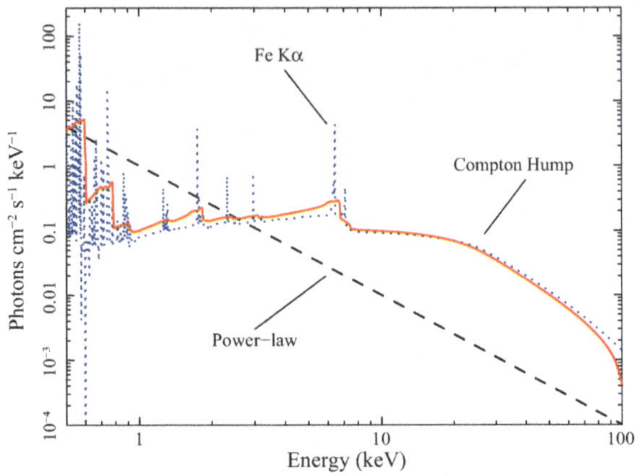

Fig. 2.14 Reflection spectrum for a neutral accretion disk irradiated by a power-law continuum with photon index $\Gamma = 2$. The dashed black line indicates the power-law continuum from the corona; the dotted blue curve is for the reflection spectrum at the emission point in the rest-frame of the gas (only atomic physics is involved); the solid red curve is for the reflection spectrum of a non-rotating black hole at the detection point and is blurred by relativistic effects (gravitational redshift, Doppler boosting, light bending). From [21], reproduced with permission

nuclear reactions) and the probability of fluorescent line emission is also high (scaling as Z^4, where Z is the atomic number). Moreover, X-ray detectors typically have high sensitivity around 6 keV and there are no other atomic lines around this energy. The iron Kα line is a very narrow feature in the rest-frame of the emitter, while the one observed in the reflection spectrum of black holes can be very broad and skewed, as the result of relativistic effects occurring in the strong gravity region of the object (gravitational redshift, Doppler boosting, light bending) [11, 21, 32, 127]. While the iron Kα line is usually the strongest feature, accurate measurements of black hole spins require a fit to the whole reflection spectrum, not just the iron line. Figure 2.14 shows the reflection spectrum for a neutral accretion disk: the unblurred reflection spectrum in the rest-frame of the gas is the dotted blue curve while the solid red curve is the blurred reflection spectrum of the accretion disk around a Schwarzschild black hole as detected by a distant observer.

Reflection models describing the reflection component of accretion disks around Kerr black holes depend on several parameters: the black holes spin a_*, the inner edge of the disk $R_{\rm in}$ (which may or may not be assumed at the ISCO radius), the outer edge of the disk $R_{\rm out}$, the inclination angle of the disk i, the metallicity (or the iron abundance), the ionization of the disk, and some parameters related to the emissivity profile of the disk. The latter is quite a crucial ingredient and depends on the geometry of the corona, which is currently unknown. A phenomenological approach is to model the emissivity profile with a power-law (is $I \propto 1/r^q$ where q is the emissivity index) or with a broken power-law ($I \propto 1/r^{q_{\rm in}}$ for $r < R_{\rm br}$, $I \propto 1/r^{q_{\rm out}}$ for $r > R_{\rm br}$). In the case of supermassive black holes, it is often necessary to take the cosmological redshift z into account. For stellar-mass black holes, their relative

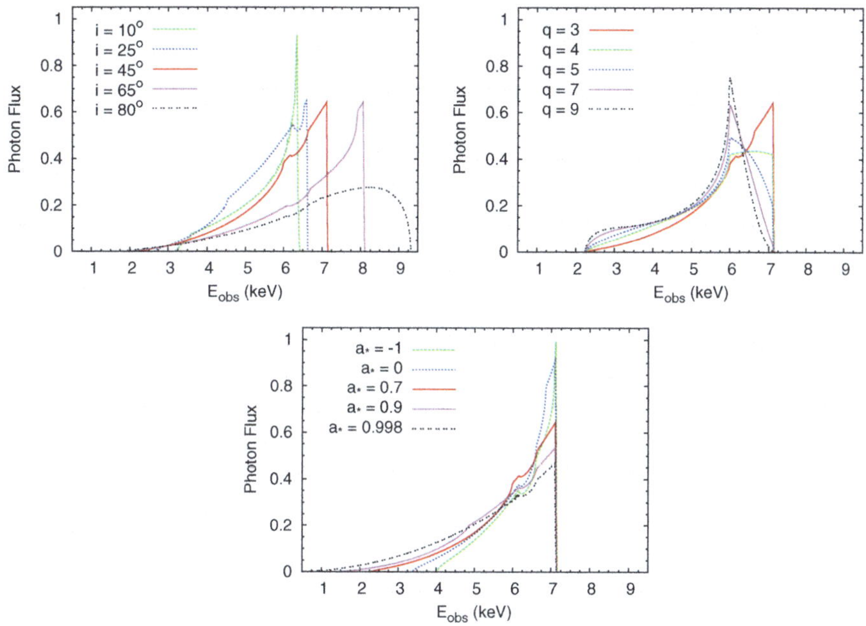

Fig. 2.15 Impact of the the inclination angle of the disk i, the emissivity index q, and the spin parameter a_* on the shape of an iron line at 6.4 keV emitted from a thin accretion disk. When not shown, the values of the parameters are: $i = 45°$, $q = 3$, $a_* = 0.7$, and $R_{\rm out} = 400\, r_{\rm g}$. From [11], reproduced with permission

motion in the Galaxy is of order 100 km/s and the redshift can be ignored. Figure 2.15 shows the impact of the the inclination angle of the disk i, the emissivity index q (assuming an emissivity profile described by a simple power-law $I \propto 1/r^q$), and the spin parameter a_* on the shape of an iron line at 6.4 keV emitted from a thin accretion disk around a black hole. A significant advantage of the iron line method is that it does not require independent measurements of the black hole mass M, the distance D, and the inclination angle of the disk i, three quantities that are required in the continuum-fitting method, are usually difficult to measure, and have large uncertainty. The reflection spectrum is independent of the former two, and can directly measure the inclination angle of the disk.

Current spin measurements of stellar-mass black holes with the iron line method are summarized in the third column in Table 2.2 (see the corresponding references in the fourth column for more details). Note that some black holes have their spin measured with both the continuum-fitting and the iron line methods. In general, the two measurements agree (GRS 1915+105, Cygnus X-1, LMC X-1, XTE J1550-564). For GX 339-4 and GRO J1655-40, the two measurements are not consistent. There can be a few reasons for this discrepancy. The iron line method is usually applied when the source is in the hard state, when the reflection spectrum is stronger, but the disk may be truncated at a radius larger than the ISCO. This would lead to an underestimation of the black hole spin, but since the iron line method provides spin values higher than the continuum-fitting method in the case of GX 339-4 and GRO J1655-40, this cannot be the reason for the discrepancy. Rather, since the continuum-fitting method crucially depends on independent measurements of the black hole mass M, the distance D, and the inclination angle of the disk i, large systematic uncertainties in these measurements may cause the continuum-fitting method to deviate. For instance, in the case of GRO J1655-40 there are a few mass measurements reported in the literature, but they are not consistent amongst each other.

A summary of spin measurements of supermassive black holes with the iron line method is reported in Table 2.3 (see the references in the last column for more details and the lists of spin measurements in [21, 127, 161] for a few more sources with a constrained spin). Note the very high spin of several objects. In part, this can be explained noting that fast-rotating black holes are brighter and thus the spin measurement is easier. If these measurements are correct, they would point out that these

Table 2.3 Summary of spin measurements of supermassive black holes reported in the literature. See the references in the last column for more details

Object	a_* (Iron)	Principal references
IRAS 13224-3809	>0.99	[167]
Mrk 110	>0.99	[167]
NGC 4051	>0.99	[115]
1H0707-495	>0.98	[167, 180]
RBS 1124	>0.98	[167]
NGC 3783	>0.98	[24]
Fairall 9	$0.973^{+0.003}_{-0.003}$	[81]
NGC 1365	$0.97^{+0.01}_{-0.04}$	[23, 130]
Swift J0501-3239	>0.96	[167]
PDS 456	>0.96	[167]
Ark 564	$0.96^{+0.01}_{-0.06}$	[167]
3C120	>0.95	[80]
Mrk 79	>0.95	[43]
NGC 5506	$0.93^{+0.04}_{-0.04}$	[147]
MCG-6-30-15	$0.91^{+0.06}_{-0.07}$	[22, 84]
Ton S180	$0.91^{+0.02}_{-0.09}$	[167]
1H0419-577	>0.88	[167]
IRAS 00521-7054	>0.84	[150]
Mrk 335	$0.83^{+0.10}_{-0.13}$	[111, 167]
Ark 120	$0.81^{+0.10}_{-0.18}$	[106, 167]
Swift J2127+5654	$0.6^{+0.2}_{-0.2}$	[97]
Mrk 841	>0.56	[167]

objects have been spun up by prolonged disk accretion and therefore would provide information about galaxy evolutions (see the discussion in Sect. 2.2.3). However, the very high spin measurements have to be taken with some caution, as they may be affected by large systematic uncertainties in the model employed to infer the black hole spin. For example, if the mass accretion rate is near the Eddington limit, which is probably the case for several sources, the spin parameter can be easily overestimated if we employ a model that assumes a thin disk [129]. More details on the possible interpretation of current spin measurements of supermassive black holes can be found in [127].

2.5.3 *Quasi Periodic Oscillations*

Quasi-periodic oscillations (QPOs) are a common feature in the X-ray power density spectrum of neutron stars and stellar-mass black holes [160]. The power density spectrum $P(\nu)$ is the square of the Fourier transform of the photon count rate $C(t)$. If we use the Leahy normalization, we have

$$P(\nu) = \frac{2}{N} \left| \int_0^T C(t) e^{-2\pi i \nu t} dt \right|^2 , \tag{2.8}$$

where N is the total number of counts and T is the duration of the observation. QPOs are narrow features in the X-ray power density spectrum of a source. Figure 2.16 shows the power density spectrum obtained from an observation of the stellar-mass black hole XTE J1550-564, where we can see a QPO around 5 Hz, one at 13 Hz, and one at 183 Hz in the inset.

In the case of black hole binaries, QPOs can be grouped into two classes: low-frequency QPOs (0.1–30 Hz) and high-frequency QPOs (40–450 Hz). The exact nature of these QPOs is currently unknown, but there are several proposals in the literature. In most scenarios, the frequencies of the QPOs are related to the fundamental frequencies of a particle orbiting the black hole [5, 6, 146]:

1. *Orbital frequency* ν_ϕ, which is the inverse of the orbital period.
2. *Radial epicyclic frequency* ν_r, which is the frequency of radial oscillations around the mean orbit.
3. *Vertical epicyclic frequency* ν_θ, which is the frequency of vertical oscillations around the mean orbit.

In the Kerr metric, we have a compact analytic form for the expression of these frequencies

$$\nu_\phi = \frac{c}{2\pi} \sqrt{\frac{r_g}{r^3}} \left[1 \pm a_* \left(\frac{r_g}{r} \right)^{3/2} \right]^{-1} , \tag{2.9}$$

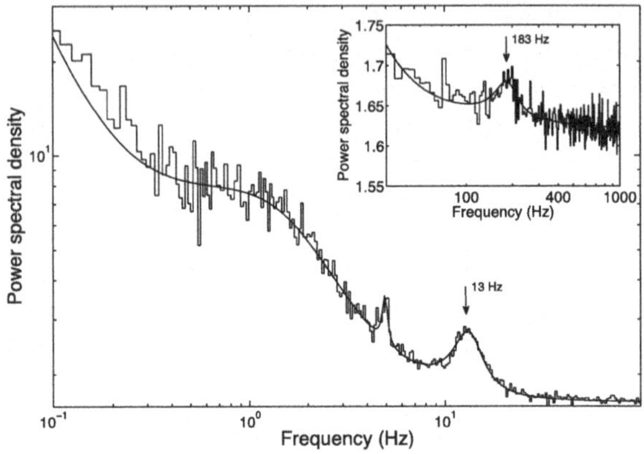

Fig. 2.16 Power density spectrum from an observation of XTE J1550-564. We see a QPO around 5 Hz, a QPO at 13 Hz (marked by an arrow), and a QPO at 183 Hz in the inset (marked by an arrow). Figure 2.1 from [102], reproduced by permission of Oxford University Press

$$\nu_r = \nu_\phi \sqrt{1 - 6\frac{r_g}{r} \pm 8a_* \left(\frac{r_g}{r}\right)^{3/2} - 3a_*^2 \left(\frac{r_g}{r}\right)^2}\,, \tag{2.10}$$

$$\nu_\theta = \nu_\phi \sqrt{1 \mp 4a_* \left(\frac{r_g}{r}\right)^{3/2} + 3a_*^2 \left(\frac{r_g}{r}\right)^2}\,, \tag{2.11}$$

where r is the orbital radius in Boyer–Lindquist coordinates. To have an idea of the order of magnitude of these frequencies, we can write the orbital frequency for a Schwarzschild black hole

$$\nu_\phi(a_* = 0) = 220 \left(\frac{10\,M_\odot}{M}\right) \left(\frac{6\,r_g}{r}\right)^{3/2} \text{Hz}\,. \tag{2.12}$$

High-frequency QPOs at 40–450 Hz are thus of the right magnitude to be associated to the orbital frequencies near the ISCO radius of stellar-mass black holes. Interestingly, we also have evidence of high-frequency QPOs in supermassive black holes (<1 mHz) [52] and intermediate-mass black holes (~1 Hz) [114].

Since it is often possible to measure the frequencies of QPOs with quite a good precision, if we knew the exact relation between QPOs and fundamental frequencies, it would be possible to measure black hole spins with high precision. For instance, in [101] the authors interpret the observed QPOs of the black hole binary GRO J1655-40 within the relativistic precession model and obtain the mass measurement $M/M_\odot = 5.31 \pm 0.07$ and the spin measurement $a_* = 0.290 \pm 0.003$.

2.5.4 Direct Imaging

Since the black hole does not allow any light to come out from inside the event horizon, and the disk outside this region is radiating, an interesting possibility is to observe the *black hole shadow*. Depending on the geometry of the accretion disk and

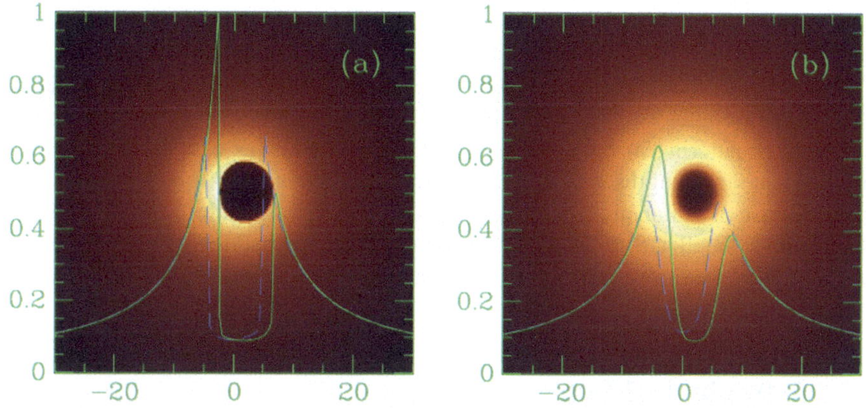

Fig. 2.17 Direct image of a black hole surrounded by an optically thin emitting medium with the characteristics of that of Sgr A*. The black hole spin parameter is $a_* = 0.998$ and the viewing angle is $i = 45°$. Panel *a*: image from ray-tracing calculations. Panel *b*: image from a simulated observation of an idealized VLBI experiment at 0.6 mm wavelength taking interstellar scattering into account. The solid green curve and the dashed purple curve show, respectively, the intensity variations of the image along the x-axis and the y-axis. From [35]. ©AAS. Reproduced with permission

on its optical properties (thin/thick), if we could image the accretion flow around a black hole with a resolution of at least some gravitational radii, we would observe a dark area in the middle of a brighter surrounding. The dark area is usually referred to as the black hole shadow (see Fig. 2.17). The shape of the shadow is determined by the bending of light in the strong gravity region [12].

Very long baseline interferometry (VLBI) uses several radio telescopes located in various continents and combines the data to mimic a single telescope of the size of the Earth. This helps achieving much smaller angular resolutions than a single telescope. The Event Horizon Telescope (EHT)[8] is an international collaboration that uses mm and sub-mm VLBI techniques to image supermassive black holes. They released the image of the supermassive black hole at the center of the galaxy M87 in April 2019 [9] and work is underway to get a similar image of SgrA*, the supermassive black hole at the center of our Galaxy.

The mass of Sgr A* is about $4 \cdot 10^6\, M_\odot$ and it is at $d \approx 8$ kiloparsec from us, so its angular size in the sky is roughly

$$\theta \sim \frac{r_g}{d} \sim 0.05 \text{ milliarcseconds}\,. \tag{2.13}$$

There are three particular conditions that make the observation of the shadow of Sgr A* achievable. (i) The angular resolution of VLBI facilities scales as λ/D, where λ is the electromagnetic radiation wavelength and D is the distance among

[8]http://www.eventhorizontelescope.org/.

different stations. For $\lambda < 1\,\text{mm}$ and stations located in different continents ($D > 10^3\,\text{km}$), it is possible to reach an angular resolution of 0.1 milliarcseconds. (*ii*) The emitting medium around the black hole at the center of the Galaxy is optically thick at wavelengths $\lambda > 1\,\text{mm}$, but becomes optically thin for $\lambda < 1\,\text{mm}$. (*iii*) The interstellar scattering at the center of our Galaxy dominates over intrinsic source structures at wavelengths $\lambda > 1\,\text{mm}$, but becomes subdominant for $\lambda < 1\,\text{mm}$.

In the case of stellar-mass black holes in our Galaxy, the angular size is 4–5 orders of magnitude smaller. Similar angular resolutions are impossible today, but they may be possible in the future with X-ray interferometric techniques [158, 171].

2.6 Astrophysical Jets

A very exciting phenomenon observed in nature are astrophysical jets. Jets are collimated streams of matter emerging from a extraterrestrial object. They are usually highly ionized and relativistic, and beamed in the direction of the rotation axis. They are a common feature of several astrophysical objects, including protostars, stars, neutron stars, and black holes. Jets are observed both from stellar-mass black holes in X-ray binaries and supermassive black holes in galactic nuclei, see e.g. [37, 98, 177].

2.6.1 Theory of Jets

The two most popular mechanisms for the formation of black hole jets are the Blandford–Znajek model [19] and the Blandford–Payne model [18], both with a number of variants and extensions. There are also proposals of hybrid models, in which the two mechanisms can coexist [92].

In the *Blandford–Znajek scenario*, magnetic fields thread the black hole horizon and can extract the rotational energy of the compact object via some version of the Penrose process [19, 117]. This mechanism exploits the existence of the ergoregion. However, strictly speaking, the extraction of the rotational energy of a compact object may be possible even in the case of neutron stars in the presence of magnetic fields anchored on the surface of the body. The paper by Blandford and Znajek derived the jet power $P_{\rm BZ}$ perturbatively, for slowly rotating black holes. In that case, one finds $P_{\rm BZ} \propto a_*^2$. A more detailed analysis provides the following formula [152]

$$P_{\rm BZ} = \frac{\kappa}{16\pi} \Phi_{\rm B}^2 \Omega_{\rm H}^2 f(\Omega_{\rm H}) \, , \tag{2.14}$$

where κ is a constant that depends on the magnetic field configuration, $\Omega_{\rm H}$ is the angular frequency at the black hole horizon and reads

$$\Omega_{\rm H} = \frac{c a_*}{2 r_{\rm H}} = \frac{c}{2 r_{\rm g}} \frac{a_*}{1 + \sqrt{1 - a_*^2}} \, , \tag{2.15}$$

$\Phi_{\rm B}$ is the magnetic flux threading the black hole horizon, and $f(\Omega_{\rm H})$ is a dimensionless function that takes into account higher order terms in $\Omega_{\rm H}$

$$f(\Omega_{\rm H}) \approx 1 + c_1 \Omega_{\rm H}^2 + c_2 \Omega_{\rm H}^4 + \cdots \, , \tag{2.16}$$

where $\{c_i\}$ are numerical coefficients and this last formula assumes units in which $M = c = 1$. For example, for a black hole with a thin accretion disk, $c_1 = 1.38$ and $c_2 = -9.2$ [152].

In the *Blandford–Payne model*, magnetic fields thread the accretion disk, corrotating with it [18]. Now the energy is provided by the gravitational potential energy of the accretion flow. The power of the jet can be written as

$$P_{\rm BZ} \sim \varepsilon \, L \, \ln \left(\frac{r_{\rm out}}{r_{\rm in}} \right) \, , \tag{2.17}$$

where ε is the efficiency of the transformation of the binding energy of the accreting matter into jet power at the inner radius of the disk $r_{\rm in}$, $r_{\rm out}$ is the outer radius of the disk, and L is the accretion luminosity.

2.6.2 Observations of Jets

We will discuss the observational aspects of astrophysical jets in the cases of black hole binaries and active galactic nuclei separately.

In the case of black hole binaries, we observe two kinds of jets [37]. *Steady jets* manifest when a source is in the hard state. The jet is steady, typically not very relativistic, and may extend up to a few tens of astronomical units (roughly 150 million kilometers). *Transient jets* are instead observed when a source switches from the hard to the soft state and crosses the "jet line" (see Fig. 2.9 and Sect. 2.4.1). These parsec-scale jets appears as blobs of plasma emitting mainly in the radio band, and are relativistic. They have features similar to the kiloparsec-scale jets observed in AGNs and for this reason the black hole binaries producing transient jets are also called microquasars [98].

If the mechanism responsible for the formation of jets were the Blandford–Znajek model, one may expect a correlation between black hole spin measurements and estimates of the jet power. Such a correlation has been found in some studies [103], while other studies did not find any correlation [38]. Presently, this is a controversial issue [88, 131]. Both studies are based on a small number of data with large uncertainty. Future observations are expected to provide a conclusive answer to this issue [88].

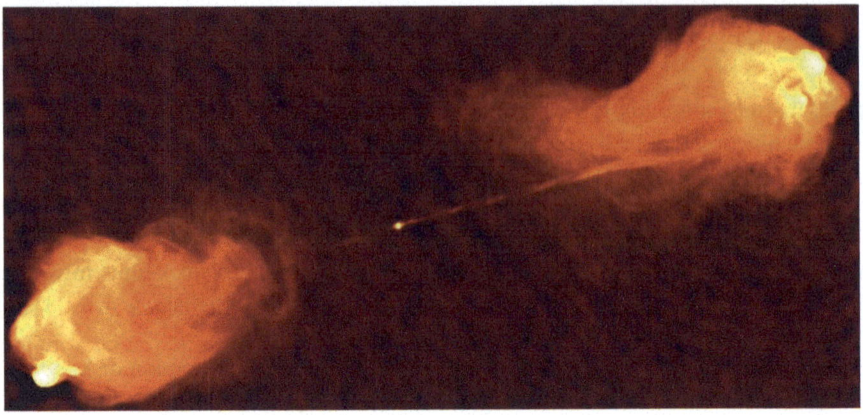

Fig. 2.18 Radio image of Cygnus A. The bright dot at the center is the location of the supermassive black hole, where the two relativistic jets are generated. The jets are stopped by the intergalactic medium, forming two giant lobes. Image courtesy of NRAO/AUI

In the case of AGNs, only a small fraction of them, around 10%, exhibit relativistic, kiloparsec-scale jets. One of the most spectacular examples is Cygnus A (see Fig. 2.18). Radio images of this object show two highly collimated jets from the very center of the galaxy, where its supermassive black hole is supposed to be located. The two jets extend well outside the galaxy, for hundreds of kiloparsec. For AGNs with an accretion luminosity above 1% of the Eddington limit, the most natural interpretation is that their jets are the counterpart of the transient jets in black hole binaries. This conclusion may be supported by the consideration that microquasars show intermittent jets for a few percent of the time, which is similar to the fraction of radio-loud AGNs [107]. The time scale of these systems is proportional to their mass, so intermittent jets in black hole binaries look like persistent jets in AGNs. However, in the case of AGNs with a luminosity below 1% of the Eddington limit, this explanation cannot work: black hole binaries with a low accretion luminosity are all radio-loud.

Jets dominate the spectrum of AGNs at radio frequencies. There are apparently two distinct populations of AGNs: radio-loud AGNs and radio-quiet AGNs (see Figs. 2.11 and 2.12). This classification is particularly evident when the optical luminosity and radio luminosity are plotted. For the same optical luminosity, radio-loud AGNs have a radio luminosity 3–4 orders of magnitude higher than that of radio-quiet AGNs. These two populations seem to follow two different tracks with a gap between them. The origin of this radio-quiet/radio-loud dichotomy is not understood [136]. One popular interpretation is that the dichotomy is determined by the black hole spin. When the accretion luminosity is low, it turns out that radio-loud AGNs are in elliptical galaxies, while radio-quiet AGNs are mainly in spiral galaxies. Galaxies with different morphology have likely a different merger and accretion history. This, in turn, may have produced two populations of black holes, with high and low values of spin [165]. A difference in radio luminosity of 3–4 orders of magnitude between the two populations is impossible to explain if the jet power is proportional to $\Omega_{\rm H}^2$, but in the case of thick disks the jet power may scale as $\Omega_{\rm H}^6$ [152]. Another interpretation is to doubt the existence of this dichotomy, attributing it to observational bias [137].

If jets are powered by the rotational energy of the accreting compact object, it is possible to extract energy and have an accretion efficiency $\eta > 1$. Some observations indicate that some AGNs may have $\eta > 1$ [51, 91]. If these measurements are correct, the jet is extracting energy from the system, and it is likely that this is the rotational energy of the black hole; some version of the Blandford–Znajek mechanism is working. While in the past general-relativistic magnetohydrodynamic simulations have not been able to find high accretion efficiency from jets, more recent simulations have achieved $\eta > 1$ [153].

References

1. B.P. Abbott et al., LIGO Scientific and Virgo Collaborations. Phys. Rev. Lett. **116**, 061102 (2016). arXiv:1602.03837 [gr-qc]
2. M.A. Abramowicz, M. Calvani, L. Nobili, Astrophys. J. **242**, 772 (1980)
3. M.A. Abramowicz, B. Czerny, J.P. Lasota, E. Szuszkiewicz, Astrophys. J. **332**, 646 (1988)
4. M.A. Abramowicz, P.C. Fragile, Living Rev. Rel. **16**, 1 (2013). arXiv:1104.5499 [astro-ph.HE]
5. M.A. Abramowicz, W. Kluzniak, Astron. Astrophys. **374**, L19 (2001). arXiv:astro-ph/0105077
6. M.A. Abramowicz, W. Kluzniak, Z. Stuchlik, G. Torok, Astron. Astrophys. **436**, 1 (2005). arXiv:astro-ph/0401464
7. M.A. Abramowicz, J.P. Lasota, Acta Astron. **30**, 35 (1980)
8. E. Agol, M. Kamionkowski, L.V.E. Koopmans, R.D. Blandford, Astrophys. J. **576**, L131 (2002). arXiv:astro-ph/0203257
9. K. Akiyama et al., Event horizon telescope collaboration. Astrophys. J. **875**, L1 (2019). arXiv:1906.11238 [astro-ph.GA]
10. M. Bachetti et al., Nature **514**, 202 arXiv:1410.3590 [astro-ph.HE]
11. C. Bambi, *Black Holes: A Laboratory for Testing Strong Gravity* (Springer, Singapore, 2017). https://doi.org/10.1007/978-981-10-4524-0
12. J.M. Bardeen, Timelike and null geodesics in the Kerr metric, in *"Black Holes"*, ed. C. DeWitt, B.S. DeWitt (Gordon & Breach, 1973), pp. 215–239
13. J.M. Bardeen, J.A. Petterson, Astrophys. J. **195**, L65 (1975)
14. J.E. Barnes, L. Hernquist, Astrophys. J. **471**, 115 (1996)
15. K. Belczynski, T. Bulik, C.L. Fryer, A. Ruiter, J.S. Vink, J.R. Hurley, Astrophys. J. **714**, 1217 (2010). arXiv:0904.2784 [astro-ph.SR]
16. T.M. Belloni, Lect. Notes Phys. **794**, 53 (2010). arXiv:0909.2474 [astro-ph.HE]
17. E. Berti, M. Volonteri, Astrophys. J. **684**, 822 (2008). arXiv:0802.0025 [astro-ph]
18. R.D. Blandford, D.G. Payne, Mon. Not. Roy. Astron. Soc. **199**, 883 (1982)
19. R.D. Blandford, R.L. Znajek, Mon. Not. Roy. Astron. Soc. **179**, 433 (1977)
20. H. Bondi, Mon. Not. Roy. Astron. Soc. **112**, 195 (1952)
21. L. Brenneman, *Measuring Supermassive Black Hole Spins in Active Galactic Nuclei* (Springer, New York, 2013). arXiv:1309.6334 [astro-ph.HE]
22. L.W. Brenneman, C.S. Reynolds, Astrophys. J. **652**, 1028 (2006). arXiv:astro-ph/0608502
23. L.W. Brenneman, G. Risaliti, M. Elvis, E. Nardini, Mon. Not. Roy. Astron. Soc. **429**, 2662 (2013). arXiv:1212.0772 [astro-ph.HE]
24. L.W. Brenneman et al., Astrophys. J. **736**, 103 (2011). arXiv:1104.1172 [astro-ph.HE]

25. J. Casares, P.G. Jonker, Space Sci. Rev. **183**, 223 (2014). arXiv:1311.5118 [astro-ph.HE]
26. Z. Chen, L. Gou, J.E. McClintock, J.F. Steiner, J. Wu, W. Xu, J. Orosz, Y. Xiang,arXiv:1601.00615 [astro-ph.HE]
27. C.Y. Chiang, R.C. Reis, D.J. Walton, A.C. Fabian, Mon. Not. Roy. Astron. Soc. **425**, 2436 (2012). arXiv:1207.0682 [astro-ph.HE]
28. E.J.M. Colbert, R.F. Mushotzky, Astrophys. J. **519**, 89 (1999). arXiv:astro-ph/9901023
29. M. Coleman Miller, E.J.M. Colbert, Int. J. Mod. Phys. D **13**, 1 (2004). arXiv:astro-ph/0308402]
30. T. Dauser, J. Garcia, J. Wilms, M. Bock, L.W. Brenneman, M. Falanga, K. Fukumura, C.S. Reynolds, Mon. Not. Roy. Astron. Soc. **430**, 1694 (2013). arXiv:1301.4922 [astro-ph.HE]
31. A.M. El-Batal et al., Astrophys. J. **826**, L12 (2016). arXiv:1607.00343 [astro-ph.HE]
32. A.C. Fabian, K. Iwasawa, C.S. Reynolds, A.J. Young, Publ. Astron. Soc. Pac. **112**, 1145 (2000). arXiv:astro-ph/0004366
33. A.C. Fabian, A. Lohfink, E. Kara, M.L. Parker, R. Vasudevan, C.S. Reynolds, Mon. Not. Roy. Astron. Soc. **451**, 4375 (2015). arXiv:1505.07603 [astro-ph.HE]
34. A.C. Fabian et al., Mon. Not. Roy. Astron. Soc. **424**, 217 (2012). arXiv:1204.5854 [astro-ph.HE]
35. H. Falcke, F. Melia, E. Agol, Astrophys. J. **528**, L13 (2000). [astro-ph/9912263]
36. W.M. Farr, N. Sravan, A. Cantrell, L. Kreidberg, C.D. Bailyn, I. Mandel, V. Kalogera, Astrophys. J. **741**, 103 (2011). arXiv:1011.1459 [astro-ph.GA]
37. R.P. Fender, T.M. Belloni, E. Gallo, Mon. Not. Roy. Astron. Soc. **355**, 1105 (2004). arXiv:astro-ph/0409360
38. R. Fender, E. Gallo, D. Russell, Mon. Not. Roy. Astron. Soc. **406**, 1425 (2010). arXiv:1003.5516 [astro-ph.HE]
39. L. Ferrarese et al., Astrophys. J. **644**, L21 (2006). arXiv:astro-ph/0603840
40. P.C. Fragile, O.M. Blaes, P. Anninois, J.D. Salmonson, Astrophys. J. **668**, 417 (2007). arXiv:0706.4303 [astro-ph]
41. T. Fragos, M. Tremmel, E. Rantsiou, K. Belczynski, Astrophys. J. **719**, L79 (2010). arXiv:1001.1107 [astro-ph.HE]
42. T. Fragos, J.E. McClintock, Astrophys. J. **800**, 17 (2015). arXiv:1408.2661 [astro-ph.HE]
43. L.C. Gallo, G. Miniutti, J.M. Miller, L.W. Brenneman, A.C. Fabian, M. Guainazzi, C.S. Reynolds, Mon. Not. Roy. Astron. Soc. **411**, 607 (2011). arXiv:1009.2987 [astro-ph.HE]
44. E. Gallo, T. Treu, J. Jacob, J.H. Woo, P. Marshall, R. Antonucci, Astrophys. J. **680**, 154 (2008). arXiv:0711.2073 [astro-ph]
45. J. Garcia, T. Dauser, C.S. Reynolds, T.R. Kallman, J.E. McClintock, J. Wilms, W. Eikmann, Astrophys. J. **768**, 146 (2013). arXiv:1303.2112 [astro-ph.HE]
46. J. Garcia et al., Astrophys. J. **813**, 84 (2015). arXiv:1505.03607 [astro-ph.HE]
47. J.A. Garcia et al., Astrophys. J. **864**, 25 (2018). arXiv:1807.01949 [astro-ph.HE]
48. K. Gebhardt, R.M. Rich, L.C. Ho, Astrophys. J. **578**, L41 (2002). arXiv:astro-ph/0209313
49. K. Gebhardt, R.M. Rich, L.C. Ho, Astrophys. J. **634**, 1093 (2005). arXiv:astro-ph/0508251
50. A.M. Ghez, S. Salim, S.D. Hornstein, A. Tanner, M. Morris, E.E. Becklin, G. Duchene, Astrophys. J. **620**, 744 (2005). arXiv:astro-ph/0306130
51. G. Ghisellini, F. Tavecchio, L. Foschini, G. Ghirlanda, L. Maraschi, A. Celotti, Mon. Not. Roy. Astron. Soc. **402**, 497 (2010). arXiv:0909.0932 [astro-ph.CO]
52. M. Gierlinski, M. Middleton, M. Ward, C. Done, Nature **455**, 369 (2008). arXiv:0807.1899 [astro-ph]
53. L. Gou, J.E. McClintock, J. Liu, R. Narayan, J.F. Steiner, R.A. Remillard, J.A. Orosz, S.W. Davis, Astrophys. J. **701**, 1076 (2009). arXiv:0901.0920 [astro-ph.HE]
54. L. Gou, J.E. McClintock, J.F. Steiner, R. Narayan, A.G. Cantrell, C.D. Bailyn, J.A. Orosz, Astrophys. J. **718**, L122 (2010). arXiv:1002.2211 [astro-ph.HE]
55. L. Gou et al., Astrophys. J. **742**, 85 (2011). arXiv:1106.3690 [astro-ph.HE]
56. L. Gou et al., Astrophys. J. **790**, 29 (2014). arXiv:1308.4760 [astro-ph.HE]
57. F. Haardt, L. Maraschi, Astrophys. J. **380**, L51 (1991)
58. F. Haardt, L. Maraschi, Astrophys. J. **413**, 507 (1993)

59. A. Heger, C.L. Fryer, S.E. Woosley, N. Langer, D.H. Hartmann, Astrophys. J. **591**, 288 (2003). arXiv:astro-ph/0212469
60. A. Heger, S.E. Woosley, Astrophys. J. **567**, 532 (2002). arXiv:astro-ph/0107037
61. J. Homan, T. Belloni, Astrophys. Space Sci. **300**, 107 (2005). arXiv:astro-ph/0412597
62. S.A. Hughes, R.D. Blandford, Astrophys. J. **585**, L101 (2003). arXiv:astro-ph/0208484
63. A. Ingram, C. Done, P.C. Fragile, Mon. Not. Roy. Astron. Soc. **397**, L101 (2009)
64. M. Jaroszynski, M.A. Abramowicz, B. Paczynski, Acta Astron. **30**, 1 (1980)
65. V. Kalogera, G. Baym, Astrophys. J. **470**, L61 (1996)
66. P.D. Kiel, J.R. Hurley, Mon. Not. Roy. Astron. Soc. **369**, 1152 (2006). arXiv:astro-ph/0605080
67. A.R. King, U. Kolb, Mon. Not. Roy. Astron. Soc. **305**, 654 (1999). arXiv:astro-ph/9901296
68. A.R. King, S.H. Lubow, G.I. Ogilvie, J.E. Pringle, Mon. Not. Roy. Astron. Soc. **363**, 49 (2005). arXiv:astro-ph/0507098
69. A.R. King, J.E. Pringle, Mon. Not. Roy. Astron. Soc. **373**, L93 (2006). arXiv:astro-ph/0609598
70. A.L. King et al., Astrophys. J. **784**, L2 (2014). arXiv:1401.3646 [astro-ph.HE]
71. M. Kolehmainen, C. Done, Mon. Not. Roy. Astron. Soc. **406**, 2206 (2010). arXiv:0911.3281 [astro-ph.HE]
72. J. Kormendy, D. Richstone, Ann. Rev. Astron. Astrophys. **33**, 581 (1995)
73. A.K. Kulkarni et al., Mon. Not. Roy. Astron. Soc. **414**, 1183 (2011). arXiv:1102.0010 [astro-ph.HE]
74. S. Kumar, J.E. Pringle, Mon. Not. Roy. Astron. Soc. **213**, 435 (1985)
75. J.P. Lasota, Black hole accretion discs. https://doi.org/10.1007/978-3-662-52859-4_1, arXiv:1505.02172 [astro-ph.HE]
76. J.M. Lattimer, Ann. Rev. Nucl. Part. Sci. **62**, 485 (2012). arXiv:1305.3510 [nucl-th]
77. L.X. Li, E.R. Zimmerman, R. Narayan, J.E. McClintock, Astrophys. J. Suppl. **157**, 335 (2005). arXiv:astro-ph/0411583
78. J. Liu, J. McClintock, R. Narayan, S. Davis and J. Orosz, Astrophys. J. **679**, L37 (2008) [Erratum: Astrophys. J. **719**, L109 (2010)] arXiv:0803.1834 [astro-ph]
79. G. Lodato, J.E. Pringle, Mon. Not. Roy. Astron. Soc. **368**, 1196 (2006). arXiv:astro-ph/0602306
80. A.M. Lohfink et al., Astrophys. J. **772**, 83 (2013). arXiv:1305.4937 [astro-ph.HE]
81. A. Lohfink et al., Astrophys. J. **821**, 11 (2016). arXiv:1602.05589 [astro-ph.GA]
82. P. Madau, F. Haardt, M. Dotti, Astrophys. J. **784**, L38 (2014). arXiv:1402.6995 [astro-ph.CO]
83. E. Maoz, Astrophys. J. **494**, L181 (1998). arXiv:astro-ph/9710309
84. A. Marinucci et al., Astrophys. J. **787**, 83 (2014). arXiv:1404.3561 [astro-ph.HE]
85. R.G. Martin, J.E. Pringle, C.A. Tout, Mon. Not. Roy. Astron. Soc. **381**, 1617 (2007). arXiv:0708.2034 [astro-ph]
86. R.G. Martin, C.A. Tout, J.E. Pringle, Mon. Not. Roy. Astron. Soc. **387**, 188 (2008). arXiv:0802.3912 [astro-ph]
87. L. Mayer, S. Kazantzidis, P. Madau, M. Colpi, T.R. Quinn, J. Wadsley, Science **316**, 1874 (2007). arXiv:0706.1562 [astro-ph]
88. J.E. McClintock, R. Narayan, J.F. Steiner, Space Sci. Rev. **183**, 295 (2014). arXiv:1303.1583 [astro-ph.HE]
89. J.E. McClintock, R. Shafee, R. Narayan, R.A. Remillard, S.W. Davis, L.X. Li, Astrophys. J. **652**, 518 (2006). arXiv:astro-ph/0606076
90. J.E. McClintock et al., Class. Quant. Grav. **28**, 114009 (2011). arXiv:1101.0811 [astro-ph.HE]
91. B.R. McNamara, M. Rohanizadegan, P.E.J. Nulsen, Astrophys. J. **727**, 39 (2011). arXiv:1007.1227 [astro-ph.CO]
92. D.L. Meier, Astrophys. J. **548**, L9 (2001). arXiv:astro-ph/0010231
93. M. Middleton, J. Miller-Jones, R. Fender, Mon. Not. Roy. Astron. Soc. **439**, 1740 (2014). arXiv:1401.1829 [astro-ph.HE]
94. J.M. Miller et al., Astrophys. J. **775**, L45 (2013). arXiv:1308.4669 [astro-ph.HE]
95. J.M. Miller et al., Astrophys. J. **799**, L6 (2015). arXiv:1411.1921 [astro-ph.HE]
96. J.M. Miller et al., Astrophys. J. **860**, L28 (2018). arXiv:1806.04115 [astro-ph.HE]

97. G. Miniutti, F. Panessa, A. De Rosa, A.C. Fabian, A. Malizia, M. Molina, J.M. Miller, S. Vaughan, Mon. Not. Roy. Astron. Soc. **398**, 255 (2009). arXiv:0905.2891 [astro-ph.HE]
98. I.F. Mirabel, L.F. Rodriguez, Ann. Rev. Astron. Astrophys. **37**, 409 (1999). arXiv:astro-ph/9902062
99. K. Mitsuda et al., Publ. Astron. Soc. Jap. **36**, 741 (1984)
100. K. Mori et al., arXiv:1910.03459 [astro-ph.HE]
101. S.E. Motta, T.M. Belloni, L. Stella, T. Muñoz-Darias, R. Fender, Mon. Not. Roy. Astron. Soc. **437**, 2554 (2014). arXiv:1309.3652 [astro-ph.HE]
102. S.E. Motta, T. Munoz-Darias, A. Sanna, R. Fender, T. Belloni, L. Stella, Mon. Not. Roy. Astron. Soc. **439**, 65 (2014). arXiv:1312.3114 [astro-ph.HE]
103. R. Narayan, J.E. McClintock, Mon. Not. Roy. Astron. Soc. **419**, L69 (2012). arXiv:1112.0569 [astro-ph.HE]
104. R. Narayan, I. Yi, Astrophys. J. **428**, L13 (1994). arXiv:astro-ph/9403052
105. R. Narayan, I. Yi, Astrophys. J. **452**, 710 (1995). arXiv:astro-ph/9411059
106. E. Nardini, A.C. Fabian, R.C. Reis, D.J. Walton, Mon. Not. Roy. Astron. Soc. **410**, 1251 (2011). arXiv:1008.2157 [astro-ph.HE]
107. C. Nipoti, K.M. Blundell, J. Binney, Mon. Not. Roy. Astron. Soc. **361**, 633 (2005). arXiv:astro-ph/0505280
108. I.D. Novikov, K.S. Thorne, Astrophysics and black holes, in *Black Holes*, ed. by C. De Witt, B. De Witt (Gordon and Breach, New York, 1973)
109. K. Ohsuga, S. Mineshige, Astrophys. J. **736**, 2 (2011). arXiv:1105.5474 [astro-ph.HE]
110. D.N. Page, K.S. Thorne, Astrophys. J. **191**, 499 (1974)
111. M.L. Parker et al., Mon. Not. Roy. Astron. Soc. **443**(2), 1723 (2014). arXiv:1407.8223 [astro-ph.HE]
112. M.L. Parker et al., Astrophys. J. **808**, 9 (2015). arXiv:1506.00007 [astro-ph.HE]
113. M.L. Parker et al., Astrophys. J. **821**, L6 (2016). arXiv:1603.03777 [astro-ph.HE]
114. D.R. Pasham, T.E. Strohmayer, R.F. Mushotzky, Nature **513**, 74 (2014). arXiv:1501.03180 [astro-ph.HE]
115. A.R. Patrick, J.N. Reeves, D. Porquet, A.G. Markowitz, V. Braito, A.P. Lobban, Mon. Not. Roy. Astron. Soc. **426**, 2522 (2012). arXiv:1208.1150 [astro-ph.HE]
116. R.F. Penna, J.C. McKinney, R. Narayan, A. Tchekhovskoy, R. Shafee, J.E. McClintock, Mon. Not. Roy. Astron. Soc. **408**, 752 (2010). arXiv:1003.0966 [astro-ph.HE]
117. R. Penrose, Riv. Nuovo Cim. **1**, 252 (1969) [Gen. Rel. Grav. **34**, 1141 (2002)]
118. M. Postman et al., Astrophys. J. **756**, 159 (2012). arXiv:1205.3839 [astro-ph.CO]
119. Y. Qin, P. Marchant, T. Fragos, G. Meynet, V. Kalogera, arXiv:1810.13016 [astro-ph.SR]
120. M.J. Reid, J.E. McClintock, J.F. Steiner, D. Steeghs, R.A. Remillard, V. Dhawan, R. Narayan, Astrophys. J. **796**, 2 (2014). arXiv:1409.2453 [astro-ph.GA]
121. R.C. Reis, A.C. Fabian, R.R. Ross, J.M. Miller, Mon. Not. Roy. Astron. Soc. **395**, 1257 (2009)
122. R.C. Reis, A.C. Fabian, R. Ross, G. Miniutti, J.M. Miller, C. Reynolds, Mon. Not. Roy. Astron. Soc. **387**, 1489 (2008). arXiv:0804.0238 [astro-ph]
123. R.C. Reis, J.M. Miller, Astrophys. J. **769**, L7 (2013). arXiv:1304.4947 [astro-ph.HE]
124. R.C. Reis, J.M. Miller, M.T. Reynolds, A.C. Fabian, D.J. Walton, Astrophys. J. **751**, 34 (2012). arXiv:1111.6665 [astro-ph.HE]
125. R.C. Reis et al., Mon. Not. Roy. Astron. Soc. **410**, 2497 (2011). arXiv:1009.1154 [astro-ph.HE]
126. R.A. Remillard, J.E. McClintock, Ann. Rev. Astron. Astrophys. **44**, 49 (2006). arXiv:astro-ph/0606352
127. C.S. Reynolds, Space Sci. Rev. **183**, 277 (2014). arXiv:1302.3260 [astro-ph.HE]
128. C.E. Rhoades, R. Ruffini, Phys. Rev. Lett. **32**, 324 (1974)
129. S. Riaz, D. Ayzenberg, C. Bambi, S. Nampalliwar, arXiv:1911.06605 [astro-ph.HE]
130. G. Risaliti et al., Nature **494**, 449 (2013). arXiv:1302.7002 [astro-ph.HE]
131. D.M. Russell, E. Gallo, R.P. Fender, Mon. Not. Roy. Astron. Soc. **431**, 405 (2013). arXiv:1301.6771 [astro-ph.HE]

132. R. Shafee, J.E. McClintock, R. Narayan, S.W. Davis, L.X. Li, R.A. Remillard, Astrophys. J. **636**, L113 (2006). arXiv:astro-ph/0508302
133. N.I. Shakura, R.A. Sunyaev, Astron. Astrophys. **24**, 337 (1973)
134. S.L. Shapiro, S.A. Teukolsky, *Black Holes, White Dwarfs, and Neutron Stars: The Physics of Compact Objects* (Wiley-VCH, New York, 1983)
135. I. Shlosman, M.C. Begelman, J. Frank, Nature **345**, 679 (1990)
136. M. Sikora, L. Stawarz, J.P. Lasota, Astrophys. J. **658**, 815 (2007). arXiv:astro-ph/0604095
137. J. Singal, V. Petrosian, A. Lawrence, L. Stawarz, Astrophys. J. **743**, 104 (2011). arXiv:1101.2930 [astro-ph.CO]
138. M. Spera, M. Mapelli, A. Bressan, Mon. Not. Roy. Astron. Soc. **451**(4), 4086 (2015). arXiv:1505.05201 [astro-ph.SR]
139. J.F. Steiner, J.E. McClintock, Astrophys. J. **745**, 136 (2012). arXiv:1110.6849 [astro-ph.HE]
140. J.F. Steiner, J.E. McClintock, J.A. Orosz, R.A. Remillard, C.D. Bailyn, M. Kolehmainen, O. Straub, Astrophys. J. **793**, L29 (2014). arXiv:1402.0148 [astro-ph.HE]
141. J.F. Steiner, J.E. McClintock, R.A. Remillard, L. Gou, S. Yamada, R. Narayan, Astrophys. J. **718**, L117 (2010). arXiv:1006.5729 [astro-ph.HE]
142. J.F. Steiner, J.E. McClintock, M.J. Reid, Astrophys. J. **745**, L7 (2012). arXiv:1111.2388 [astro-ph.HE]
143. J.F. Steiner, D.J. Walton, J.A. Garcia, J.E. McClintock, S.G.T. Laycock, M.J. Middleton, R. Barnard, K.K. Madsen, arXiv:1512.03414 [astro-ph.HE]
144. J.F. Steiner et al., Mon. Not. Roy. Astron. Soc. **416**, 941 (2011). arXiv:1010.1013 [astro-ph.HE]
145. J.F. Steiner et al., Mon. Not. Roy. Astron. Soc. **427**, 2552 (2012). arXiv:1209.3269 [astro-ph.HE]
146. L. Stella, M. Vietri, S. Morsink, Astrophys. J. **524**, L63 (1999). arXiv:astro-ph/9907346
147. S. Sun, M. Guainazzi, Q. Ni, J. Wang, C. Qian, F. Shi, Y. Wang, C. Bambi, Mon. Not. Roy. Astron. Soc. **478**, 1900 (2018). arXiv:1704.03716 [astro-ph.HE]
148. R.A. Sunyaev, J. Truemper, Nature **279**, 506 (1979)
149. R.A. Sunyaev, L.G. Titarchuk, Astron. Astrophys. **86**, 121 (1980)
150. Y. Tan, J. Wang, X. Shu, Y. Zhou, Astrophys. J. **747**, L11 (2012). arXiv:1202.0400 [astro-ph.HE]
151. L. Tao, J.A. Tomsick, J. Qu, S. Zhang, S. Zhang, Q. Bu, arXiv:1910.11979 [astro-ph.HE]
152. A. Tchekhovskoy, R. Narayan, J.C. McKinney, Astrophys. J. **711**, 50 (2010). arXiv:0911.2228 [astro-ph.HE]
153. A. Tchekhovskoy, R. Narayan, J.C. McKinney, Mon. Not. Roy. Astron. Soc. **418**, L79 (2011). arXiv:1108.0412 [astro-ph.HE]
154. K.S. Thorne, Astrophys. J. **191**, 507 (1974)
155. F.X. Timmes, S.E. Woosley, T.A. Weaver, Astrophys. J. **457**, 834 (1996). arXiv:astro-ph/9510136
156. J.A. Tomsick et al., Astrophys. J. **780**, 78 (2014). arXiv:1310.3830 [astro-ph.HE]
157. C.M. Urry, P. Padovani, Publ. Astron. Soc. Pac. **107**, 803 (1995). arXiv:astro-ph/9506063
158. P. Uttley et al., arXiv:1908.03144 [astro-ph.HE]
159. E.P.J. van den Heuvel, Endpoints of stellar evolution: The incidence of stellar mass black holes in the galaxy, in *"Environment Observation And Climate Modelling Through International Space Projects"*, vol. 29 (1992)
160. M. van der Klis, arXiv:astro-ph/0410551
161. R.V. Vasudevan, A.C. Fabian, C.S. Reynolds, J. Aird, T. Dauser, L.C. Gallo, Mon. Not. Roy. Astron. Soc. **458**, 2012 (2016). arXiv:1506.01027 [astro-ph.HE]
162. M. Volonteri, Astron. Astrophys. Rev. **18**, 279 (2010). arXiv:1003.4404 [astro-ph.CO]
163. M. Volonteri, F. Haardt, K. Gultekin, Mon. Not. Roy. Astron. Soc. **384**, 1387 (2008). arXiv:0710.5770 [astro-ph]
164. M. Volonteri, G. Lodato, P. Natarajan, Mon. Not. Roy. Astron. Soc. **383**, 1079 (2008). arXiv:0709.0529 [astro-ph]

165. M. Volonteri, M. Sikora, J.P. Lasota, Astrophys. J. **667**, 704 (2007). arXiv:0706.3900 [astro-ph]
166. D.J. Walton, R.C. Reis, E.M. Cackett, A.C. Fabian, J.M. Miller, Mon. Not. Roy. Astron. Soc. **422**, 2510 (2012). arXiv:1202.5193 [astro-ph.HE]
167. D.J. Walton, E. Nardini, A.C. Fabian, L.C. Gallo, R.C. Reis, "Mon. Not. Roy. Astron. Soc. **428**, 2901 (2013). arXiv:1210.4593 [astro-ph.HE]
168. D.J. Walton et al., Astrophys. J. **826**, 87 (2016). arXiv:1605.03966 [astro-ph.HE]
169. D.J. Walton et al., Astrophys. J. **839**, 110 (2017). arXiv:1609.01293 [astro-ph.HE]
170. J.M. Wang, Y.M. Chen, L.C. Ho, R.J. McLure, Astrophys. J. **642**, L111 (2006). arXiv:astro-ph/0603813
171. N. White, Nature **407**, 146 (2000)
172. S.E. Woosley, J.S. Bloom, Ann. Rev. Astron. Astrophys. **44**, 507 (2006). arXiv:astro-ph/0609142
173. Y. Xu et al., Astrophys. J. **852**, L34 (2018). arXiv:1711.01346 [astro-ph.HE]
174. Y. Xu et al., arXiv:1805.07705 [astro-ph.HE]
175. S.C. Yoon, N. Langer, C. Norman, Astron. Astrophys. **460**, 199 (2006). arXiv:astro-ph/0606637
176. L.R. Yungelson, J.-P. Lasota, G. Nelemans, G. Dubus, E.P.J. van den Heuvel, J. Dewi, S. Portegies Zwart, Astron. Astrophys. **454**, 559 (2006). arXiv:astro-ph/0604434
177. J.A. Zensus, Ann. Rev. Astron. Astrophys. **35**, 607 (1997)
178. S.N. Zhang, W. Cui, W. Chen, Astrophys. J. **482**, L155 (1997). arXiv:astro-ph/9704072
179. V.V. Zhuravlev, P.B. Ivanov, P.C. Fragile, D.M. Teixeira, Astrophys. J. **796**, 104 (2014). arXiv:1406.5515 [astro-ph.HE]
180. A. Zoghbi, A. Fabian, P. Uttley, G. Miniutti, L. Gallo, C. Reynolds, J. Miller, G. Ponti, Mon. Not. Roy. Astron. Soc. **401**, 2419 (2010). arXiv:0910.0367 [astro-ph.HE]

Chapter 3
How to Detect X-Rays and Gamma-Rays from Space: Optics and Detectors

Valentina Fioretti and Andrea Bulgarelli

3.1 Observing the Universe in Energy, Space, and Time

The observation techniques at high energies can be based on direct photo-detection or indirect reconstruction. In photo-detection systems the photon is totally or in part absorbed in the material and its energy frees an atomic electron. The ionized electron collides with other atoms and generates a cascade of secondary electrons, then collected with direct (e.g. proportional counters) or indirect (e.g. scintillators) methods to produce an electric signal. The signal is proportional to the energy amount deposited in the sensitive region of the detector. Photo-detectors can be used as stand-alone imaging and/or spectrometry instruments from X-ray to soft gamma-ray energies (<hundreds of keV) depending on the material's photo-electric attenuation efficiency and thickness. At higher energies, where only a fraction of the photon energy is deposited, they still can be used as stand-alone rate-meters (e.g. the AGILE/MCAL scintillator operating as gamma-ray burst detector up to 50 MeV [1]).

The use of combined photon and particle detectors allows for the exploitation of Compton scattering or pair production interactions to build gamma-ray telescopes based on indirect reconstruction of the photon properties. Their sensitivity ranges from hundreds of keV to hundreds of GeV, well above the domain of stand-alone photo-detectors.

V. Fioretti (✉) · A. Bulgarelli
INAF OAS Bologna, Via P. Gobetti 93/3, 40129 Bologna, Italy
e-mail: valentina.fioretti@inaf.it

A. Bulgarelli
e-mail: andrea.bulgarelli@inaf.it

C. Bambi (ed.), *Tutorial Guide to X-ray and Gamma-ray Astronomy*,
https://doi.org/10.1007/978-981-15-6337-9_3

3.1.1 Principles of Radiation Interaction with Matter

The main physics processes behind the radiation interaction with matter, from X-rays to gamma-rays, are:

- photoelectric absorption (until few hundreds of keV);
- Compton scattering (from around 100 keV up to 30 MeV);
- pair production (>few MeVs).

The cross-section of each physics process expresses its probability of occurrence in terms of the area of interaction and the properties of the material. Depending on the science goals of the mission, the observations will be focused in a certain energy band: this is the first driver in detector selection since it defines the main interaction process to be used.

3.1.1.1 Photoelectric Absorption

The interaction occurs between the X-ray photon and the atoms in the material. In photoelectric absorption, or photo-ionization, the photon transfers all of its energy to an electron in one of the atomic shells, that is then freed from its atomic bond (Fig. 3.1). The kinetic energy of the resulting free electron E_{e^-} (the photo-electron) is given by the difference between the photon energy E_γ and the electron binding energy E_b:

$$\mathrm{E_{e^-}} = \mathrm{E}_\gamma - \mathrm{E_b} \ . \tag{3.1}$$

This interaction is possible only when the photon has sufficient energy to overcome the binding energy and remove the electron from the atom. Photo-electrons rapidly lose their energy in the surrounding matter resulting in a deposit not far from the photoelectric interaction, hence within the detector material.

Most X-ray photo-absorptions occur in the inner K or L shells. Once a vacancy is created, an electron from the outer shells moves to fill in the vacancy. This process is followed by one of the following phenomena:

- the emission of an X-ray fluorescence photon, with energy equal to the difference between the vacancy site and the binding energy of the outer shell electron filling the vacancy;
- the emission of an electron (Auger electron) from an outer shell, with kinetic energy given by the difference between the energy released in the vacancy filling and the bonding energy of the outer shell.

The probability of the X-ray fluorescence or Auger electron emission (the fluorescence yield) depends on the detector atomic number (Fig. 3.2). Photoelectric absorption is the dominant process at low energies but its cross section σ_{ph} increases rapidly with atomic number (Z):

$$\sigma_{\mathrm{ph}} \propto \frac{\mathrm{Z^n}}{E_\gamma^3}, \tag{3.2}$$

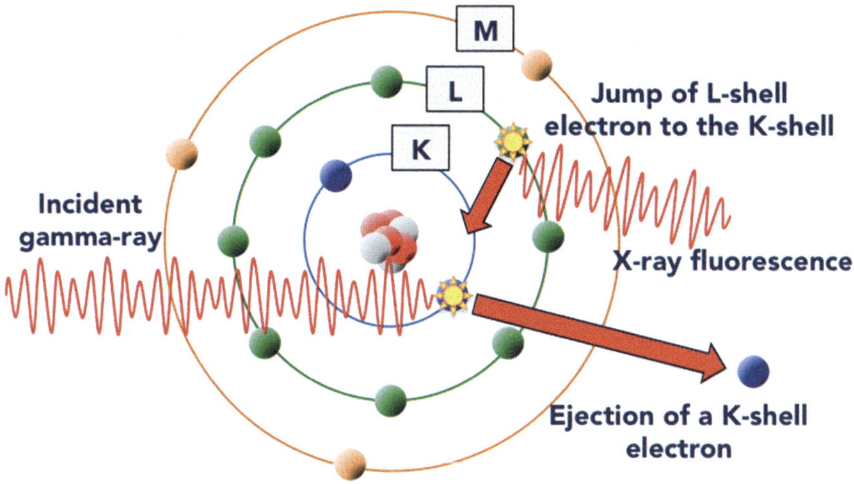

Fig. 3.1 Schematic view of photoelectric absorption and emission of an X-ray fluorescence photon

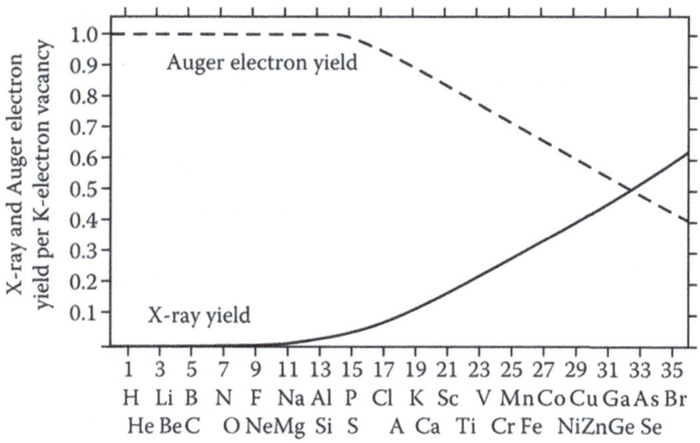

Fig. 3.2 X-ray fluorescence and Auger electron yields per K-shell electron vacancies as functions of the atomic number [2]

where n varies between 4 and 5. The cross section is also characterized by discontinuities (absorption edges) as thresholds for photo-ionization, or for production of photo-electrons, of various atomic levels are reached. Each edge occurs when the photon energy is equal to the binding energy of electrons in the K, L, M, etc., shells of the absorbing elements.

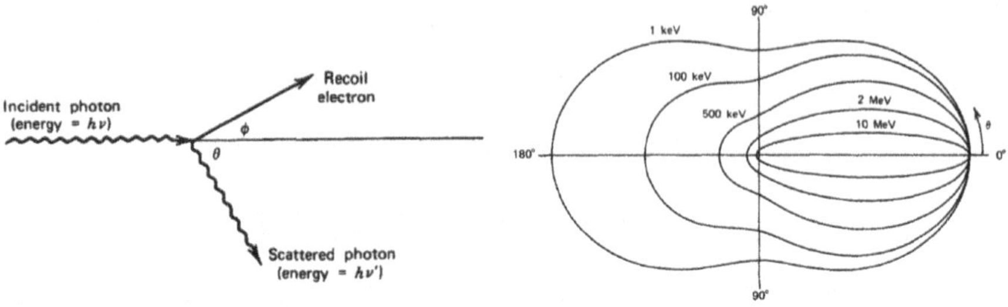

Fig. 3.3 *Left panel:* schema for the Compton scattering interaction. *Right panel*: Klein–Nishina cross-section as a function of the scattering angle θ for different energies (photon incident from the left) [3]

3.1.1.2 Scattering

In the Thomson (or Rayleigh) scattering the X-ray photon interacts with the whole atom and its direction is changed, but not its energy nor the energy of the atom. The probability of coherent scattering is much higher than Compton scattering at low energies (<tens of keV) but still orders of magnitude lower than photoelectric absorption, which remains the dominant process in the soft X-ray band.

From hundreds of keV up to several tens of MeV, Compton scattering becomes the dominant interaction mechanism. X-rays at these energies can scatter off atomic electrons during their passage through matter and a part of the photon energy is transferred to the recoil electron (Fig. 3.3, left panel). The photon is deflected by an angle θ from its path. If the electron is assumed to be free and at rest, the photon energy after scattering (E'_{γ}) as a function of the scattering angle and the initial energy (E_{γ}) can be derived from the energy and momentum conservation laws:

$$E'_{\gamma} = \frac{E_{\gamma}}{1 + \alpha(1 - \cos\theta)} \; , \tag{3.3}$$

where $\alpha = E_{\gamma}/\mathrm{m_0c^2}$ and $\mathrm{m_0c^2}$ is the rest energy of the electron (511 keV). The differential cross-section for Compton scattering is described by the Klein–Nishina formula (for unbound electrons) and it predicts the angular distribution of photons after scattering (Fig. 3.3, right panel):

$$\frac{d\sigma_c}{d\Omega} = \mathrm{Z}r_0^2 \left(\frac{1}{1+\alpha(1-\cos\theta)}\right)^2 \left(\frac{1+\cos^2\theta}{2}\right)\left(1 + \frac{\alpha^2(1-\cos\theta)^2}{(1+\cos^2\theta)[1+\alpha(1-\cos\theta)]}\right) , \tag{3.4}$$

where r_0 is the classical electron radius.

When a photon Compton scatters within a detector, the recoil electron is absorbed in a region close to the interaction. Depending on the detector design, the electron can be absorbed locally or its scattering angle and energy can be tracked by a series of thin particle detectors. A second detection system can collect the energy and position of the scattered photon. If the energy and scattering angles of both the photon and electron are measured, the energy and position of the incident photon can be reconstructed with good accuracy. Compton telescopes (see Sect. 3.2.4) are applications of this indirect method of X-ray and γ-ray detection in space. Partial or missing measurements decrease the detected energy and angular resolution, because of the difficulty in tracking the recoil electron together with the unknown momentum of the electron, which in reality is bound to the atom.

3.1.1.3 Pair Production

When the photon energy starts exceeding twice the rest mass of the electron (1022 keV), pair production comes into play, where the gamma-ray is converted to an electron-positron pair in the coulomb field of the nucleus. Pair production is the dominant interaction process at high energies, starting from a few MeVs. The excess energy carried by the photon is transferred to the pair kinetic energy, while a small portion—undetectable—is released to the nucleus. If the photon energy is more than four times the electron rest energy a fraction of pair creations can happen in the field of an electron, and a triplet, the pair and the recoil electron, is created. The recoil electron energy can be measured as the Compton recoil electron. The conversion cross-section in the nucleus $\sigma_{p,n}$ field scales with Z^2, with Z the atomic number of the target, while if a triplet is generated the cross-section $\sigma_{p,e}$ is proportional to Z [4]. The larger the Z, the lower the probability of a conversion in the electron field:

$$\sigma_{\mathrm{p,e}} \propto \frac{1}{\mathrm{Z}} \sigma_{\mathrm{p,n}}. \tag{3.5}$$

The measurement of the pair direction, by means of tracking devices, and energy, by means of the tracker and absorbers placed at the bottom of the conversion, allows the reconstruction of the primary photon energy with a certain level of accuracy. Pair production telescopes use this indirect detection method to perform observations above tens of MeV.

3.1.2 Linear Attenuation

The relative dominance of the photoelectric effect (σ_{ph}), Compton scattering (σ_{c}), and pair production (σ_{p}) for different detector materials and photon energies is shown in Fig. 3.4 (left panel). The lines on the left and right refer to the energies and atomic number values where photo-absorption and Compton scattering or Compton scattering and pair production have the same cross-section.

The linear attenuation coefficient $\mu = \sigma n$, where σ is the cross-section and n is the number of atoms per cm^3, is the probability per unit length of a photon to interact with the absorber and it is expressed in cm^{-1}. The inverse of the linear attenuation, $1/\mu$, is the mean free path λ, defined as the average distance traveled in the absorber before an interaction takes place.

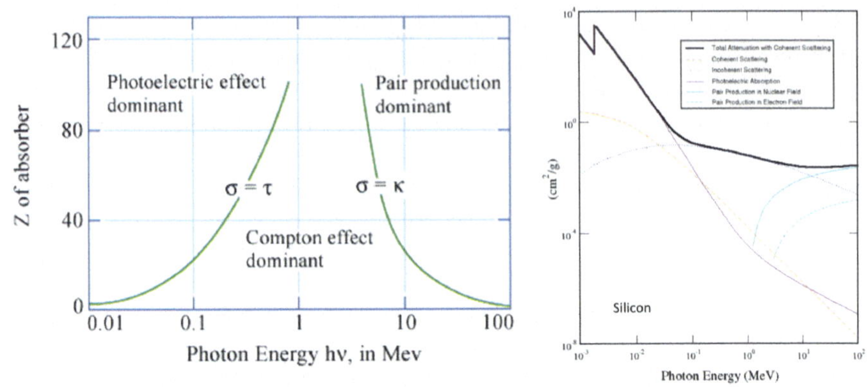

Fig. 3.4 *Left panel*: Probability of different interactions of radiation with matter as a function of the atomic number Z and photon energy [CREDIT: MIT OpenCourseWare]. *Right panel*: The total mass attenuation coefficient and the contribution from each process for Silicon, tabulated by the National Institute of Standards and Technology (NIST)

Given t the thickness of the material and I_0 in the incident radiation, the Beer–Lambert law describes the intensity of the attenuated radiation after crossing the material:

$$\mathrm{I} = \mathrm{I_0 e}^{-\mu t} \quad . \tag{3.6}$$

The transmission and attenuation efficiency of the material is then given by I/I_0 and $1 - I/I_0$. The mass attenuation coefficient, μ/ρ with ρ the material density, allows for the expression of the attenuation with no dependence on the density. Its values for Silicon, in cm2/g , are shown in Fig. 3.4 (right panel). Figure 3.5 shows the transmission efficiency of different configurations of the optical filters in front the ATHENA wide field detector [5]. These filters, while blocking the UV and visible light entering the field of view, must remain transparent to the X-ray radiation. The visible scales are caused by the absorption edges of the photo-electric cross-section (Sect. 3.1.1.1). The total attenuation of a material is the sum of the probability of all the processes listed before:

$$\mu = \mu_{ph} + \mu_c + \mu_p = n(\sigma_{ph} + \sigma_c + \sigma_p) \ . \tag{3.7}$$

3.1.3 Passage of Electrons Through Matter

The generation of electrons, as charge clouds within the detector, or as recoil particles, or in conversion pairs for indirect signal reconstruction, is the key in X-ray and γ-ray detection (see [7] for a full review). Electrons and positrons lose energy mainly

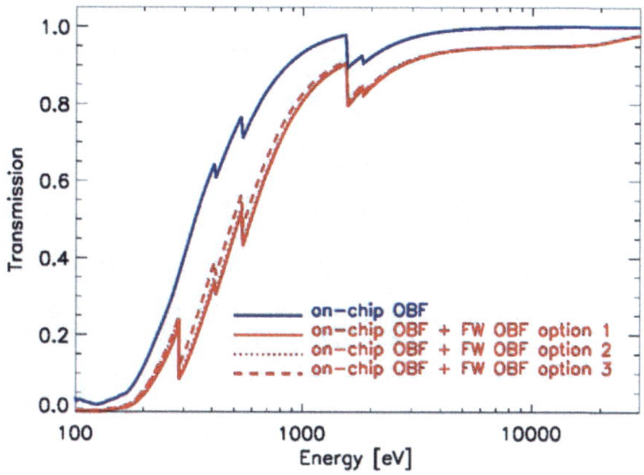

Fig. 3.5 X-ray transmission of the optical blocking filter (OBF) directly deposited (on-chip) on the ATHENA wide field detector and the total transmission including three configurations for the Filter Wheel (FW) [6]

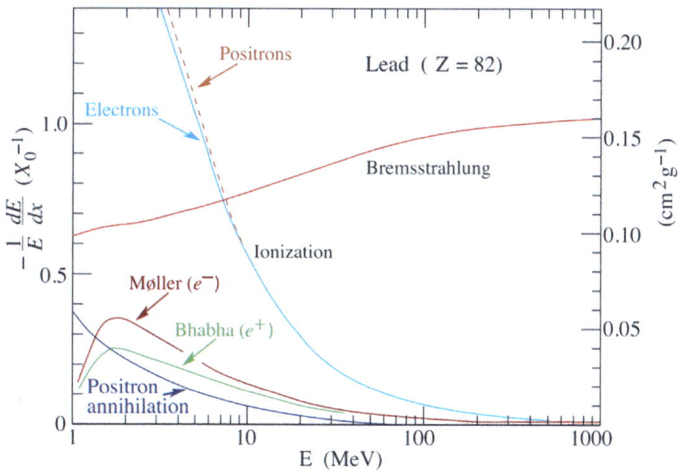

Fig. 3.6 Fractional energy loss per radiation length in Lead as a function of electron or positron energy [CREDIT: http://pdg.lbl.gov/]

by ionization and bremsstrahlung, with the first dominating at low energies and the latter dominating above a few tens of MeV in most materials (Fig. 3.6). Also, Møller scattering, Bhabha scattering, and positron annihilation add a small contribution at low energies.

In photo-detectors operating in the X-ray band, from the photo-electric effect to the Compton scattering regime the electrons lose energy by ionization. In tracking devices for pair production the probability of bremsstrahlung production can limit the angular resolution of the telescope [8]. If the electron/positron tracker is composed

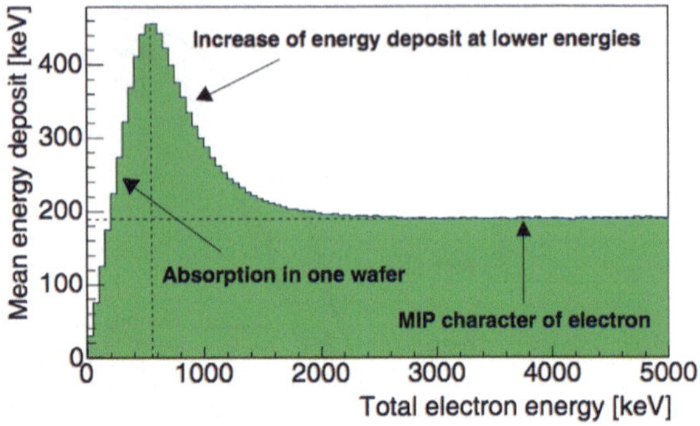

Fig. 3.7 Mean energy deposit in 500 μm of Silicon as a function of electron energy [9]

of Silicon strips (see Sect. 3.6.7) with a thickness of 500 μm, electrons are totally absorbed up to ∼400 keV, then the mean energy deposit decreases and reaches a constant minimum where electrons behave as minimum ionizing particles (MIP, Fig. 3.7). Electrons and positrons from pair production and Compton scattering can be considered as MIPs when interacting with Silicon trackers, i.e. their energy losses are nearly constant, on average, despite the variable energy of the particles. Energy losses for single particles in thin layers can vary a lot (the *straggling* effect) and follow a Landau (or Landau–Vavilov) distribution. It is characterized by a narrow peak with a long tail towards larger energy losses due to the small number of individual collisions, each with a small probability of transferring large amounts of energy.

The direction of the electrons and positrons crossing the detector layer changes when interacting with the Coulomb potential of the atomic nuclei, resulting in multiple low angle scattering described by Molière theory. It predicts an angular distribution roughly gaussian for small deflection angles but with larger tails at larger angles. Its width projected on a scattering plane, defined as θ_0, is given by:

$$\theta_0 = \frac{13.6\ \mathrm{MeV}}{\beta \mathrm{cp}} \sqrt{\frac{\mathrm{x}}{\mathrm{X}_0}} \left(1 + 0.038 \ln \frac{\mathrm{x}}{\mathrm{X}_0}\right), \tag{3.8}$$

where βc and p are the velocity and momentum of the electron (or positron), x/X_0 is the projected thickness of the scattering medium (e.g. the detector) along the particle path in radiation lengths. The Molière scattering limits the accuracy in measuring the electron (or positron) direction, and this limitation increases with higher thicknesses of the tracker layer and lower particle energies, i.e. the angular spread increases as the particle loses energy in the layers.

The radiation length is a property of the material and it can be described as both the mean distance over which a high-energy electron loses all but $1/e$ of its energy by bremsstrahlung, or 7/9 of the mean free path of a γ-ray before converting into a pair. It can be computed analytically, with an accuracy of a few percent, with [7]:

$$\mathrm{X}_0 = \frac{716.4\ \mathrm{g\,cm^{-2}A}}{\mathrm{Z(Z+1)\ln(287/\sqrt{Z})}}. \tag{3.9}$$

By dividing by the material density X_0 is expressed in cm, and it represents the reference scale length for describing high-energy electromagnetic cascades. In Silicon, $X_0 = 9.35\,\text{cm}$ and a layer of $500\,\mu\text{m}$ gives $\sim 0.005\,X_0$. The Earth's atmosphere is $28\,X_0$.

3.1.4 Detection Efficiency

The detection efficiency, often expressed as a percentage or fraction of the incident radiation, depends on the material linear attenuation coefficient, the detector thickness, the energy of the incident photons, and the signal read-out technology. While the total efficiency considers all counts no matter the energy released in the material, the photo-peak efficiency assumes that only those interactions that deposit the full energy of the incident radiation are counted. If the detection is aimed at measuring the energy of the incoming photons, the latter should be considered. The efficiency can also be absolute or intrinsic, depending if the amount of incoming radiation is counted from the source or onto the detector. The absolute efficiency at the photo-peak $\varepsilon_{p,abs}$ is the ratio of the number of counts at the photo-peak $C(E)$ over the total number of photons $N_s(E)$ with energy E emitted by the source. For an isotropic source, it is related to the intrinsic efficiency by:

$$\varepsilon_{\text{p,abs}} = \frac{\text{C(E)}}{\text{N}_\text{s}\text{(E)}} = \frac{\Omega}{4\pi}\varepsilon_{\text{p,in}} = \frac{\Omega}{4\pi}\frac{\text{C(E)}}{\text{N}_\text{in}\text{(E)}}, \tag{3.10}$$

where $N_{\text{in}}(E)$ is the number of photons hitting the detector surface and Ω the solid angle subtended by the detector.

In astronomy the term quantum efficiency QE, which is the fraction of incident photons that produce events in the detector (e.g. photo-electrons), is used to generally refer to the fraction of incident photons registered by the detector. The QE of current photo-detection technologies such as CCDs (Sect. 3.4.1) and active pixel sensors (Sect. 3.4.2) can reach more than 90%.

3.1.5 Energy and Angular Resolution

The energy and angular resolutions of an instrument define the ability to discriminate photons of different energy and at different positions.

A detector performing radiation spectroscopy measures the energy distribution of the incoming photons, a property that gives access to temperature and acceleration processes of all kinds of astrophysical objects. Given a mono-energetic source of radiation, the energy response function is the energy distribution of the pulses generated in the detector. Once the background or any continuum is subtracted, and defining the full width half maximum (FWHM) of the response function as the width

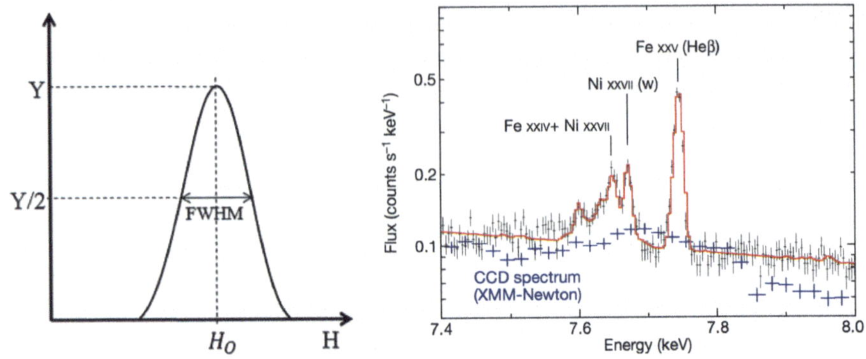

Fig. 3.8 Gaussian energy response function for a mono-energetic source (left panel [10]) and the X-ray spectrum of the Perseus cluster (right panel [11]) detected with the HITOMI micro-calorimeter, in red, and the XMM-Newton CCD, in blue

of the distribution at the half of its maximum, the energy resolution $\Delta E/E$ or R (see Eq. 3.34) is defined as the FWHM divided by the peak centroid (Fig. 3.8, left panel), and it is usually expressed as a percentage.

Semi-conductor based photo-detectors have energy resolutions tens of times better than scintillators, but current thermistors applied to micro-calorimeters (Sect. 3.5.2) can reach resolutions up to ~100 times better than semi-conduction photo-detectors. Figure 3.8 (right panel) compares the X-ray spectrum of the Perseus cluster obtained by the HITOMI Soft X-ray Spectrometer (SXS), based on Silicon thermistors, and the XMM-Newton Silicon-based CCDs. The high-resolution spectroscopy achieved by the micro-calorimeter unveiled properties of the hot plasma permeating galaxy clusters that were inaccessible with previous instruments.

The point spread function (PSF) describes the response of any imaging system to a point source and it is given by the distribution of the detected photon position against the real one. For an ideal imaging system, all detected or reconstructed photon positions in the sky would be the same as the real position of the astrophysical source. However, each design of choice introduces an error in the final position, which depends on the energy and position of the source in the field of view (Fig. 3.9). For focusing X-ray optics (Sect. 3.2.1), the PSF is worsened by fabrication errors, the accuracy in the mirror alignment, and micro-roughness of the mirror surface. For Compton and pair production telescopes based on electron-tracking devices, the electron and positron multiple scattering in the tracker layers deviates the particle path reducing the accuracy of the reconstruction algorithm.

If the PSF is assumed to be symmetric with respect to the source's real position, it can be expressed as a surface brightness radial profile, i.e. the count density for different radial bins of spherical distance from the target position in the sky. Otherwise, a two-dimensional surface brightness profile is required. Common ways to measure the PSF are as containment radius of the detected photons, e.g. the radius within which 68% (one standard deviation) of the X-ray or γ-ray photons fall, or as

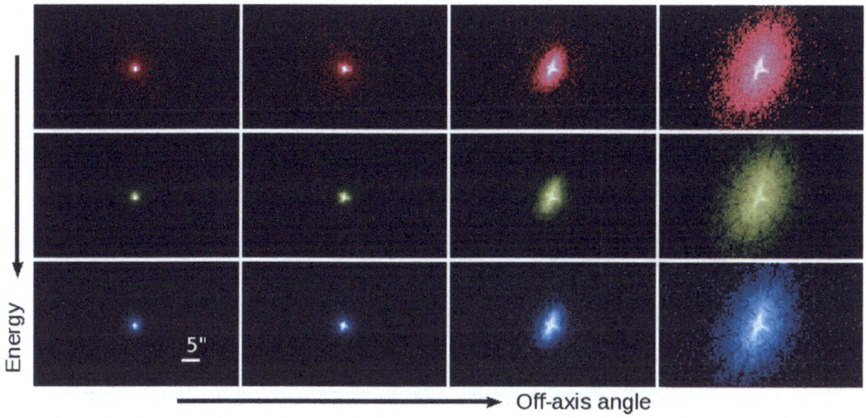

Fig. 3.9 PSF of the Chandra X-ray optics simulated at a set of off-axis angles (0', 2.4', 4.7' and 9.6') and mono-chromatic energies (0.92, 1.56 and 3.8 keV) [CREDIT: http://cxc.harvard.edu]

the FWHM of the radial profile, or as half energy width (i.e. the angular diameter enclosing 50% of the photons), or by fitting the profile with dedicated models. In both X-rays (e.g. XMM-Newton [12]) and pair conversion telescopes (e.g. AGILE [13]), the PSF is usually fitted by a King function or double King functions to account for tails in the distribution. The King function K(r, σ, γ) describing the PSF surface brightness profile takes the form:

$$K(r,\sigma,\gamma) = \frac{1}{2\pi\sigma^2}\left(1-\frac{1}{\gamma}\right)\left(1+\frac{r^2}{2\sigma^2\gamma}\right)^{-\gamma} , \tag{3.11}$$

where r is the radial distance and σ and γ characterize the size of the angular distribution and the weight of the tail, respectively. The 68% containment radius CR68 is derived by the best-fit values of σ and γ with this relation [14]:

$$\mathrm{CR68} = \sigma\sqrt{2\gamma\left((1-0.68)^{\frac{1}{1-\gamma}}-1\right)} . \tag{3.12}$$

3.1.6 *Effective Area*

The effective area A_{eff} of a telescope is the geometrical area A_{geom} exposed to the incoming flux convoluted with the efficiency ε of this area in collecting and detecting such photons:

$$A_{\mathrm{eff}} = \varepsilon \times A_{\mathrm{geom}} . \tag{3.13}$$

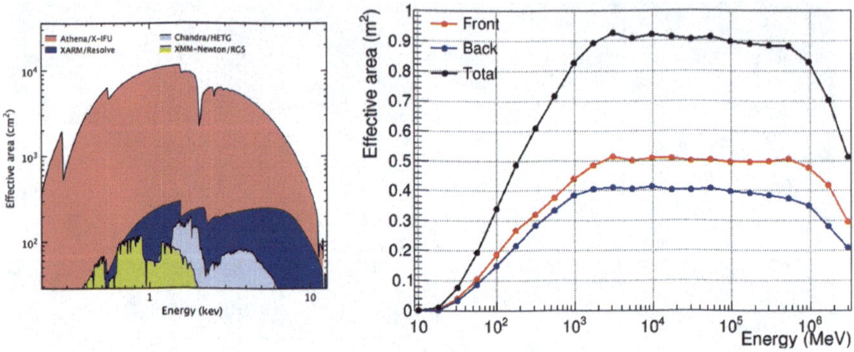

Fig. 3.10 *Left panel*: Effective area in the 0.2–12 keV energy range of the future ATHENA and XARM micro-calorimeters and current Chandra and XMM-Newton grating spectrometers [15]. *Right panel*: Effective area of the Fermi pair conversion detector from 10 MeV to 1 TeV given by the front, in red, back, in blue, tracker sections and the total value, in black [CREDIT: https://www.slac.stanford.edu]

The larger the effective area, the higher the ability to observe photons emitted by an astrophysical source. In some missions, the term exposure is used to indicate the observation time multiplied by the effective area. The factors that reduce this efficiency are mission-dependent, ranging from the reflectivity and vignetting of the mirrors in grazing angle focusing telescopes to the detector quantum efficiency and the reconstruction efficiency in indirect detection systems. Figure 3.10 shows the effective area of present and future X-ray telescopes (left panel) and the Fermi/LAT gamma-ray tracker [8].

3.1.7 Sensitivity and Background

The sensitivity, or the minimum detectable flux, of a telescope is a key science requirement of a telescope, since it defines its ability to attain the science goals of a mission. Both the telescope (e.g. effective area, energy and angular resolution) and observation (e.g. background level, observation time, required significance for a detection) parameters play a role in the evaluation of the sensitivity, which can be for point or extended sources, for continuum or line emission, and differential (as a function of incoming energy) or integral (above an energy threshold). We will focus here on the definition of continuum sensitivity for point sources, but the derived formulae can be easily extended to other cases.

In an integration time T, the detector collects a total number of counts that is the sum of counts from the source N_s and the background N_b, with the latter including all the events not generated by the astrophysical source. Various sources that affect the measurement of the signal contribute to the noise, that can be generally described as the error in the measurement in standard deviation. The total number of counts

N_{tot}, including its error, is:

$$N_{tot} = (N_s + N_b) \pm \sigma(N_s + N_b) \quad . \tag{3.14}$$

The source counts are obtained by subtracting the background from the total, which removes the mean background contribution but not its noise:

$$N_s = (N_s + N_b) \pm \sigma(N_s + N_b) - (N_b \pm \sigma(N_b)) = N_s \pm \sigma(N_s + 2N_b) \quad . \tag{3.15}$$

The signal-to-noise ratio, often abbreviated as SNR, measures the level of excess—the signal of the source—above the total noise. In statistics, the SNR also expresses the mean, the signal excess in this case, divided by the standard deviation. From Eq. 3.15, the signal-to-noise ratio can be written as:

$$SNR = \frac{N_s}{\sigma(N_s + 2N_b)} = n_\sigma \ , \tag{3.16}$$

and it gives the significance of the detection in standard deviations. For example, an SNR of 3 means the ability to detect an excess above the background with a significance of 3σ.

For a focusing X-ray telescope, defining $f_s(E, A, t)$ and $f_b(E, A, t)$ as the source and background flux in units of energy, area, and time, the number of photons generated from the source and the background can be written as:

$$N_s = \int_{E_0}^{E_1} \int_{t_0}^{t_1} f_s(E, A, t) A(E) \, dE \, dt = T \, A_{eff} \int_{E_0}^{E_1} f_s(E) \, dE \tag{3.17}$$

and

$$N_b = \int_{E_0}^{E_1} \int_{t_0}^{t_1} f_b(E, A, t) A(E) \, dE \, dt = T \, A_{PSF} \int_{E_0}^{E_1} f_b(E) \, dE \ , \tag{3.18}$$

where T is the integration time, A_{eff} is the mirror effective area, and A_{PSF} is the area subtended by the source PSF on the detector surface[1] and E_0 and E_1 are the energy integration limits.

Assuming the photon counting a random process, with independent arrival times, the standard deviation of a measure can be expressed as the square root of the detected counts:

$$N_s = n_\sigma \sqrt{N_s + 2N_b}. \tag{3.19}$$

The minimum observable flux to reach a significance of n_σ, integrated over the energy channel of interest, takes the form:

[1] Both the effective area and the PSF depend on the source energy and should be integrated, but for simplicity they are assumed to be constant in energy.

$$F_s = \int_{E_0}^{E_1} f_s(E)dE = n_\sigma \frac{\sqrt{N_s + 2N_b}}{A_{eff}T}. \tag{3.20}$$

For a faint source, where $N_s << N_b$, the formula for the sensitivity, in units of photons cm^{-1} s^{-1} for a given significance n_σ, can be written as:

$$F_s = n_\sigma \frac{\sqrt{2N_b}}{A_{eff}\ T} \tag{3.21}$$

$$= n_\sigma \frac{\sqrt{2\int_{E_0}^{E_1} f_b(E)dEA_{PSF}T}}{A_{eff}T} \tag{3.22}$$

$$= \frac{n_\sigma}{A_{eff}}\sqrt{\frac{2F_bA_{PSF}}{T}}, \tag{3.23}$$

with F_b the energy integrated background flux in counts cm^{-2} s^{-1}.

An increase in the effective area or integration time results in a better sensitivity of the instrument, while an increase in background flux lowers the sensitivity (or increases the minimum detectable flux). One important feature of X-ray and gamma-ray missions is high efficient background prevention and rejection techniques (e.g. the use of dedicated data analysis techniques or shielding design) to ensure a good sensitivity throughout the mission life and a minimum loss of observation time. (see Sect. 3.7 on current methods to reduce or prevent the background of high energy missions).

From Eq. 3.21 it is clear that for a collimated or coded X-ray telescope (see Sect. 3.1), where the area of detection of photons is the same as the area of collection, i.e. $A_{PSF} \sim A_{eff}$, the background has a greater impact on the final sensitivity with respect to focusing telescopes.

An alternate way to assign a significance to a detection, not requiring approximations in its analytical form, is by using the method of hypotheses test, the null and alternate hypotheses, and their maximum likelihood ratio on the basis of Wilks's theorem [16]. By using this theorem, and considering the detection as performed with two observations (*on* and *off* source), the work by [17] provides the following formula for the significance:

$$n_\sigma = \sqrt{2}\left\{N_{on}\ln\left[\frac{1+\alpha}{\alpha}\left(\frac{N_{on}}{N_{on}+N_{off}}\right)\right] + N_{off}\ln\left[(1+\alpha)\left(\frac{N_{off}}{N_{on}+N_{off}}\right)\right]\right\}^{1/2}, \tag{3.24}$$

where α is the ratio of effective area, integration time, and observation region size of the *on* among the *off* observation, and N_{on} and N_{off} are the number of counts collected in the two observations. N_{off} measures the background to be subtracted from the *on* detection. This method is particularly useful for low statistics detections, where the background fluctuations can not be ignored.

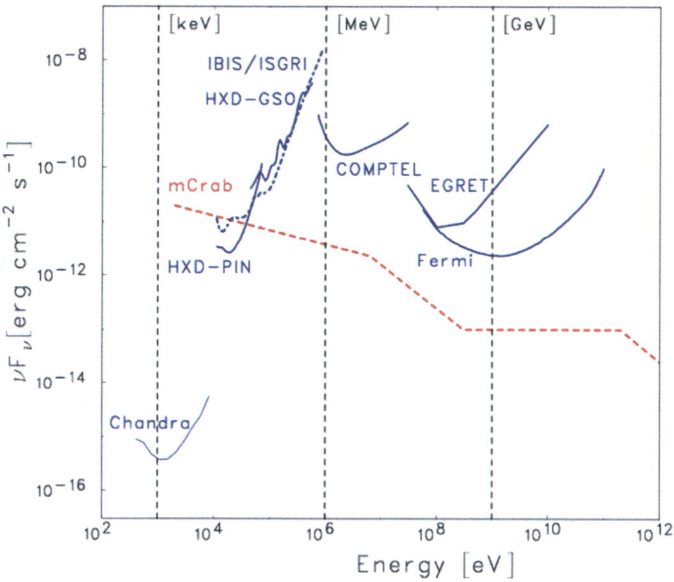

Fig. 3.11 Point source differential sensitivity (3σ detection, 100 ks exposure) of different X-ray and γ-ray telescopes based on [18]. A spectral bin of $\Delta E/E = 1$ and 0.5 is assumed for Chandra and all the other instruments

The point source differential sensitivity for a selection of X-ray and γ-ray satellites is shown in Fig. 3.11 as determined from their respective effective areas, imaging performance, and background minimization.

3.2 Imaging in X-Rays and Gamma-Rays

The imaging capability of a detection system is mandatory to identify the source of X-rays or gamma-rays in the sky and a key parameter to increase the sensitivity of the telescope (see Eq. 3.21). Its angular resolution and collection area is defined by the applied imaging technology, hardware or software, which in turn depends on the energy range of interest.

3.2.1 Grazing Incidence X-Ray Optics

In the X-ray band, from below 1 keV to ~80 keV, the introduction of X-ray focusing mirrors (Sect. 3.2.1) with the Einstein Observatory, or HEAO–2, in 1978 marked the birth of a new era in X-ray astronomy thanks to the increase in sensitivity of about a factor of 1000 with respect to collimated proportional counters (see [19, 20] for a review). The performance of an X-ray telescope is characterized by (i) the image quality (the PSF), (ii) the effective area, and (iii) the stray light rejection efficiency.

3.2.1.1 X-Ray Reflection

The refractive index n of a material takes the form:

$$\mathrm{n} = 1 - \delta + i\beta \ , \tag{3.25}$$

where the real part δ accounts for the refraction effect and the imaginary part $i\beta$ represents the absorption of the medium. In the X-ray band both terms are very small ($10^{-4} - 10^{-6}$), e.g. for Iridium the values of δ and β at 2 keV are 4.2×10^{-4} and 1.4×10^{-4} respectively, and the refractive index is always slightly smaller than 1. Contrary to optical light, where refractive indices in transparent media are larger than 1 and refractive lenses are a feasible focusing solution, the principle of reflection is an effective method of choice up to tens of keV. From Snell's law (Fig. 3.12, left panel), the incident and refraction/reflection angle depend on the material refractive index as follows:

$$\frac{\sin\alpha_\mathrm{i}}{\sin\alpha_\mathrm{r}} = \frac{\mathrm{n}_2}{\mathrm{n}_1}. \tag{3.26}$$

A total reflection $\alpha_r > 90°$ from vacuum ($n_1 = 1$) to any medium translates into a maximum incident angle $\theta_c = 90 - \alpha_i$, usually referred to as the critical angle, for which reflection occurs:

$$\cos\theta_\mathrm{c} = \mathrm{n}_2 \ . \tag{3.27}$$

Given n_2 close to 1, X-ray reflection can only happen at small, or grazing, angles of incidence. Assuming total reflection ($\beta \to 0$) and linking the δ value to the atomic properties of the material, the critical angle takes the form:

$$\theta_\mathrm{c} \approx \sqrt{2\delta} = \sqrt{\frac{\mathrm{N}_a \rho r_e \mathrm{h}^2\mathrm{c}^2\mathrm{f}_1}{\mathrm{M}_a \pi}\frac{1}{\mathrm{E}_\gamma}} \propto \frac{\sqrt{\rho}}{\mathrm{E}_\gamma}, \tag{3.28}$$

where N_a is the Avogadro's constant, M_a is the molar mass, ρ is the medium density, r_e the classical electron radius, h the Plank constant, E_γ the photon energy, and f_1 the first component of the scattering factor. In a real case, only a near-total reflection is possible because of scattering caused by micro-roughness on the mirror surface [21] and the photo-electric absorption—included in the β term—then can not be neglected as the incident angle increases.

From Eq. 3.28, the critical angle increases with the material density, hence high-Z metals (Au, Ir, Pt) are used as coating layers for X-ray grazing mirrors on top of lighter supporting plates (e.g. glass, Ni). At the same time, a higher photon energy

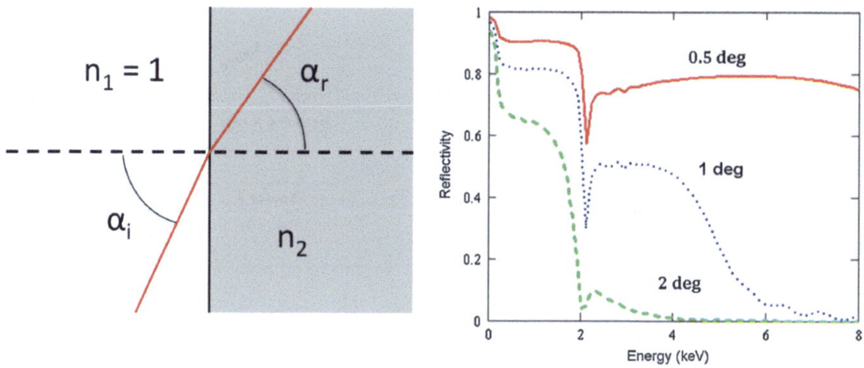

Fig. 3.12 *Left panel*: schema for Snell's refraction law from vacuum ($n_1 = 1$) to a medium with refractive index n_2. *Right panel*: Reflectivity of 30 nm of Iridium as a function of incident photon energy for three angles (0.5°, 1° and 2°) [20]

leads to a lower critical angle: for Platinum ($\rho = 21.4$ g/cm^3), $\theta_c < 0.15°$ at 30 keV and 0.8° at 5 keV. For a fixed incident angle, only photons below a certain cut-off energy can be reflected. The reflectivity probability of 30 nm of Iridium (Fig. 3.12, right panel) as a function of energy confirms this behaviour, approaching unity only for <0.5° incident angles. Reflecting materials are the result of a trade-off analysis between a high critical angle and a low photo-absorption cross-section.

3.2.1.2 Multi-layer Coatings

The use of multiple layers [22, 23] of alternate high (Tungsten, Platinum) and low (Silicon, Carbon) atomic number allows for the extension of the reflection efficiency for X-rays towards higher energies (>10 keV). The single reflection efficiency of each high-Z layer is small at high energies, and a great fraction of the light continues along its path. Multi-layer coatings are a stack of hundreds of coating pairs, few nm thick, with each pair (e.g. Ni/C, W/Si, Pt/C) having different density and refractive index. From Bragg's law, the distance d required between the reflection layers to produce constructive interference depends on the photon wavelength λ and the grazing angle θ as follows:

$$2\mathrm{d} \sin\theta = \mathrm{n}\lambda. \quad (3.29)$$

Reflection above 10 keV is achieved by choosing the layer optical thickness in order to produce constructive interference between each coating pair and summing the reflection amplitude of all the stacks. The low density layer acts as a mechanical spacer between the reflecting layers. If the distance between each layer is fixed, the system will create constructive interference only at some wavelengths. Graded multilayer coatings, with varying (graded) spacing between layers, ensure a broad band reflection efficiency that is mandatory for astronomy applications.

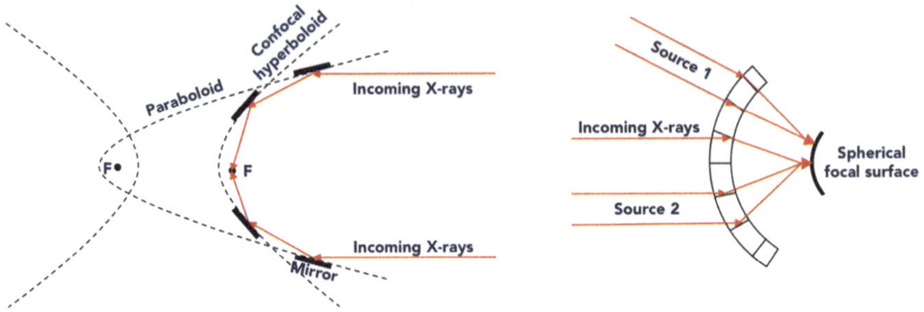

Fig. 3.13 Schematic view of the Wolter-I (left panel) and the lobster-eye (right panel) configuration in X-ray grazing angle telescopes

Multi-layer coatings make it possible to build X-ray mirrors up to ~80 keV (e.g. the NuSTAR mission). Above this energy the required spacing between layers becomes too small, around the scale of the surface micro-roughness. Mirrors at higher energies can still be possible by means of Laue lenses or Bragg diffracting crystals (see Sect. 3.2.5).

3.2.1.3 X-Ray Mirrors

Grazing angle reflection results in a very small projected collection area. A very diffuse technique is the use of confocal and coaxial mirror shells of varying size nested together to increase the effective area.

A paraboloid produces a perfect focus for on-axis photons but a coma blur for off-axis photons, with a size proportional to the off-axis angle [25]. In 1952 Hans Wolter [26] proposed three configurations (Wolter I, II and III) where an even number of reflections from confocal conic-like optics avoids coma by ensuring the same optical path for all incident X-ray photons (the *Abbe sine condition*). The Wolter-I type (Fig. 3.13, left panel) is the most used mirror configuration. It consists of a paraboloid as primary mirror and a confocal coaxial hyperboloid as secondary mirror: X-ray photons strike first the paraboloid inner surface, then the hyperboloid and the detector is placed at the focus of the latter. X-ray telescopes such as XMM-Newton and Chandra are Wolter-I mirrors. A perfect coma free condition is obtained with more complex mirror geometries. The Wolter–Schwarzschild lay-out [27], chosen for the ATHENA X-ray telescope, improves the off-axis angular resolution by placing the mid-planes of the Wolter-I mirror modules on a spherical plane with a radius equal to the focal length of the telescope. The advantages of Wolter-I like mirrors are: a short focal length (from few to 12 m) to aperture ratio, maximising the collecting area despite the restrictions on diameter and length imposed by space vehicles; the ability to concentrate photons on smaller detector surfaces, which reduces the impact of the non-X-ray background.

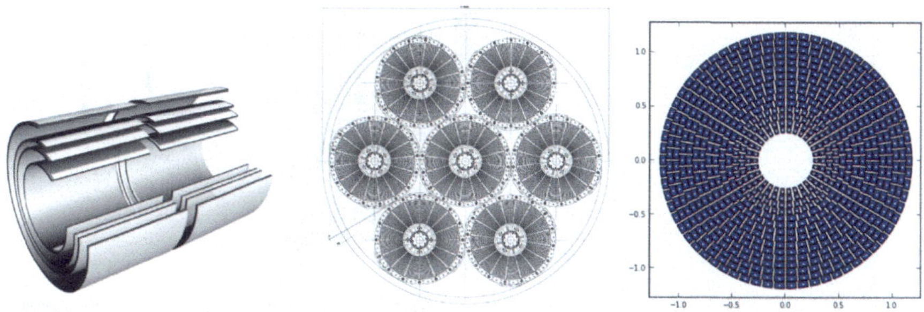

Fig. 3.14 X-ray grazing angle mirror design of the Chandra [CREDIT: Raytheon Co] (left), eRosita [28] (center) and ATHENA [CREDIT: https://sci.esa.int] (right) telescopes

The way Wolter-I X-ray mirrors are fabricated and assembled together can largely vary the final scientific performance of the telescope. Mirror shells can be obtained through (i) direct polishing, (ii) electroformed-nickel replication and (iii) epoxy-replicated thin foils.

Monolithic shells of glass that are figured and directly polished achieve a high angular resolution (less than few arcsec Half Power Diameter (HPD)), but with relatively large mass penalty, hence a reduced number of shells can be nested together with a decrease in effective area (e.g. the Chandra High Resolution Mirror Assembly, Fig. 3.14, left panel). X-ray shells can instead be replicated by electro-forming a thin layer of nickel over a mandrel that is the inverse of the desired mirror. With this technique, used for example for the XMM-Newton telescope, X-ray shells are more cost-effective and lighter than polished glass without, however, the same figure error. More shells can be nested together to increase the telescope throughput and effective area, or more than one optics system (Fig. 3.14, central panel) can be mounted on the telescopes because of the lower mass budget (1/4 the weight per unit axial length [19]), but the angular resolution is of the order of 10–20 arcsec (HPD). In epoxy replication a gold mandrel is coated with >100 nm of Gold and a thin ($\sim 100\,\mu$m) Aluminum foil acting as a mirror substrate. Hundreds of replicated mirror foils can be nested together thanks to the very low mass budget. Since thin mirrors are prone to deform, they ensure high throughput X-ray telescopes with low cost but with a modest angular resolution ($>$arcmin HPD).

A completely different mirror configuration, the lobster eye optics (see [30] for a review) emulating the arrangements of eyes of lobsters (Fig. 3.13, right panel), is currently being proposed in, e.g., the SVOM/MXT [42] and THESEUS/SXI [24] telescopes. The first lobster-eye prototype telescope, based on multi-foil optics, was launched on the CZECH VZLUSAT-1 nano-satellite in 2017 [31].

Two arrangements exist for lobster-eye mirrors: Angel's [32] and Schmidt's [33]. In the first one, grazing angle reflections occur within a system of square-sided tubes (1–2 mm thick) arranged over a spherical surface with a radius twice the focal length of the telescope. X-ray photons are focused on a spherical detection surface. In the second, flat mirror surfaces, with both sides reflecting, are arranged around the

Fig. 3.15 Micro-pore applied to the SVON lobster eye (left panel [29]) and ATHENA Wolter-type (right panel, [CREDIT: cosine Research BV, DTU space]) X-ray grazing angle optics

perimeter of a cylinder of radius *R* and photons are focused on a cylindrical surface of radius *R/2*. Placing two of such systems orthogonal to each other creates a double reflection and a bi-dimensional imaging system.

The lobster eye mirror can cover large field of views ($\sim$1 sr) with homogeneous sensitivity and it is the perfect solution for X-ray sky monitors. The only drawbacks are the still limited angular resolution (few arcmins) and energy band (<few keV).

Kirkpatrick-Baez mirrors (see [34] for a review), comprised of a pair of perpendicular reflecting mirrors, have yet to be proposed for an actual X-ray mission despite many works describing their application in X-ray reflection.

The use of compact micro-pores, with a size less than 1 mm, has been proposed to minimise mass and volume of the optics system while increasing the effective area. Since the walls of the pore structure are very thin the reflecting surfaces can be stacked very densely, effectively leading to small pores and short optics. They are currently being proposed with a Wolter-Schwarzschild configuration for the ATHENA mission (Fig. 3.14, right panel). In this case the micro-pores are obtained by stacking together Silicon plates with ribs on one side. Glass plates are instead slumped over a spherical surface to build a lobster eye micro-pore mirror. Figure 3.15 shows the micro-pore technique applied to both a lobster eye (left panel) and Wolter-Schwarzschild (right panel) X-ray mirror, resulting in a compact size for the whole mirror system.

X-ray mirrors are usually coupled with thermal baffles at the sky-side, preventing any thermal gradient across the surface that could decrease the image quality, and

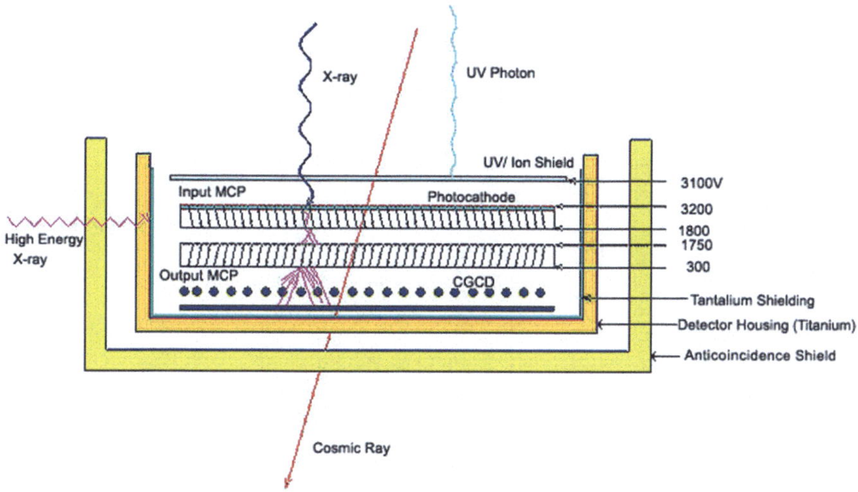

Fig. 3.16 The micro-channel plate detection system on board Chandra, consisting of two clusters of lead oxide glass tubes and a cross-grid detector at the bottom [CREDIT: https://chandra.harvard.edu]

X-ray baffles, consisting of concentric annular plates placed at the mirror exit for Wolter-type nested X-ray shells, to block off-axis photons and reduce stray light contamination. The stray light is caused by X-ray photons coming from outside the field of view but still able to reach the focal plane by means of single reflections on the hyperbola section of the Wolter-I mirror.

As it should be clear from the listed mirror configurations, improving the angular resolution while increasing the effective area and keeping under control the mass envelope of the system are conflicting goals in most cases and the chosen mirror design depends on the mission science goals. Table 3.1 summarizes the design of a selection of current and proposed X-ray telescopes representing the state of the art of X-ray mirror technologies (Fig. 3.16).

3.2.1.4 Micro-channel Plates

A micro-channel plate (see [43] for a review) is an array of tiny tubes (~10 μm in diameter and few mm long) with walls coated by high photoelectric yield material that acts as particle amplifiers. X-ray photons hitting the channels release electrons that are accelerated down the tubes at high voltage. The resulting charge is then amplified by orders of magnitude by a cascade effect within the tube, which is held at an electrical potential. The electronic cascade at the exit strikes a cross-grid detector and the locations of the charges on the grid are processed to create a high spatial resolution map of the source. Once placed at the focus of an X-ray mirror system, the combination of the effective area and spatial resolution of the optics with the size

Table 3.1 List of X-ray grazing angles technologies applied to launched (XMM-Newton [35], Swift/XRT [36], eRosita [28], Chandra [37], Suzaku/XRT [38], HITOMI/SXT [39], NuSTAR [40]) and proposed (ATHENA [41], SVOM/MXT [42], THESEUS/SXI [24]) X-ray missions

X- RAY GRAZING MIRROR TECHNOLOGIES

Telescope type	X-ray missions	Design	Energy range	Angular resolution (HPD)	Effective area	Focal length
Electroformed Wolter I	XMM-Newton	3 × 58 shells, Ni + Au coating, 420 kg/mod.	0.15–15 keV	15″	1500 cm^2 at 1 keV, 900 cm^2 at 7 keV	7.5 m
	Swift/XRT	12 shells, Ni + Au coating, 225 kg/mod.	0.2–10 keV	18″	110 cm^2 at 1.5 keV	3.5 m
	eRosita	7 × 54 shells, Ni + Au coating, 50 kg/mod.	0.2–10 keV	16.1″	1400 cm^2 at 1keV	1.6 m
Polished Wolter I	Chandra	58 shells, Glass + Ir coating, 18500 kg/m^2	0.1–10 keV	0.5″	800 cm^2 at 0.25 keV, 400 cm^2 at 5 keV	10 m
Thin foil double cones	Suzaku/XRT	175 shells, Al + Au coating, 20 kg/mod.	0.2–12 keV	2′	450 cm^2 at 1.5 keV, 250 cm^2 at 7 keV	4.75 m
	HITOMI/SXT	203 shells, Al + Au coating, 43 kg/mod.	0.3–12 keV	1.3′	390 cm^2 at 6 keV	5.6 m
Multi-layer double cones	NuSTAR	2 × 133, Glass + Pt/SiC and W/Si multilayers, 37 kg/mod.	5–80 keV	58″	200 cm^2 at 20 keV	10 m
Silicon Pore Optics Wolter–Schwarzschild	ATHENA	15 rows of modules, Si + Ir/SiC coating, 299.2 kg (total)	0.1–15 keV (WFI)	5″	1.4 m^2 at 1 keV	12 m
Micro-channel plate lobster eye	SVOM/MXT	Glass + Ir coating, 35 kg (total)	0.3–6 keV	4.2′ (FWHM)	27 cm^2 at 1 keV	1 m
	THESEUS/SXI	4× units, Glass + coating, 40 kg/mod.	0.3–6 keV	4.5′ (FWHM)	4 cm^2 at 1 keV	0.3 m

of the microchannel plate and the grid spacing of the wires dramatically increases the spatial resolution of the system (0.5 arcsec for the Chandra High-Resolution Camera).

3.2.2 Coded Masks

At higher energies, from tens of keV to tens of MeV, coded aperture telescopes are the best solution for ensuring a fine angular resolution while providing a large field of view (Fig. 3.17, left panel). A mask composed of opaque and transparent elements made of high-Z material (e.g. Tungsten) is placed at a distance H over a position sensitivity detector with the same spatial resolution d for the mask (few mm). The radiation is modulated differently according to the attenuation efficiency of the mask elements. Each source casts a unique shadowgram on the detector and the final image is an overlap of the shadowgrams of all the sources in the field. In coded aperture imaging applied to X-ray telescopes, uniformly redundant arrays (URA) of elements, or modifications of it, are the most used family of mask patterns [44]: reconstruction of the sky image is obtained using deconvolution algorithms that correlate the recorded image with the decoding array derived from the mask pattern. Coded aperture telescopes are able to perform simultaneous observations of the source and background fluxes, with an angular resolution of a few arcmins defined by the angle subtended by one hole at the detector ($\sim d/H$). The effective area depends on the number of transparent elements viewed by the detector. Reducing d, the hole size, or increasing H, the mask distance, while increasing the number of elements improves the angular resolution without loss in sensitivity [45]. For a fully coded mask, the field of view (FOV) is given by:

$$\mathrm{FOV} = \arctan\left(\frac{\mathrm{D_m} - \mathrm{D_d}}{\mathrm{H}}\right), \tag{3.30}$$

where D_m and D_d are the size of the mask and the detector, respectively. Masks of dimensions greater or equal to the detector dimensions are employed to reach large field of views. Coded masks are usually employed in wide field survey instruments.

The IBIS coded mask on board INTEGRAL (Fig. 3.17, right panel), with a total coded area of 1064×1064 mm, covers the energy range from 20 keV up to 10 MeV, with a 12 arcmin angular resolution and a FOV of 9×9 degrees. The Swift/BAT coded mask, operative in the 14–195 keV energy range, reaches a field of view of 1.4 sr (partially coded) with a PSF of 22 arcmin (FWHM).

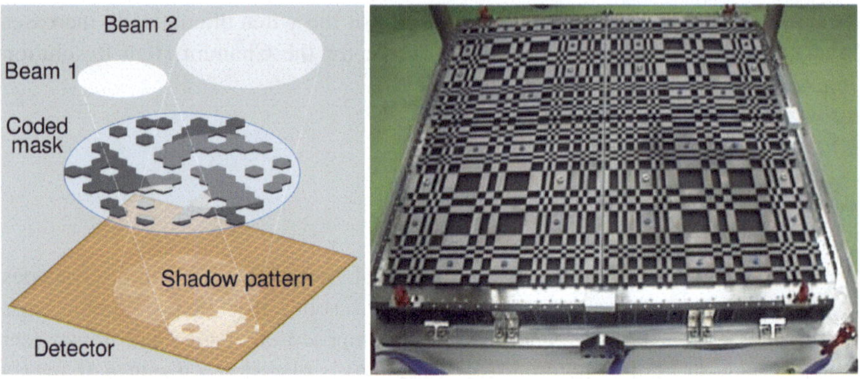

Fig. 3.17 Principle of operation of the Hexagonal Uniformly Redundant Array (HURA) coded mask used for the INTEGRAL/SPI instrument (left panel) [CREDIT: Cmglee] and picture of the Modified Uniformly Redundant Array (MURA) of the INTEGRAL/IBIS [CREDIT: https://ipl.uv.es]

3.2.3 Collimated Systems

A collimator is a mechanical device that narrows the field of view of an instrument. All early X-ray astronomy satellites, from Uhuru (1970–1973) to HEAO–1 (1977–1979), made use of collimators on top of proportional counters or scintillators to locate the source position. At present, collimated instruments are still used for observations with single photo-detectors from the hard X-ray to the soft gamma-ray band (from tens to hundreds of keV). If the Compton scattering or the pair production mechanisms, or coded patterns, are not exploited in such bands only collimated observations can provide some information on the source position. The angular resolution, expressed as FWHM, takes the form:

$$\theta_{\mathrm{FWHM}} = \arctan\left(\frac{\mathrm{d}}{2\mathrm{H}}\right), \tag{3.31}$$

where H is the collimator height and d the distance between the walls. In the case of the Suzaku Hard X-ray Detector [46], covering a wide energy range (10–600 keV) by combination of silicon PIN diodes and GSO scintillators, the BGO anti-coincidence system walls surround the four sides of the detector and limit its field of view, operating as an active collimator. The field of view ranges from half to $\sim 5^{\circ}$ (Fig. 3.18).

Collimated systems have large drawbacks: they provide small field of views, the effective area is reduced with respect to a focused system, and the background determination requires an off-source pointing that reduces the observation time. If the collimator is not coupled with a coded mask system (Sect. 3.2.2) or a position sensitive detector, the angular resolution almost coincides with the field of view.

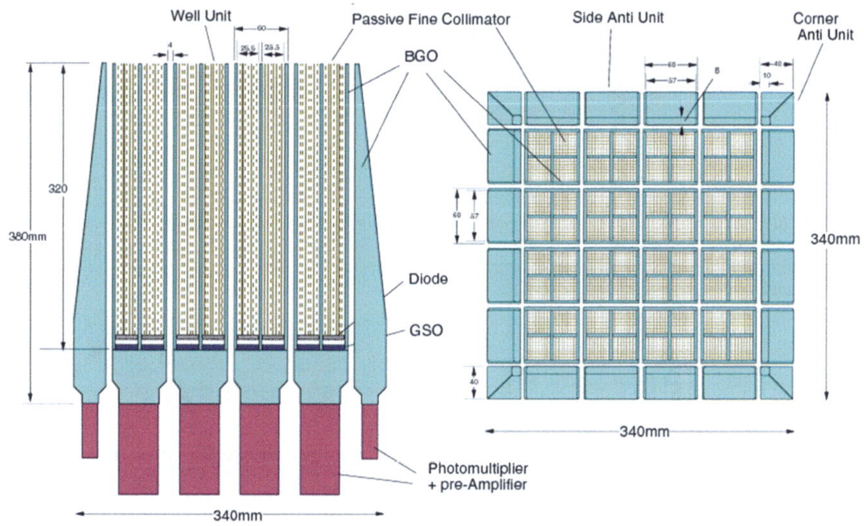

Fig. 3.18 Schematic view of the HXD instrument on board the Suzaku satellite [CREDIT: https://heasarc.gsfc.nasa.gov]

3.2.4 Tracking Telescopes

A different way of reconstructing the origin of photons becomes feasible at energies above a few hundred of keV where Compton scattering or pair production become dominant. In this energy range single event detector systems must be developed, i.e. telescopes that record the interaction of a single photon within the telescope, recording the track of the secondary/scattered particles to retrieve origin, energy, and, sometimes, polarisation of the original photon. Two main types of telescopes, for this purpose, have been developed: (i) Compton telescopes, (ii) pair-tracking telescopes.

Figure 3.19 shows two concepts of Compton telescopes. In the left panel, two planes of detectors act as scatterer and absorber, separated by a large distance to measure the time-of-flight of the scattered gamma-ray and and reduce the contamination by background events. An example of this concept is the COMPTEL instrument on board CGRO ([47], see Chap. 4), where the two planes are made of low Z (a liquid scintillator) and high Z (NaI(Tl) scintillator) material. Right panel shows an example of electron tracking Compton telescope. The tracker consists of several layers, thin enough to track the recoil electron. The scattered photon is stopped in a second detector around or below the tracker. The track of the electron determines the direction of motion of the photon. The Advanced Compton Telescope [48] proposed

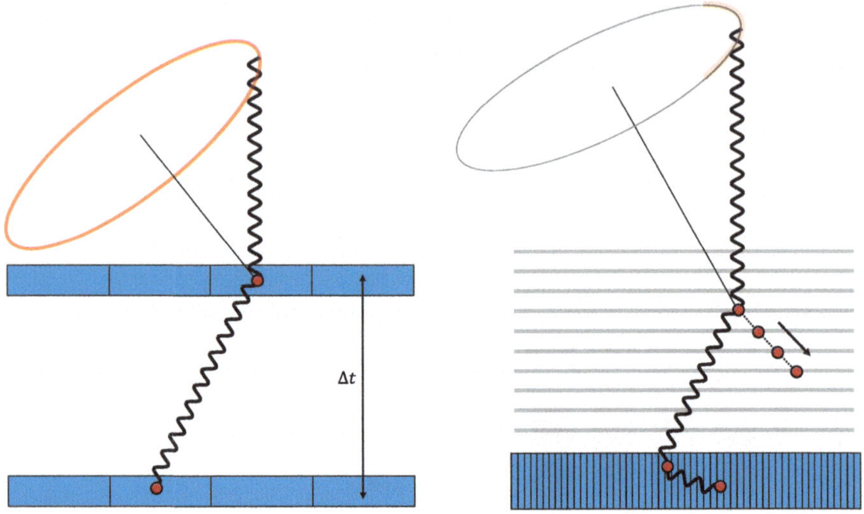

Fig. 3.19 *Left panel*: The COMPTEL instrument concept with two detector planes, the upper scatterer and the lower absorber. *Right panel*: an electron tracking Compton telescope where the track of the electron determines the direction of motion of the photon

instead the use of two arrays of low-Z (Si) and high-Z (Ge) detectors for scatterer and absorber, respectively.

Pair tracking telescopes consist of a pair-tracking device coupled with a gamma-ray absorber (a calorimeter that could be a scintillator detector) and an anti-coincidence system to reject background. The pair-tracking device is an instrument made of a vertical series of layers, with converter layers interleaved with tracking material layers. The converter is typically a heavy metal (such as Tungsten or Lead) which provides the target for creating the initial electron-positron pair while the tracking material detects the electron and positron tracks, which are recorded by an electronic readout system. The absorber device (called calorimeter) is used to measure the energy of particles produced within the tracking device. The tracks are used to determine the direction and energy of the primary photon. A plastic scintillator (see Sect. 3.6.2) surrounds the instruments and acts as anti-coincidence shielding system for background rejection. Figure 3.20 (left panel) shows the basic design of a pair-tracking telescope. More details in Chap. 4.

3.2.5 *Diffracting X-Ray Telescopes*

The use of crystals as X-ray concentrators by means of Bragg's diffraction (Eq. 3.29) is a possible solution to extend X-ray focusing techniques to the hard X-ray and soft gamma-ray bands, since the maximum incident angle for constructive interference

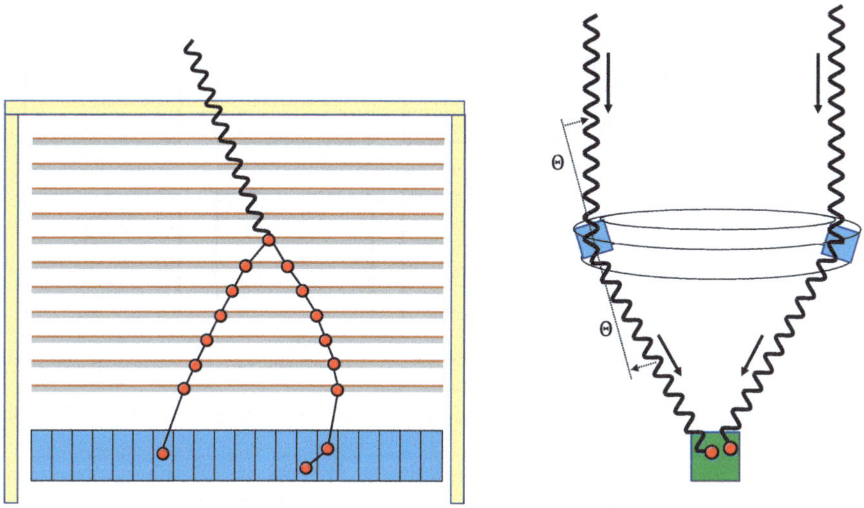

Fig. 3.20 A pair-tracking telescope (left panel), with the calorimeter in blue and the tracker planes in red, and a schematic view (right panel) of gamma-ray focusing obtained by Bragg's diffraction through Laue lenses

is larger than the angular range allowed in total external X-ray reflection. The use of Mosaic crystal mirrors composed by many, almost parallel, crystals, enabling the focusing of X-ray photons over a wide energy band, have been proposed in Schmidt's lobster eye configuration [49] for hard X-ray telescopes (>10 keV).

Focusing diffraction lens telescopes featuring Laue lenses [50] make use of mosaic or curved crystals to diffract the radiation in a focus point over a certain energy range (Fig. 3.20, right panel). Technological progress in the field of diffracting crystals for Laue lenses have increased the sensitivity and signal-to-noise of such systems such that Laue lenses for astrophysical applications have been proposed in recent years to image the Universe in the 100–1000 keV energy range. See, e.g., [51–53] for a review of Laue lenses for astrophysics.

3.3 Semiconductor Detectors

Detection systems based on semiconductors or solid state materials (e.g. Silicon, Germanium, CdTe, CdZnTe) have largely replaced scintillation detectors (See Sect. 3.6.2) in many applications and they are currently the most used detection devices in astronomy (see [54] for a review). The basic information carriers are the electron-hole pairs generated by ionizing radiation. Their motion in an applied electrical field creates the detector electrical signal.

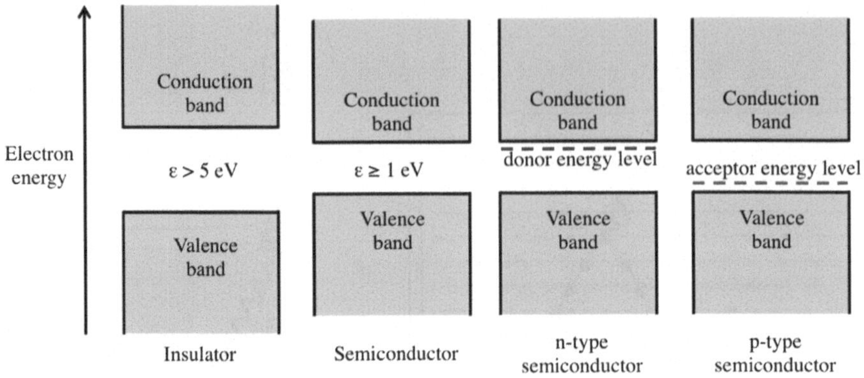

Fig. 3.21 Schematic view of energy bands in, from left, insulators, pure, and n-type and p-type semiconductors

3.3.1 Principles of Operation

In solids, the discrete energy levels of atomic electrons become a set of energy bands. At zero temperature, the highest energy band containing electrons is called the valence band. The electrons within the crystal almost completely fill the electron states within the valence band and electrons are unable to move between atoms in the solid. The conduction band lies at the top of the valence band and is separated from it by a band gap. The size of the band gap (Fig. 3.21) defines the solid as an insulator (>5 eV) or a semiconductor (≥1 eV). In the absence of thermal excitation, the conduction band is empty and there is no electrical conductivity. Electrons need to gain an amount of energy sufficient to cross the band gap and reach the conduction band.

In metals, the valence and conduction bands overlap and the latter is easily populated by electrons free to move, hence the high electrical conductivity of such materials.

In semiconductors, the band gap is small enough to only allow conductivity after an excitation process, e.g. an increase in temperature or the absorption of a photon, where an electron-hole pair is generated: an electron moves to the conduction band leaving a vacancy (a *hole*) in the valence band. The electron-hole pairs are the solid state analog of electron-ion pairs in gas-filled ionization chambers. The electron is able to drift through the detector under an applied electric field and the vacancy, representing a positive charge, moves to an opposite direction until they are collected at an electrode creating an electric signal. Without an applied electric field, the electron-hole pairs recombine after a certain amount of time.

The probability $p(T)$ to thermally generate an electron-hole pair at a give time is given by:

$$\mathrm{P(T)} = \mathrm{C\,T}^{3/2} \exp\left(-\frac{\mathrm{E}_g}{2\mathrm{kT}}\right), \tag{3.32}$$

where T is the absolute temperature, E_g is the band gap energy, k the Boltzmann constant, and C a constant depicting the properties of the material. The smaller the band gap energy and/or the temperature, the lower the probability for electrons to move to the conduction band from thermal excitation only. Under an electric field, the current of the radiation-induced free electrons is superimposed by the temperature-dependent free charge density (leakage current) and random thermal motion resulting in charge diffusion from the point of origin.

3.3.2 Impurities in Semiconductors

The diffusion, that can be described as a Gaussian spread in the charge position, is generated by impurities present at different levels in all semiconductor materials. Donor (*n-type*) and acceptor (*p-type*) impurities increase the electrical conductivity of the material by additional conduction electrons or valence holes, respectively. Deep impurities, where materials such as Gold, Zinc, Cadmium, or other metallic atoms, substitute lattice positions, can trap charge carriers or accelerate the recombination reducing their lifetime in the crystal from 1 to 10^{-5} s. The charge collection time in a semiconductor detector must be around 10–100 ns to collect a large fraction of all the carriers. Impurity-induced effects can be important for large volume detectors and high precision position or time measurements, since the mobility of electron-hole pairs should be high and the lifetimes long for fast signal rise time and efficient charge collection.

In an n-type semiconductor, impurities have one more electron than the host material and its electrons can more easily be excited towards the conduction band than valence electrons. On the contrary, p-type semiconductor impurities have an electron less than the host in the outer electron structure and the valence electrons can easily migrate to fill the vacant energy level leaving a hole. The presence of both donor and acceptor can compensate for the impurity induced imbalance, resulting in conduction properties close to pure or intrinsic semiconductors. This is the case in Lithium drifted detectors (SiLi, GeLi) where p-type Silicon or Germanium acceptor impurities are compensated by Lithium atoms.

In a so-called *p-n junction*, where a p-type and n-typed doped regions within the same semiconductor material are placed in contact, conduction electrons from the n-type side diffuse towards the p-type region where they combine with holes while holes diffuse across the junction in the opposite direction (e.g. in Silicon diodes). The combined effect is the creation of a negative charge in the p-side and a positive charge in the n-side and the generation of an electric field that prevents additional diffusion through the junction. The region along the p-n junction where a charge imbalance exists is defined as the *depletion region*. The electric field extending in the depletion region causes any electron and hole generated within or near the junction to move to the n-type and p-type region, respectively. The region is then *depleted* of electrons and holes.

3.3.3 Semiconductors as Detection Devices

The depletion region in semiconductors, having the concentration of charge carriers greatly suppressed, is characterized by a very high resistivity. An electron-hole pair generated by ionising radiation would drift away because of the electric field generating an electrical signal. The role of the depletion region as active detection area is increased in the presence of a reverse biased junction, where a negative voltage is applied to the p-side and a positive one to the n-side. In this case, the resistivity is enhanced and the thickness of the depletion region increased, extending the active area where charge carriers produced by radiation can be created and collected while lowering the leakage current.

Semiconductor materials are a thousand times denser than gases, hence they have a much higher stopping power (see Eq. 3.6). Thin layers with few μm precision can be produced at low cost while ensuring the same, or higher, attenuation efficiency at X-rays and gamma-rays. The most important advantage of semiconductor devices with respect to scintillators in high energy astronomy is, however, the smaller ionization energy ε, ranging from about 3 eV to create an electron-hole pair for Silicon and Germanium to about 30 eV to create an ion pair in gas-filled detectors. The increased number of charge carriers generated by the same energy deposit reduces the statistical fluctuations, which limit the energy resolution of a detector, and increases the signal-to-noise ratio. As result, the energy resolution obtained in semiconductors is close to the intrinsic limit of the Fano factor (2% at 6 keV for Si, 1% at 60 keV for Cd(Zn)Te) while for NaI scintillators it is limited by the photoelectron statistical fluctuations to about 6% at 0.662 MeV. The Fano correction factor F is a property of the material and defines a better energy dispersion than the one predicted by Poisson statistical fluctuations, since the creation of charge carriers is not entirely a random process because of discrete electron shells:

$$\mathrm{F} = \frac{\sigma^2}{\mathrm{E}_\gamma/\varepsilon}\,, \tag{3.33}$$

where σ^2 is the observed variance of carriers and E_g is the photon energy. F is equal to 1 for Poisson statistics but lower than 1 for observations. Given $N_c = E_\gamma/\varepsilon$ the mean number of charge carriers, the energy resolution (Sect. 3.1.5) can be linked to the Fano factor as follows:

$$\mathrm{R} = \frac{\mathrm{FWHM}}{\mathrm{E}_\gamma} \approx \frac{2.355\sigma}{\mathrm{N}_c} = 2.355\sqrt{\frac{\mathrm{F}\varepsilon}{\mathrm{E}_\gamma}}\,. \tag{3.34}$$

Si, Ge, CdTe, CdZnTe detectors are tailored for both high resolution imaging and spectroscopy observations with versatile readout, compact dimensions, and high granularity. One major disadvantage in applying semiconductor materials for X-ray and gamma-ray detectors is their greater sensitivity towards radiation damage induced by particles hitting the material, with consequent loss in detection performance (e.g. increase in charge transfer inefficiency and leakage current). Dedicated studies of the space radiation environment and consequent application of shielding solutions must be in place (see Sect. 3.7).

3.3.4 Silicon, Germanium, CdTe and Cd(Zn)Te

Silicon (Si) has a lower atomic number and density (Table 3.2) than Germanium (Ge), Cadmium Telluride (CdTe) and Cadmium Zinc Telluride (CZT), and, therefore, a lower photon detection efficiency above 40 keV. For this reason, Si-based devices are used for detection of X-rays below tens of keV while Ge, CdTe, and CZT are applied at higher energies.

The width of the band gap defines the feasibility of a semiconductor material for operations at room temperature. The large band gap of Si, CdTe, and CZT make them suitable for observing at ~300 K, while cooling at liquid nitrogen temperature is required for Ge-based devices. However, the smaller band gap of Ge facilitates the production of free carriers and increases its spectral resolution.

Si is the semiconductor material of choice in the production of position sensitive detectors (e.g. particle trackers) thanks to its abundance, bandgap, the possibility to fully deplete the volume by means of reverse biased junctions and the microscopic structuring in strips and pixels [55]. The EPIC detectors on-board XMM-Newton and the electron-positron tracking planes in Fermi/LAT and AGILE/GRID are made of Si.

High purity Germanium detectors are instead applied in gamma-ray instruments that do not require a fine spatial resolution but, rather, a high resolution spectroscopy capability, which can be achieved in a broad energy range from tens of keV to tens of MeV. Their depleted region can reach centimeters, contrary to the Si that has a few mm thickness at maximum, allowing Ge detectors to fully absorb gamma-rays up to a few MeVs. The spectrometer on-board INTEGRAL is based on Ge.

Table 3.2 Atomic number, density (in g/cm^3), energy of the band gap between valence and conduction band (in eV), the ionization energy required to move an electron to the conduction band (in eV), and energy resolution in FWHM of Silicon, Germanium, CdTe, and Cd(Zn)Te [3, 55]

Properties of common semiconductor materials

Type	Atomic number	Density	Bandgap energy (300° K)	Ionization energy	Energy resolution (FWHM)
Si	14	2.33	1.14	3.62	550 eV at 122 keV (77 K)
Ge	32	5.33	0.67	2.96	400 eV at 122 keV (77 K)
CdTe	48–52	6.20	1.54	4.43	3.5 keV at 122 keV (300 K)
Cd(Zn)Te	48–30–52	5.78	1.58	4.64	3.2 keV at 122 keV (300 K)

CdTe semiconductor devices combine a high atomic number, which results in a high quantum efficiency for observations in the 10–500 keV energy range, with a wide band gap making their operations possible at room temperature [56]. The CdTe photoelectric absorption per unit path is 4–5 times higher than in Ge and 100 times higher than in Si, allowing for a compact detector size. CdTe is the material of choice for hard X-ray detectors coupled with X-ray focusing mirrors and coded-aperture telescopes. In comparison with Ge they show a modest spectroscopy performance because of poor charge collection efficiency, especially for holes, an issue that limits their dimension in size. By increasing the Zinc (Zn) concentration, CdZnTe semiconductors have a larger band gap than CdTe resulting in a higher resistivity and lower leakage current, which increases the energy resolution and allows for larger sizes. The hard X-ray detector on board NuSTAR is made of CdZnTe.

3.4 X-Ray Wide Field Imagers

If performing large surveys of the high energy Universe or continuously scanning the sky for transient sources are the main objectives of the mission, wide field imagers covering large portions of the sky are in place based on the use of pixelated imaging detector arrays.

3.4.1 Charge-Coupled Devices (CCD)

Charge-coupled devices (CCDs) constitute one of the most used sensors for X-ray astronomy imaging since the 1970s because they combine large sizes and efficiency with low noise (see [57] for a review). Astronomers rapidly adopted CCDs for optical imaging in ground-based telescopes. Conventional devices consist of small and thin electrodes of metal-oxide-semiconductor (MOS) capacitor structures on top of a p-type silicon wafer. A positive bias applied to the electrodes creates depletion regions used to store the signal charge (the electrons) in potential wells underneath the p-type silicon oxide, and then the charge is transferred to a collecting electrode by a suitable change of the gate potentials. All CCDs integrate two features to limit the charge transfer inefficiency. A thin n-doped layer between the silicon and the oxide (the n-buried channel) prevents the trapping of electrons by the surface crystal defects. The p-type strips implanted perpendicular to the gate electrodes (the channel stops) prevent charge spreading towards the adjacent rows during transfer.

A strip of an insulator is implanted to create an array of CCD pixels, in structures known as channel stops, to isolate the charge packets held under each pixel. In most CCDs, each pixel has three electrodes attached to it, to hold in place the charge packet, with the central electrode held at a more positive potential than the two on either side. Together, the channel stops and the electrode triplet define the pixel. Figure 3.22 shows the schematic view of a portion of CCD.

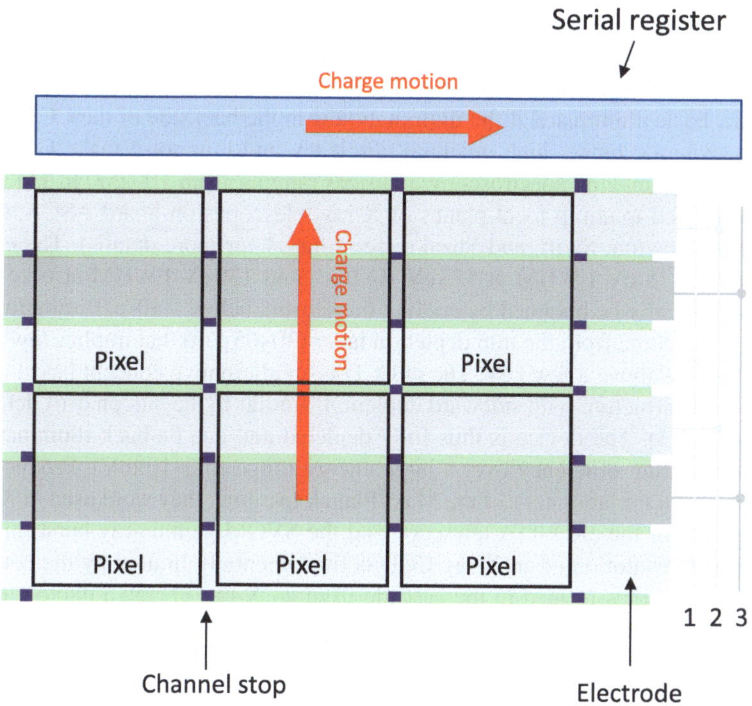

Fig. 3.22 Schema of a portion of a CCD. The combination of channel stops and triplet electrode structure define the CCD pixel and is used to confine the charge packets. Charges are moved in the CCD circuit by manipulating the voltages on the gates of the electrodes in sequence (electrodes 1, 2 and 3 in the figure, see text for more details). The charge is moved to a horizontal serial register, is amplified, and acquired by an electronic readout system

The method of signal readout gives CCDs their name. CCDs are shift registers, in which the collected charge is shifted across the device in a "bucket brigade" technique towards the readout amplifier. Once the CCD has been exposed to light for the required amount of time, each pixel will contain a charge packet of size proportional to the number of incident photons. To measure this charge, it is necessary to move these charge packets, one by one, off to the chip. Usually, three electrodes are used to transfer charge between pixels. During the exposure time, a large positive voltage applied to one of these electrodes will attract the photo-electrons (electrode 1 in Fig. 3.22). To move the charge, the voltage in an electrode adjacent to the electrode holding the charge packet is raised to the same level (electrode 2 in Fig. 3.22); this allows the charge to flow between the two electrodes. The voltage of the first electrode is decreased, and the charge transfer is complete. The same operation is also repeated for electrode 3 to move the charge packet by one pixel. The charge of a sequence of pixels (the vertical register in 3.22) is then acquired by a horizontal serial register, amplified, and read by an electronic readout system.

The detection characteristics of a CCD depends on the method of illumination (in addition to the physical characteristics of the device). CCDs are divided into front and back-illuminated. For front-illuminated CCDs, the radiation passes through the electrode, back-illuminated if the illumination is in the backside of the CCD.

High dynamic range, high quantum efficiency and low noise make CCD suitable for X-ray imaging spectroscopy; matrices ranging from 100,000 to 1M pixels were developed to equip focal planes of X-ray telescopes on-board ASCA, Chandra, XMM-Newton, Swift, and Suzaku (see Chap. 4 for more details). The energy resolution of 75 eV FWHM at 523 eV (O line) and 130 eV FWHM at 6 keV (Fe line) can typically be obtained by cooling the sensors below −50°c. Limitations for a MOS CCD come from the thin depletion layer (30–65 μm) that implies low quantum efficiency above a few keV. The pnCCD is an alternative concept based on the double diode structure with sideward depletion, similar to the silicon drift detectors (see Sect. 3.4.3). The device is thus fully depleted and can be back-illuminated to improve quantum efficiency over a large energy range (0.2–10 keV). Produced by the semiconductor laboratory of the Max–Planck Institute, they were used in XMM-Newton and for the eROSITA telescope and the SVOM gamma-ray burst mission. The spectral resolution of an X-ray CCD is fundamentally limited by the counting statistics of charges formed in the cascade from an X-ray liberated photo-electron to carriers at the band-gap of the material. This gives them moderate spectroscopic resolving power of about 50 at 6 keV.

3.4.2 Active Pixel Sensors

Active pixel sensors (APS) combine high resistance Silicon absorbers with doped Silicon front-end electronics, keeping the first stage of amplification of the signal within the pixel. With respect to CCDs, active pixel sensors are characterized by a higher charge transfer efficiency, since the charge is collected within the pixel, thus increasing the energy resolution while allowing larger array sizes and higher frame rates. The Wide Field Imager on board ATHENA will use active pixel sensors: a DEPFET (Depleted P-channel field effect transistor, i.e. a p-channel MOS-transistor) which combines sensor and amplifier in each pixel by a MOSFET integrated onto fully depleted n-doped silicon bulks [58]. The deep-n implantation below the MOS-gate forms a potential minimum for electrons. The X-ray photon is absorbed in the bulk and the electron cloud drifts towards this so-called internal gate changing the transistor current proportionally to the number of charges. By detecting the transistor modulation, the amount of collected electrons and consequently the photon energy can be measured. The charge is stored until applying a positive voltage on the clear structure to remove it from the internal gate.

3.4.3 *Silicon Drift Detectors*

In Silicon Drift Detectors (SDD), high resistivity n-type silicon wafers have p+ semiconductor junctions at both sides and a small n+ electrode on one side that are reverse biased until the Silicon bulk is fully depleted (principle of side-ward depletion [59]). The positive voltage applied to the n+ electrode (the depletion voltage) is four times lower than the voltage needed to deplete a pn diode of similar thickness (see [60] for a review). In an SDD, the p+ junctions, segmented in strips or rings, generate an electric field parallel to the surface. Ionizing radiation absorbed in the depleted region creates electrons that drift in a direction parallel to the wafer surface towards the collecting anode placed near the edge of the wafer (see Fig. 3.23). The n+ anode is then connected to an amplifier. The holes move instead towards the p+ junctions.

Thanks to the specific electrode geometry, the anode is characterized by a very low capacitance almost independent of the detector area. For this reason, SDDs have a very low noise and are well suited for low energy X-ray spectroscopy while covering large areas. Initially used in particle physics as position sensitive particle detectors [62], SDDs are today applied in X-ray astronomy both as read-out detectors in scintillator bars (e.g. in the CsI(Tl) calorimeter proposed for the e-ASTROGAM mission [63]), sensors in active pixels [64], and stand-alone X-ray imaging and spectroscopy detectors at soft X-rays (e.g. the XGIS instrument in the proposed THESEUS mission [65]).

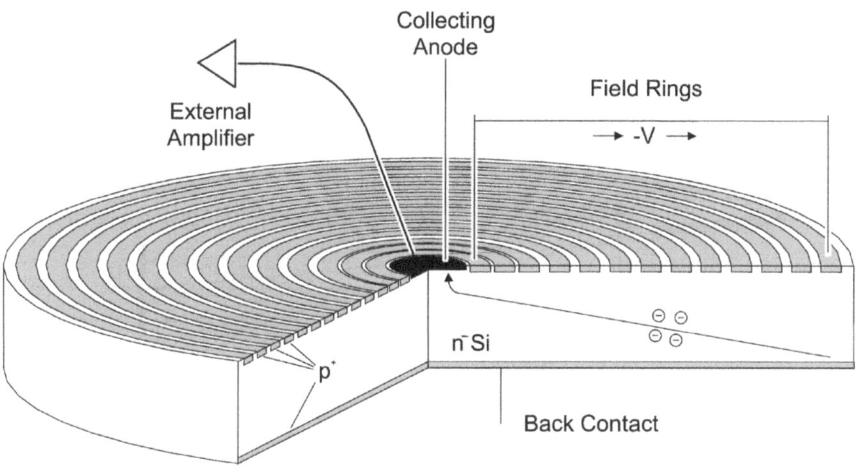

Fig. 3.23 An example of a cylindrical Silicon Drift Detector. An electric field moves the electrons to the collecting anode in the center [61]

3.5 Pushing the Resolution Limit of X-Ray Observations

The need for disentangling spectral lines from absorption and ejection processes in a large variety of stellar, interstellar, and extragalactic ionized regions, or to accurately measure their energy to derive the velocity of plasma in winds and jets, fueled the development of dedicated technologies to lower the energy resolution down to a few eVs in the soft X-ray band.

3.5.1 X-Ray Diffraction Gratings

X-ray diffraction gratings placed at the focus of the focusing system are a set of reflective stacks or transmission slits that disperse the X-rays by means of diffraction according to their wavelength. An array of photo-ionization detectors, usually CCDs, collect the energy and position of the diffracted photons to create a high resolution X-ray spectrum of the source while separating the contribution from the various overlapping grating orders and reduce the background. Gratings are on board of both the XMM-Newton (Reflection Grating Spectrometer, RGS) and Chandra (the High Energy Transmission Grating Spectrometer, HETGS) X-ray telescopes. Reflection, instead of transmission, gratings are used in XMM-Newton because of the lower angular resolution of its mirror with respect to the Chandra telescope. The XMM-Newton RGS provides high spectral resolution ($E/\Delta E$ from 200 to 800 for point sources) X-ray spectroscopy over the 0.35–2.5 keV energy range [66] while the Chandra HETGS reaches a resolution of $E/\Delta E$, for point sources, up to 1000 between 0.4 and 10 keV (Fig. 3.24).

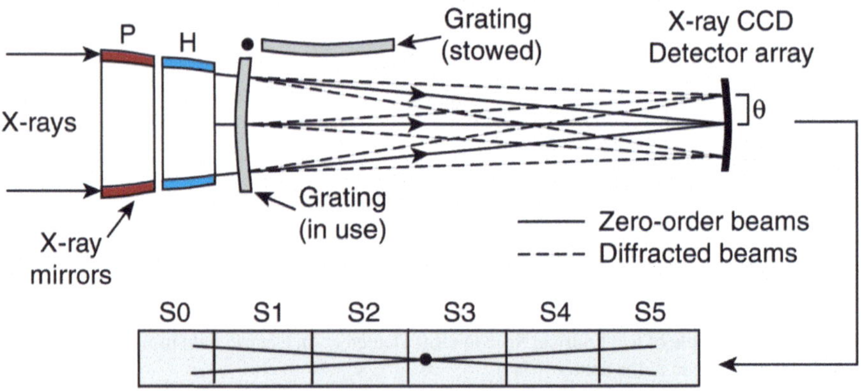

Fig. 3.24 Operating scheme of the X-ray transmission grating system on board the Chandra X-ray telescope [67]

3.5.2 X-Ray Micro-calorimeters

In X-ray micro-calorimeters the photon energy is measured as the temperature rise of an absorber material at very low temperatures, then converted to an electrical pulse. They reach an energy resolving power of over 3000, ~50 times better than solid-state detectors, over a large bandpass while ensuring a near unity efficiency. The ability to build arrays of micro-calorimeters adds imaging capability to higher resolution spectroscopy and the possibility to study extended sources (see [68] for a full review). Below 2 keV, in the very soft X-ray band, gratings still ensure a higher energy resolution ($E/\Delta E > 5000$ for the Lynx mission proposal [69]).

The incident X-ray photo-ionizes the absorber, which releases a primary photo-electron and consequent additional electrons. These electrons thermalize in the absorber converting the photon energy to temperature, causing a rise in temperature of a few milli-Kelvin (Fig. 3.25, left panel). The absorber must have a high photo-electric cross-section and a low heat capacity. A thermistor is a device that changes its electrical resistance dramatically with a small change in temperature. The thermistor, placed at good thermal contact with the absorber, rises in temperature in few milliseconds (Fig. 3.25, right panel), and the temperature gradient ΔT is proportional to the photon energy E and the absorber heat capacity C:

$$\Delta \mathrm{T} = \frac{\mathrm{E}}{\mathrm{C}} \ . \tag{3.35}$$

A weak thermal link connects a heat sink to the detector. The heat leaks to the heat bath, kept at constant temperature, and the detector returns to its base temperature ready for the next photon interaction. The accuracy in the photon energy measurement

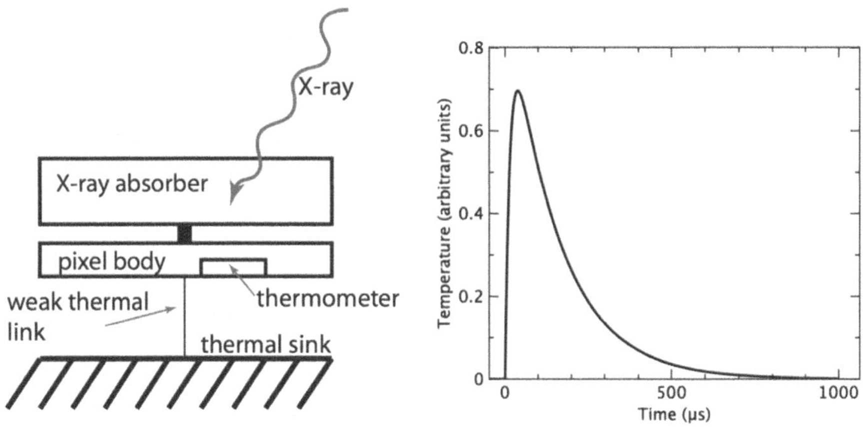

Fig. 3.25 *Left panel*: schematic view of an X-ray micro-calorimeter detector: the incoming X-ray photon is photo-absorbed in the absorber and a thermal signal is registered with a sensitive thermometer [68]. *Right panel*: the fast rise and slower decay of the temperature (or resistance) with time for the ATHENA X-IFU [5]

is limited by thermodynamic fluctuations associated with energy exchange across the weak thermal link. The energy resolution depends on the temperature T_0 of the heat sink, the heat capacity C at that temperature, and the thermistor responsivity α, taking the form:

$$\Delta \mathrm{E} \propto \sqrt{\frac{\mathrm{k_B}\ \mathrm{T_0}\ \mathrm{C}}{\alpha}} \tag{3.36}$$

where k_B is Boltzmann's constant. Cryogenic bath temperatures ($T_0 < 100$ mK) are required to reach high energy spectroscopy, which makes the realization and the readout of matrices extremely challenging.

At the time of writing, X-ray micro-calorimeters have shortly operated in space on board the Suzaku and HITOMI JAXA/NASA missions, because of cryogenic system and attitude control failures, respectively. They both used semi-conductor (Si) transistors and an absorber made of HgTe (Mercury-Telluride), a compromise between good thermalization properties and low heat capacity. They X-ray calorimeter covered the 0.3–12 keV energy band, with an efficiency >80% below 8 keV and an energy resolution <7 eV.

The use of superconductor-based transistors, commonly known as transition-edge sensors (TES), as in the case of the X-IFU instrument on board the proposed ATHENA mission, will achieve a spectral resolution of 2.5 eV up to 7 keV on $\sim 5''$ pixels. [70]. Superconductors behave like normal electrical conductors at high temperatures, but below a certain energy threshold they reach perfect conduction with zero electrical resistance. TES arrays are superconductors operated at the edge of the transition, with an electrical resistance extremely sensitive to temperature variation. The very high responsivity ($\alpha = 100-1000$) permits a wider range of absorber materials because of the greater range of allowed heat capacities. In addition, it is easier to produce large arrays of sensors with respect to semi-conductor transistors.

3.6 Hard X-Ray and γ-Ray Detectors

Moving towards higher energies—above tens of keV—photo-ionisation attenuation is reduced, hence materials with higher cross-sections and/or densities, or larger volumes, are required for photo-counting detectors. Starting from hundreds of keV, particle detectors to reconstruct the primary photon energy and position are a valid solution to overcome the decrease in detection efficiency. The most common concepts used for X-ray and gamma-ray detection and imaging are briefly described in the following sections.

3.6.1 Proportional Counters

A proportional counter is a gas-filled X-ray detector based on the principle of electron cascade. An array of wires may be used to make an imaging proportional counter. An energetic X-ray induces ionisation of the gas (typically Neon or Argon with traces of other gases) that fills the detector volume, producing photo-electrons. These energetic photo-electrons ionise a number of atoms creating a cloud of secondary electrons that are accelerated towards a thin wire anode. The number of secondary electrons is proportional to the number of the primary electrons, and consequently the initial photon energy. The energy resolution of a proportional counter is limited by the statistical fluctuations in the multiplication process, resulting in values higher than what is predicted by the Fano factor (using Eq. 3.34, more than 10% [3] at ∼6 keV). The operative energy range of proportional counters depends on the interaction efficiency of the X-ray and gamma-ray photons with the counter gas, decreasing below 10% at 10–20 keV (Argon), ∼40 keV (Krypton) and ∼90 keV (Xenon). Thus, proportional counters can be used for detection and spectroscopy of soft X-rays or gamma-rays whose energy is low enough to interact with reasonable efficiency in the counter gas.

The ROSAT satellite [71] was equipped with two redundant multi-wire position-sensitive proportional counters (PSPC) filled with a mix of about 65% argon, 20% Xeon, and 15% methane. Photons entered through a window and the area of the detector was crossed with a grid of cathode wires to determine the event location.

3.6.2 Scintillators

A scintillation detector (a.k.a scintillation counter) is a detector with a scintillator material coupled to an electronic light sensor. Examples of light sensors are photomultiplier tubes (PMT), photodiodes, silicon photomultipliers, or silicon drift detectors (see Sect. 3.4.3).

Scintillators are one of the oldest types of g-ray detectors. The incoming photon interacts within the material via the photoelectric effect, Compton scattering, or pair production, producing energetic electrons that lose energy via bremsstrahlung and ionisation, causing atomic electrons to be raised to an excited state which then return to a stable state by producing scintillation light in the visible or near-visible energy range. Charged particles can also produce scintillation light by ionisation along their track.

Figure 3.26 reports a schematic view of a scintillation detector comprising a scintillation material coupled with a photomultiplier tube. The photomultiplier is a photoemissive device in which the absorption of a photon results in the emission of a cascade of electrons. It consists of a glass tube, in which there are several electrodes that make up the dynodes. Photons hit a surface—the photocathode—through an input window, covered by a layer of material that favours the photoelectric effect. Photo-electrons are accelerated by a voltage potential and focused so that they strike the first dynode with enough energy to release additional electrons, amplifying the electrons generated by the photocathode exposed to the photon flux. At the end of the dynode chain the anode collects the electrons.

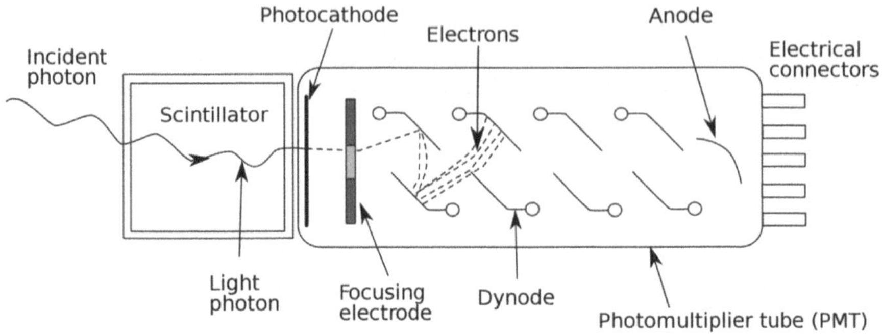

Fig. 3.26 Schematic view of a scintillation detector comprised of a scintillation material coupled with a photo-multiplier tube [CREDIT: https://web.stanford.edu/]

Two types of scintillators are commonly used, inorganic crystals and organic scintillators. The scintillation mechanism is different and depends on the structure of the crystal lattice.

In organic scintillators the emission process is based on the transition within the energy-level structure of the molecule and where optical photons are emitted via fluorescence and phosphorescence after an external excitation by gamma-ray or ionising radiation. These materials have small cross-sections and, for this reason, they are poor gamma-ray absorbers but have the fastest response time.

In a pure inorganic crystal material such as NaI, electrons are only allowed to occupy selected energy bands. Under normal conditions, the electrons are bound to the atoms of the lattice, and the conduction band, where the electrons can move freely, is empty. A charged particle moving through a crystal can excite an electron through absorption of energy that elevates electrons from the valence band to the conduction band leaving a gap in the valence band. After some time the return of an electron to the valence band occurs with the emission of a photon, but this is an inefficient process because few photons are released per decay; in addition, band gap widths in pure crystals are such that the resulting emitted photon is too high to lie within the visible range.

The discovery that small amounts of impurities can be added to the crystal to make the process more efficient changed the use of these materials. The impurities create special sites in the lattice structure at which the band gap is modified. They are called activators. Note that the energy structure of the overall crystal is not changed, just the energy band gap at the activator sites. The excitation of these activator states occur more efficiently and, with the right material, the photons emitted from the de-excitation process lie in the visible wavelength region. The energy levels created by the activator's presence within the crystal are narrower than in the pure crystal. Also, with these impurities, self-absorption is suppressed.

The main properties of a scintillator are: (i) high light output per energy unit deposited, (ii) transparency to its own fluorescence radiation; (iii) high stopping power for the type of radiation studied; (iv) good linearity over a wide energy range; (v) short rise and decay times; (vi) temperature independent.

Inorganic scintillation crystals are efficient solutions for energy measurement in gamma-ray experiments above 10 MeV because they can be implemented in large volumes (e.g. the AGILE and Fermi calorimeters). Plastic scintillators are in fact commonly applied as active shielding systems in high-energy experiments for background particle detection and rejection against incident charged particles. Surrounding the main detection instrument, they generate triggers in coincidence with the main instrument if the event is caused by a charged particle and veto the false detection, while being transparent to gamma-rays.

Some of the first detectors using scintillators include OSO-7 and the BATSE and OSSE instruments on CGRO (see next chapter for a full description of this topic).

3.6.3 Silicon Photomultipliers

Silicon photomultipliers (SiPM) are solid-state single-photon-sensitive devices based on a single-photon avalanche diode (SPAD) implemented on a common silicon substrate. A SPAD is a solid-state photo-detector in which a photo-generated carrier (via the photoelectric effect) can trigger a massive avalanche current of short duration. The creation of this avalanche through impact ionization generates electrons and holes (the carriers) accelerated to high kinetic energies through a large potential gradient (voltage). If the kinetic energy of a carrier is sufficient, further carriers are released from the lattice. In this way, the number of carriers increases exponentially. The dimension of every SPAD can vary from 10 to 100 μm.

Silicon photomultipliers (SiPMs) are an option for space-based high-energy detectors because of their requirements on the small size, weight, and power consumption, coupled with reproducibility of single elements of an array. Many experiments are adopting SiPMs for existing detectors or current developments, with different requirements in terms of performances and operating conditions of these devices. One of these examples is the proposal AMEGO (see Chap. 4).

3.6.4 Pixel CZT Detectors

Cadmium Telluride and Cadmium Zinc Telluride Cd(Zn)Te based detectors have become the best prospect for room-temperature X-ray detectors. The band gap spans from 1.4 to 2.2 eV (depending on the material). The advantages of CZT detectors include high sensitivity for X-rays and gamma-rays, due to the higher atomic numbers of Cd and Te, and better energy resolution than scintillator detectors. They achieve good spatial and energy resolutions without the need for cryogenic cooling over the broad energy range from 10 keV to $\sim$600 keV.

INTEGRAL, launched in 2002, carries the Imager on Board the Integral Satellite (IBIS) that uses CdTe detectors with a detector area of $\sim$2600 cm^2. The SWIFT satellite carries the Burst Alert Telescope (BAT) with a CZT detection area of $\sim$5240 cm^2. Both IBIS and BAT use the approach of coded mask imaging (see Sect. 3.2.2).

For hard X-ray focusing optics, new concepts of Cd(Zn)Te hybrid detectors have been developed. For the NuSTAR hard X-ray telescope, launched in 2012, the California Institute of Technology designed a 2D ASIC with front-end channels in 498 μm pixel sizes [72]. The CZT pixel sensor is connected to 2 ASICs with conductive epoxy and gold stud using a flip-chip bonding technique. The ASICs are glued and connected by wire bonding to an interface board holding the analogue to digital converters. Energy resolution better than 0.8 keV FWHM at 60 keV and 1.5 keV low-energy threshold can be obtained over the array in both designs.

3.6.5 *Cherenkov Detectors*

When a charged particle moves through a transparent medium faster than the local speed of light within the medium Cherenkov photons are emitted. These photons can be measured with normal photo-tubes and, for this reason, so-called Cherenkov detectors are similar to scintillators detectors. The satellites described in Chap. 4 and equipped with a Cherenkov detector were Explorer XI, OSO-3, and SAS-II (see Chap. 4).

3.6.6 *Spark Chambers*

The basic principle of spark chambers is the detection of a track left by the passage of a charged particle. When passing through matter, it interacts and knocks out electrons from the atoms of the material, disturbing the structure of the material and creating free electrons. These electrons, moved from their position, can be collected and tracked.

Spark chambers were used in high-energy astrophysics for $E > 30$ MeV photons, where the dominant interaction process is pair production. The initial photon interaction occurs in a high-Z material (e.g. lead or tungsten), producing a e^+e^- pair. These particles produce ionization tracks in the spark chamber, filled with gas (Neon) located between a pair of electrodes and composed of several layers of thin metal plates or metal wires. A high voltage pulse, triggered by a scintillator, is applied to the electrodes, producing a visible track of the ionization trail. Since the direction and range of the pair are recorded, it is possible to determine the initial energy of the gamma-ray. A scintillator can be used as an anti-coincidence system.

A trigger telescope is mounted below the spark chamber; it consists of two thin plastic-scintillator plates separated by several tens of cm and viewed by photomultipliers. The electrons and positrons move through the spark chamber, ionizing the gas along their flight path, then they penetrate the two plastic scintillators of the trigger telescope, where photons are produced and registered by the phototubes. A trigger pulse is applied with a coincidence measurement. This pulse is used to fire the spark chamber by applying a high voltage to its metal plates or wires, and a spark is broken along the ionization path. The position of the spark is recorded via optical or electronic readout. To reconstruct the tracks of e^+e^- in three dimensions, the readout is made from two orthogonal directions. The initial direction of the incoming gamma-ray is determined from the bisector of the angle of the V-shape tracks.

The spark chambers have a limited lifetime when used on satellites. The sparks degrade the gas, i.e. the performance of the detector. As a consequence, during the mission the tracks become less well-traced and the reconstruction more difficult and less accurate. Spark chamber applications in astrophysics are described in Sect. 4.10.2 and 4.10.2.3.

3.6.7 Single and Double-Sided Silicon Strip Detectors

A micro-strip detector is an instrument made of identical detection structures (strips) usually in a plane, i.e. a compound of fragmented metal electrodes on a common support, typically an insulator or a semiconductor, where each of them can be considered as stand-alone detectors. The main types of microstrip detectors are silicon microstrip detectors (described in this section) and microstrip gas chambers.

A silicon microstrip detector is an arrangement of strip-like shaped implants acting as charge collecting electrodes. The main geometrical parameter of a microstrip detector is the pitch, i.e. the distance between two strips. A silicon detector is thinner than gaseous detectors, with a typical thickness for high-energy physics between 100 and 500 μm. In intrinsic silicon, there are $\sim 10^9$ free charge carriers at room temperature. Still, only about 10^4 e^- are induced by a MIP traversing a thickness of 300 μm, and this means that a MIP signal is lost among a large number of free charge carriers. To operate as a silicon detector these sensors must be fully or partially depleted of free charge carriers using p-n junctions operating in reverse bias (e.g. of about 100 V) such that only thermally-generated currents contribute to the leakage current.

3.6.7.1 Single Side Strip Detector

A single side strip detector (SSSD) consists of a silicon wafer that forms a p-n junction that is segmented into strips or pixels using the planar technique. Silicon sensors are often fabricated on n-type bulk by adding a donor impurity like Phosphorus to Silicon to provide an excess of electron charge carriers. Similarly, a p-type material can be realised by adding acceptor impurities like boron (type III material) that yields an excess of holes as the majority charge carriers. Figure 3.27 shows an SSSD, which is a p-n junction with electrodes divided into individual strips. Since the mobility

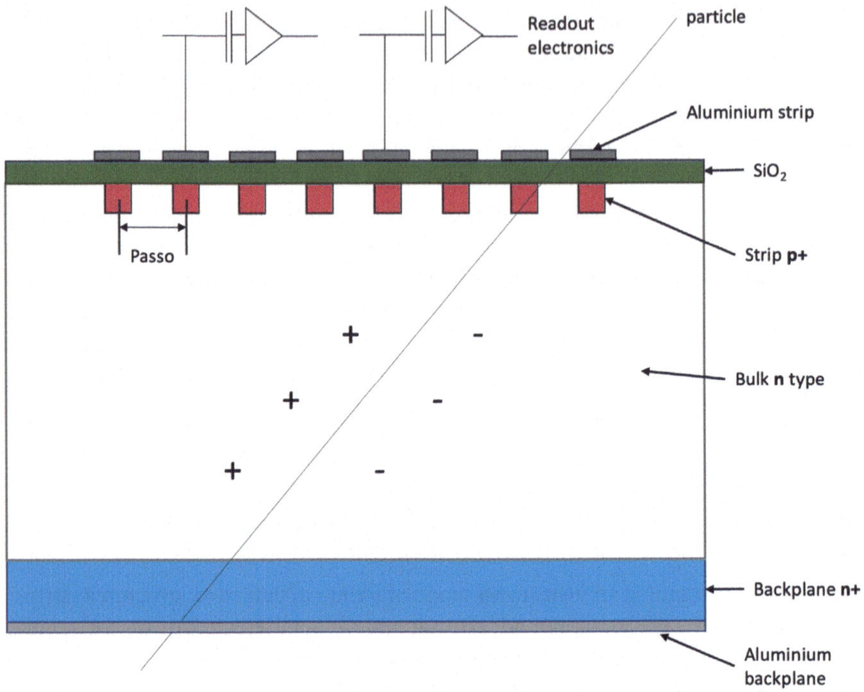

Fig. 3.27 Schematic view of a silicon microstrip detector

of electrons and holes in Silicon differs only by a factor of 3, the collection time (the time required for all charges to reach their respective electrodes) of both types of carriers are of the same order of magnitude. Furthermore, the average lifetime of the carriers (10^{-3} s for the e^- and 2×10^{-3} s for holes) is much longer than the collection time so that each electrode is sensitive to the motion of electrons and holes. The SSSD can determine the arrival position thanks to the use of the strips arranged parallel to each other and connected independently to the electronics for reading. The readout strips are connected to the electronics, while those that are not connected are called floating strips. If analog read-out is applied, the energy deposited in the strips can be used to weight the position of the hit among a group of contiguous strips, hence improving the spatial resolution. For more details, see Sect. 4.10.3.

3.6.7.2 Double Side Silicon Strip Detector

Two-dimensional position measurements can be achieved by applying additional strip-like doping on the wafer backside to build double-sided silicon strip detectors (DSSDs). Double-sided strip detectors have the advantage of having fewer read-out channels than a pixel array, possibly a more manageable connection scheme (at the sensor periphery), and thus the possibility of layer stacking. For the Hard X-ray Imager (HXI) of Astro-H integrating both double-sided Silicon strips and CdTe detectors, the strips are connected to a ceramic board by gold-stud bonding and then connected to a multi-channel front-end ASIC. Future gamma-ray missions are proposing DSSD (see Sect. 4.12) for the tracking of electrons and positrons in Compton and pair production telescopes.

3.7 Monte Carlo Simulations for Background Evaluation

The instrumental background is the sum of the residual counts surviving all shielding and filtering techniques (e.g. the vetoing of an anti-coincidence detector) and not produced by the science target. For X-ray focusing telescopes, the instrumental background is usually referred to as Non X-ray Background (NXB). The NXB is is divided in *not focused*, induced by both particles and photons interacting with the material (payload and spacecraft) surrounding the instruments and *focused*, induced by particles entering the field of view. In gamma-ray telescopes, the instrumental background is usually divided into galactic, because of the bright emission in gamma-rays of our galaxy, and isotropic, including both particles and the extra-galactic gamma-ray photons coming from all directions.

We only report the radiation effects that increase the level of photon-like events, reducing the sensitivity towards scientific observations. However, the potential degradation of the detector sensors and electronics induced by charged particles (e.g. lattice displacement, single event upsets) is always part of the engineering feasibility studies of the mission.

The action chain for the background simulation of a high energy space telescope can be summarized as:

1. environment modeling, depending on the mission orbit and science goals;
2. design of the shielding system surrounding the focal plane instruments;
3. Monte Carlo simulation of the interaction of the space particle population with the spacecraft instruments and payload;
4. processing of the simulation scientific output (e.g. background spectra production);
5. comparing the background level with the mission requirement;
6. if step 5 fails, shielding optimization or the implementation of new solutions for background minimization;
7. back to step 3.

3.7.1 Mission Orbit

The background level of a mission depends first on the spacecraft orbit, because it exposes the telescope to a specific radiation environment. All space telescopes must minimize the time spent in the Earth's radiation belts, which pose a significant radiation threat to the instruments.

First discovered by Van Allen and his collaborators on Explorer I [73], the Van-Allen belts are composed of energetic plasma (mainly protons and electrons) powered by cosmic ray particles and confined by the geomagnetic field in a toroidal region, the magnetosphere, ranging from 10^3 up to 6×10^4 km of altitude. The interaction of the magnetized solar wind with the Earth's magnetic field compresses the magnetosphere at the Sun side while generating a long tail (up to 300 Earth radii) of trapped particles, the magnetotail, at the opposite side.

High-energy space missions are operating in the following orbits:

- Low Earth Orbit (LEO): the spacecraft follows a nearly equatorial path below the radiation belts, with an apogee below 1000 km above the Earth's surface and an orbital period of about 90 minutes (see Fig. 3.28 for an example of). The Earth's magnetosphere shields most of the cosmic rays, with an efficiency depending on the orbit inclination with respect to the magnetic axis. This orbit ensures a

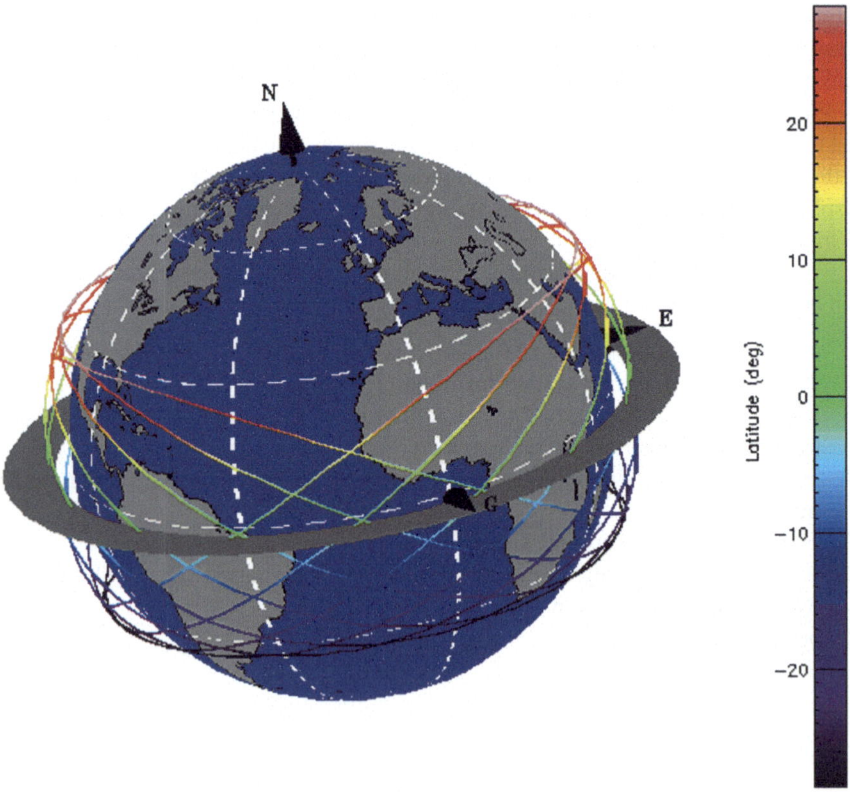

Fig. 3.28 Trajectory, integrated over 1 day, of the Fermi gamma-ray mission orbiting in LEO with an altitude of 550 km and an inclination of 28.5°, obtained with the ESA Spenvis tool

low background environment despite the satellite being exposed to low energy albedo radiation by the Earth's atmosphere, produced by both cosmic ray protons and photons interacting with its surface. The drawbacks of this orbit are the Earth shadowing a large portion of the sky ($\sim$1.2 sr for a 550 km altitude) and the exposure to the South Atlantic Anomaly (SAA). The SAA is a depression of the radiation belt due to an off-set of the geomagnetic dipole center with respect to the Earth's center that causes the trapped particles to reach lower altitude regions in the South East of Brazil [74]. Both X-ray (e.g. Swift, Suzaku) and g-ray (e.g. AGILE, Fermi) missions are operating in LEO. The 1-day Fermi orbit in LEO, with an altitude of 550 km and an inclination of 28.5; is shown in Fig. 3.28.

- Highly Elliptical Orbit (HEO): the spacecraft reaches extreme apogees ($>10^5$ km), with an orbital period of a few days, and spends most of the operating time beyond the radiation belts and without the frequent interruption of the Earth's shadowing. At the same time, the geomagnetic field does not shield the spacecraft from the charged particle environment. The Chandra X-ray Observatory (CXO) can perform observations for $\sim$85% of an orbiting cycle [75]. During the perigee passage at an altitude of $5-10 \times 10^3$ km, however, the spacecraft is forced to cross the trapped region and the instrumentation is usually switched off and protected by additional filters to avoid damages. The minimum required satellite elevation to conduct science observations for the XMM-Newton mission is currently 46000 km. Figure 3.29 shows the XMM-Newton altitude along the orbit and the corresponding count rate detected by the Low Energy (LE) proton (1-4.5MeV) and electron (0.13-15MeV) unit on board the mission radiation monitor. A sudden increase in the particle rate appears when entering and exiting the radiation belts. The INTEGRAL space observatory is also operated in HEO.

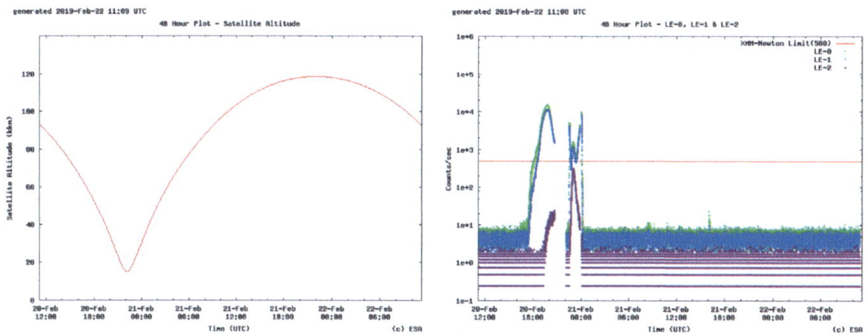

Fig. 3.29 XMM-Newton altitude (left panel) and related count rate detected by the Low Energy (LE) proton (1–4.5 MeV) and electron (0.13–15 MeV) unit on board the mission radiation monitor over 48 h of orbit. Both plots are taken from the XMM live radiation monitor feed [CREDIT: ESA]

- L1 or L2 orbit: the five Lagrangian points are regions of gravitational equilibrium where the combined gravitational forces of the Earth and the Sun combined with a small body's—the spacecraft's—centripetal force causes the latter to maintain an almost constant position relative to the two larger bodies, in the Earth orbital plane. The L1 and L2 points are located 1.5 million kilometers from the Earth towards and opposite the Sun, respectively. Lagrangian orbits, because of the constant Sun direction and the lack of passages in the Earth's shadow, ensure a stable thermal environment for the instruments, while achieving an almost constant view of the sky. Missions in L1 and L2 are fully exposed to cosmic rays and the flaring activity of the Sun (e.g. solar energetic particles and coronal mass ejection). The L2 orbit, placed at the Earth's magnetotail, is in part shielded by the solar emission but also provides a highly variable environment because of the different plasma regimes in the tail. At the time of writing, the Spectrum-Roentgen-Gamma (SRG) satellite carrying on-board the German eROSITA X-ray telescope, launched in 2019, is the only X-ray mission operating in L2. No high energy telescopes are currently operating in L1.

3.7.2 Space Radiation Environment

Evaluating the spectral, temporal, and spatial distribution of the particles hitting the spacecraft requires the collection of all data available from both particle experiments and space telescopes flown in close proximity to the orbit under study. For each particle population, the differential flux, in particles $cm^{-2}\ s^{-1}\ MeV^{-1}\ sr^{-1}$, is extracted in the energy range of interest. The sources of background can be both steady throughout the mission or highly variable. For the latter case, the maximum flux expected in a certain fraction of the observation time is usually taken into account. For each particle type, the Monte Carlo simulator randomly generates their energy according to a given spectral model (see Sect. 3.7.4 for a description of the source geometry and its conversion to a flux). For two magnetic latitudes the complex and rich radiation environment for a LEO X-ray telescope, the HITOMI mission in this case during the equatorial passages, is reported in Fig. 3.30.

3.7.2.1 Cosmic Rays

Cosmic rays (CR) are the primary source of charged particles in the space environment outside the radiation belts.

Below tens of MeV, protons, electrons, and alpha particles of solar origin dominate the CR composition, while at higher energies cosmic rays have galactic origin [76]. Solar CRs are both composed by the steady stream of solar plasma wind from the Sun's corona and highly variable outbursts of solar energetic particles [77] and coronal mass ejections (CMEs), as well as interstellar plasma charged in the Sun's proximity and carried by solar wind to the inner heliosphere (anomalous CRs). When

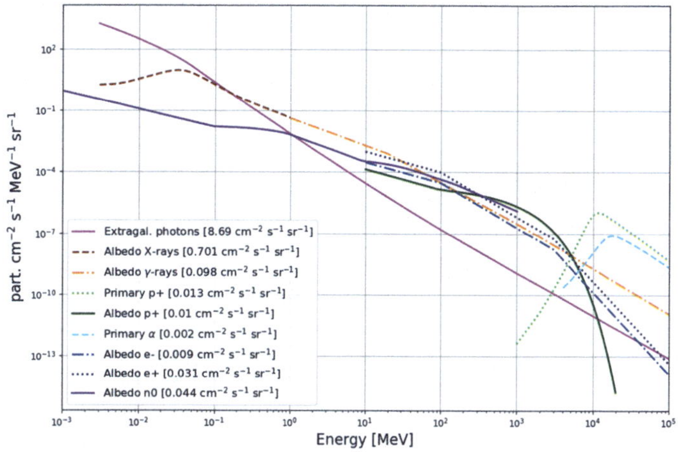

Fig. 3.30 The LEO space radiation environment modelled for the HITOMI in the equatorial passages. Mission [63]

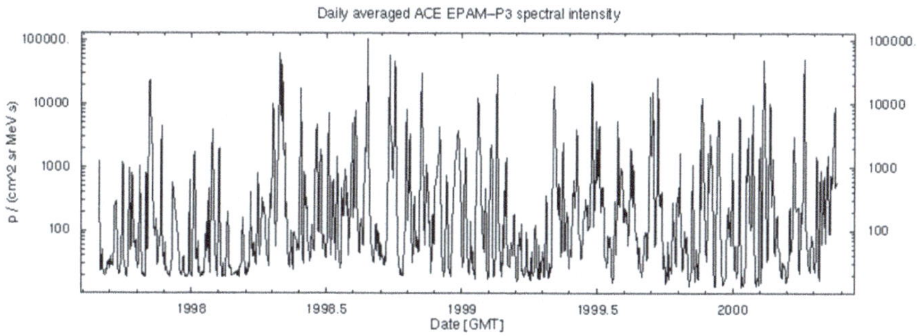

Fig. 3.31 Daily averaged proton flux detected in L1 orbit by the ACE/EPAM P3 channel, centered at ∼0.14 MeV [CREDIT: https://wwwastro.msfc.nasa.gov/ACIS/fluence/]

the solar activity reaches its maximum, the proton intensity in solar flares and mass ejections can vary by orders of magnitude. This flaring component can compromise X-ray observations and it must be continuously monitored. The Chandra X-ray Observatory Center, for example, uses a near-real-time solar-proton alert system, using ACE/EPAM data [78] from the NOAA (National Oceanic and Atmospheric Administration) Space Environment Center (Fig. 3.31), to hide the ACIS detectors during extreme solar events.

In terms of NXB contribution, the monitoring of the proton environment above 30 MeV by the GOES satellite, at $\sim 40 \times 10^4$ km of altitude, has shown that solar protons exceed the galactic component only 20% of the time [79]. This, coupled with the low energies of the solar CR component, allows us to neglect the solar protons as a source of *not focused* NXB.

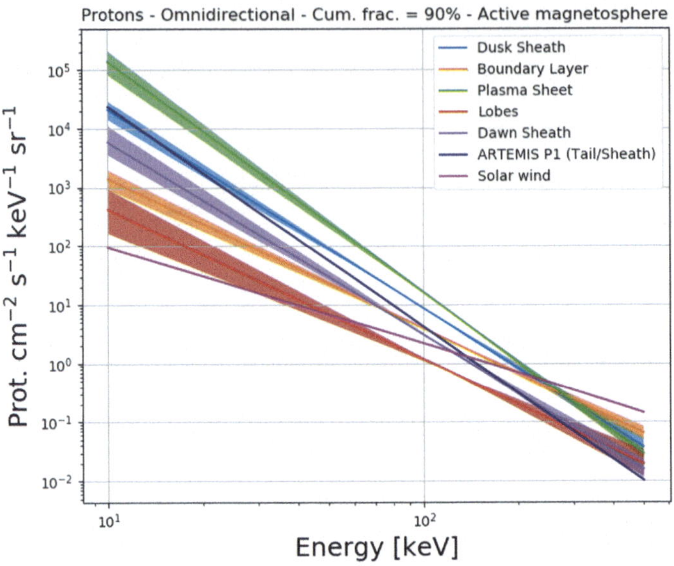

Fig. 3.32 Soft proton spectral models for each plasma regime potentially encountered in L2, with the maximum intensity expected during 90% of the operation time and for an active magnetosphere [81]

On the contrary, low energy protons and electrons (<100 keV) can enter the field of view of X-ray grazing angle optics, scatter with the mirror surface, and reach the focal plane. Low energy, or *soft*, protons populate both the interplanetary space, through the solar wind, and the trapped particle regions of the Earth magnetosphere (Fig. 3.32), while low energy electrons can also be found in LEO as a result of Earth albedo and SAA contamination [80].

Low energy particles can dramatically increase the *focused* NXB level, reducing the sensitivity or, in the most extreme cases, compromising the observation itself. Current X-ray telescopes such as XMM-Newton, Chandra, Swift, and eROSITA carry on-board an electron magnetic diverter, i.e. an array of magnets between the optics and the focal plane that can deflect charged particles away from the instrument's field of view (see, e.g. [82]). The ability of protons to scatter with the X-ray optics towards the focal plane had been discovered only after one month of operation of the Chandra satellite. Launched in 1999, after a few orbits the Chandra/ACIS front-illuminated CCDs suffered from degradation of the charge transfer efficiency when crossing the radiation belts [83]. Switching off and moving them from the focal position [84] minimized the damage, while blocking filters protect the XMM-Newton EPIC instruments below an altitude of 40×10^3 km. Above this limit and throughout the orbit, XMM-Newton scientific observations are still contaminated by sudden flares, the so-called *soft proton flares*, in the background, lasting from hundreds of seconds to hours. These events can hardly be disentangled from X-ray photon induced counts, causing the loss of large amounts (30–40%) of observing time [85]. Future high throughput X-ray grazing angle telescopes (e.g. ATHENA [81] and the Lynx [86] mission) foresee an on-board proton diverter for the shielding of the soft proton induced background. Because of the ~2000 times higher proton mass with respect to the electron mass, the design of a proton diverter poses several challenges in terms of higher required field density, higher magnet mass budget, and impact to neighbouring instruments.

First discovered in the early 900s by V. Hess with ionization chambers on balloon flights, galactic cosmic rays (GCRs) originate in particle accelerators (e.g. supernova remnants [87] and blazars [88]) in our galaxy and their energy distribution peaks in the 100 MeV–1 GeV band. GCRs are composed of about 98% nuclei (87% Hydrogen, 12% Helium and 1% heavier nuclei [89]) and 2% leptons (>90% electrons [90]). The differential energy spectrum for a given species i is well described by a power law in the form:

$$J_i(E) \propto E^{-2.7}, \tag{3.37}$$

where E is the total energy per nucleon (Fig. 3.33).

The deviation from a pure power-law and the low energy cut-off visible in Fig. 3.33 are the product of the solar modulation induced by the solar wind and the geomagnetic cut-off rigidity, respectively.

The solar activity follows a cycle, with an averaged period of 11.5 years [92] that can be divided into a 7 year maximum phase and a relatively quiet 4 year minimum phase. We are currently at the ending tail of the 24th solar cycle (since 1755) and the next is expected to continue through 2030. The interplanetary magnetic field (IMF) is generated by the Sun's magnetic field lines dragged from the Sun's corona to interplanetary space by the solar wind. GCR particles entering the heliosphere are scattered by IMF irregularities and undergo convection and adiabatic decelerations [93]. As the solar activity reaches its maximum, the solar wind and the derived IMF intensity increase. As a result the GCR particles, which interact with the IMF, decrease in rate and anti-correlate with the solar activity. This effect is clearly visible in Fig. 3.34, where the neutron flux detected on-ground, generated by GCR induced showers in the atmosphere, reaches its maximum at solar minimum.

The energy loss of cosmic rays due to diffusional, convectional, and adiabatic interactions with the solar wind is integrated into a single solar modulation parameter Φ, in units of megavolts (MV) [94]. The quantity $Ze\Phi$, where e is the electron charge and Z is the atomic number of the particle, is the loss of the particle potential energy when moving from the solar wind boundary to the Earth. From GCR measurements [95] of the ACE Cosmic Ray Isotope Spectrometer (CRIS), the modulation parameter varies from ~300 MV at solar minimum up to ~1200 MV at solar maximum, about a factor 3 of variability. If the minimization of the NXB is required, a high energy space mission should always be planned during the solar maximum period, where the shielding of GCRs by the solar wind is at its maximum. At the same time, operating at solar maximum decreases the satellite's life because of solar wind drag and increases the occurrence of solar flares, which are potentially dangerous for the instrumentation.

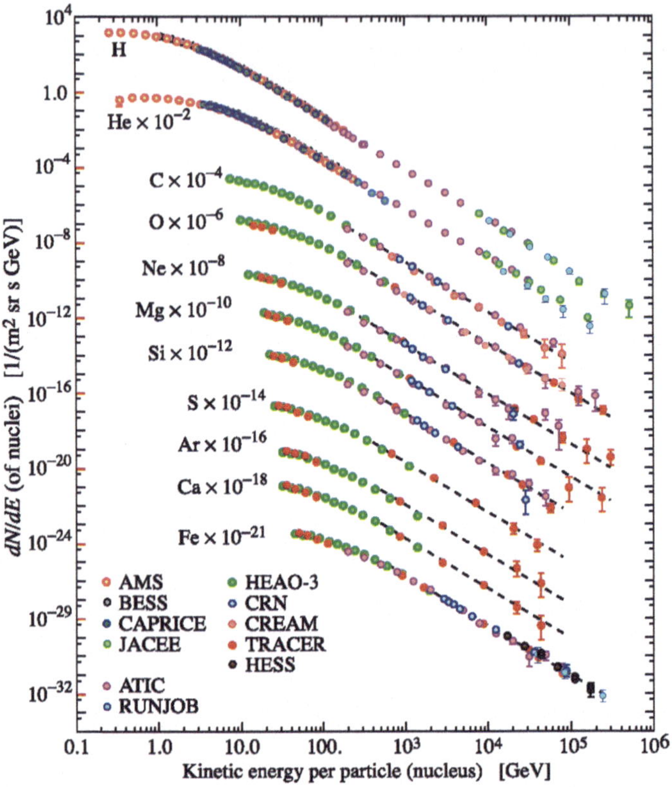

Fig. 3.33 Differential spectra of the nuclei composing the primary galactic cosmic rays obtained from direct measurements by Boyle and Mueller in [91]

Another effect to take into account when modeling the GCR flux is the shielding of the Earth's magnetic field. The propagation of a charged particle through a magnetic field depends on its rigidity, or its momentum per unit of charge (measured in Volt):

$$\mathrm{R} = \frac{\mathrm{pc}}{\mathrm{Ze}}, \tag{3.38}$$

where p is the particle momentum, c the speed of light, e is the electron charge, and Z is the atomic number of the particle. The number of nucleons composing a nucleus, or ion, is defined as A. For light nuclei, $\mathrm{A}/\mathrm{Z} \sim 2$. If we define $E_{0,n}$ and $E_{k,n}$ as the rest and kinetic energy per nucleon and E as the particle total energy, from the energy-momentum relation:

$$\mathrm{E}^2 = \mathrm{p}^2\mathrm{c}^2 + \mathrm{A}^2\mathrm{E}_{0,\mathrm{n}}^2 = \mathrm{A}^2\left(\mathrm{E}_{0,\mathrm{n}} + \mathrm{E}_{\mathrm{k},\mathrm{n}}\right)^2 \tag{3.39}$$

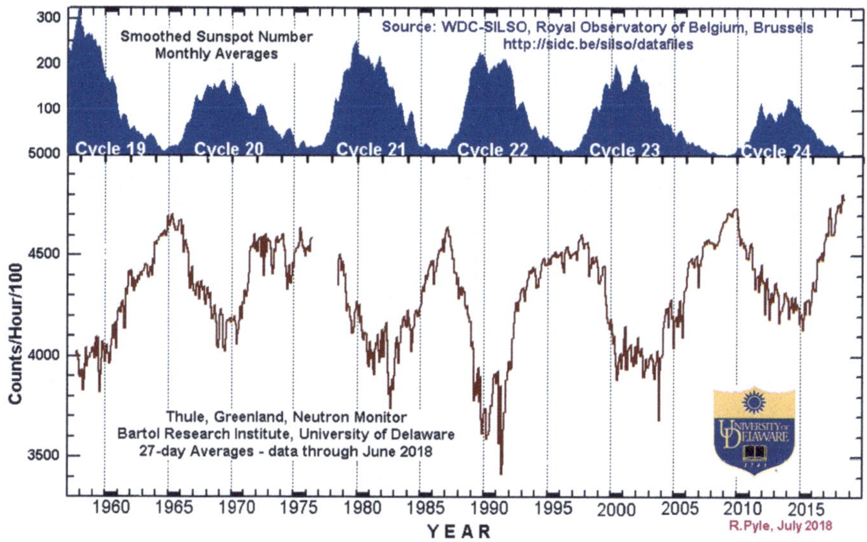

Fig. 3.34 Effect of the solar activity cycle (top panel) on the count rate recorded by a neutron monitor (bottom panel) in Thule (Greenland), directly linked to the flux of galactic cosmic rays reaching the Earth [CREDIT: Bartol Research Institute, University of Delaware (http://neutronm.bartol.udel.edu)]

we can link the rigidity to the kinetic energy per nucleon:

$$\mathrm{R} = \frac{\mathrm{A}}{\mathrm{ze}}\sqrt{\mathrm{E}_{\mathrm{k,n}}\left(\mathrm{E}_{\mathrm{k,n}} + 2\mathrm{E}_{0,\mathrm{n}}\right)}\,. \tag{3.40}$$

A proton, or an Hydrogen nucleus, with 1 GeV kinetic energy has from the previous relation a rigidity of $\sim$1.7 GV.

When interacting with the Earth's magnetic field, GCR particles follow a trajectory around the field lines. In the approximation of a locally uniform field, the motion is spiral with a gyroradius, or Larmor radius, $\rho = R/B_{\perp}$, where $B_{\perp}$ is the Earth's magnetic field component perpendicular to the particle momentum. At a given $B_{\perp}$, the higher the particle rigidity the larger is the trajectory's curvature. At each position around the Earth, the geomagnetic field defines a rigidity cut-off, i.e. the minimum particle rigidity that the particle must hold in order to avoid being shielded by the magnetic field. As a direct consequence, the energy spectra of GCRs present a low energy cut-off up to tens of GeV if the mission operates in LEO at low inclinations. From Störmer's equation in the approximation of a dipolar field geometry [96], the vertical rigidity cut-off depends only on the magnetic latitude λ_m and the spacecraft altitude r in Earth radii. For the 2000 International Geomagnetic Reference Field (IGRF), the formula, in units of GV, takes the form [97]:

$$\mathrm{R}_{\mathrm{c,v}} = \left(14.5\cos^4\lambda_{\mathrm{m}}\right)/\mathrm{r}^2\,. \tag{3.41}$$

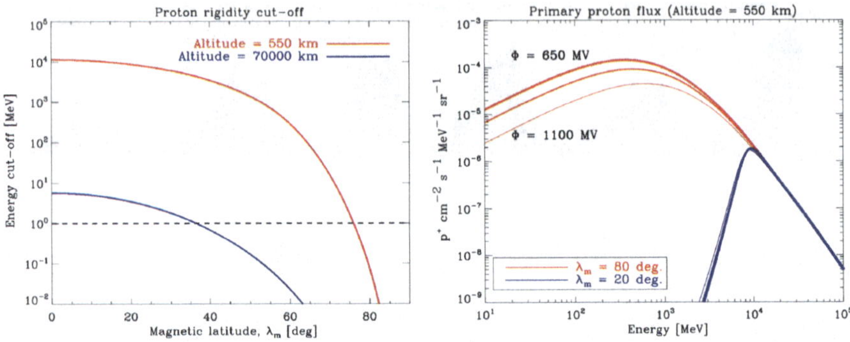

Fig. 3.35 Rigidity cut-off for two spacecraft altitudes at different magnetic latitudes (left panel) and the effect of solar modulation and rigidity cut-off (right panel) on the GCR protons flux in LEO as a function of the particle kinetic energy

Figure 3.35 (left panel) shows the $R_{c,v} - \lambda_m$ relation for a spacecraft operating in LEO and HEO. For detailed studies of the GCR environment encountered by the spacecraft, tabulated rigidity cut-off data are obtained from particle transport simulations with a real magnetosphere model.

Both the solar modulation induced by the solar wind and the geomagnetic cut-off shape the spectrum of GCRs reaching the spacecraft, depending on its orbit. In Fig. 3.35 (right panel) we use the GCR proton spectral model of [98], based on BESS and AMS measurements, to compute the differential flux expected in LEO. The solar modulation, ranging here from 650 to 1100 MV, only affects the low energy part of the spectrum and for magnetic inclinations below 20° (e.g. AGILE, Swift) it is neglectable. On the contrary, the geomagnetic field is highly effective in shielding protons below ~10 GeV. Towards the polar region, the solar activity impacts the proton flux by up to a factor of 3.

3.7.2.2 Diffuse Photons

The space radiation environment is also populated by a diffuse emission of X-ray and gamma-ray photons, the Cosmic X-ray Background (CXB), and the cosmic gamma-ray background radiation, common to all the orbits listed in Sect. 3.7.1. These high energy photons are the sum of a galactic and an extragalactic, nearly isotropic, component and contribute to the NXB by interacting with the materials surrounding the detectors and generating secondaries and/or leaking through the focal plane. Diffuse photons collected in the instrument FoV and contaminating the astrophysical observation are part of the sky background. They require discrimination techniques specific to the type of mission and detector to be applied during the scientific analysis.

In the X-ray regime, the galactic emission dominates below 1 keV and it is both thermal, originated from hot plasma in the Local Hot Bubble, the galactic disk, and the galactic halo [99], and induced by charge transfers between solar wind ions and the interstellar gas [100]. Above 1 keV most of the CXB emission is accounted for by the integrated emission of unresolved Active Galactic Nuclei (AGN) over the cosmic time, with a high fraction of highly absorbed AGN contributing to the hard part (20–40 keV) of the spectrum [101].

At higher energies (>1 MeV) our Galaxy becomes a bright emitter of gamma-rays because of the interaction of cosmic rays with the interstellar medium and local low energy radiation [102]. The galactic gamma-ray emission is superimposed over a weaker, isotropic, extragalactic gamma-ray background (EGB) mostly, but not entirely, produced by blazars and star-forming and radio galaxies [103].

The broad band CXB and EGB spectrum can be described in the 3 keV–100 GeV energy range by the model of [104], obtained by fitting together data from the HEAO-1, COMPTEL, and EGRET missions.

3.7.2.3 Earth Albedo Emission

Cosmic-rays interact with the Earth's atmosphere and produce a multitude of low energy secondaries by means of nuclear interaction cascades. A fraction of these particles, mainly hadrons, leptons, and photons, escape upward towards the sky and populate the environment of space missions orbiting in LEO. This so-called Earth albedo emission [105] strongly varies for each position of the spacecraft because of its dependence on the Earth's magnetic field and the rigidity cut-off modulating the input GCR flux. For this reason, despite the low energy of its components, albedo particles represent an important source of background uncertainty in LEO.

Spectral models for the albedo protons, electrons, and positrons at different magnetic latitudes can be derived from measurements of the AMS [106, 107] and PAMELA [108] experiments flown in near-Earth orbits. The albedo high energy radiation is the sum of CXB reflection, dominating below 50 keV, and GCR-induced photons, mainly bremsstrahlung and positron-electron annihilation emission up to 1 MeV and decay of neutral pions and kaons at higher energies. Atmospheric neutrons are also produced by GCRs through mainly spallation of nitrogen and oxygen nuclei and neutron evaporation in intranuclear cascades. Their energy distribution extends from thermal energies (few keV) to hundreds of MeV. Examples of differential fluxes for albedo photons and neutrons in LEO are shown in Fig. 3.36.

While the photonic and neutron albedo component originate from the Earth's surface, charged particles bounce in the near-Earth space following the magnetic field lines. As a first approximation, and considering the fast orbital period of a spacecraft in LEO, they can be considered uniformly distributed despite their East-West asymmetry [109].

3.7.3 *Monte Carlo Simulations*

Predicting with the required accuracy the scientific performance (e.g. effective area, background level) of a space telescope before its launch is mandatory for the success

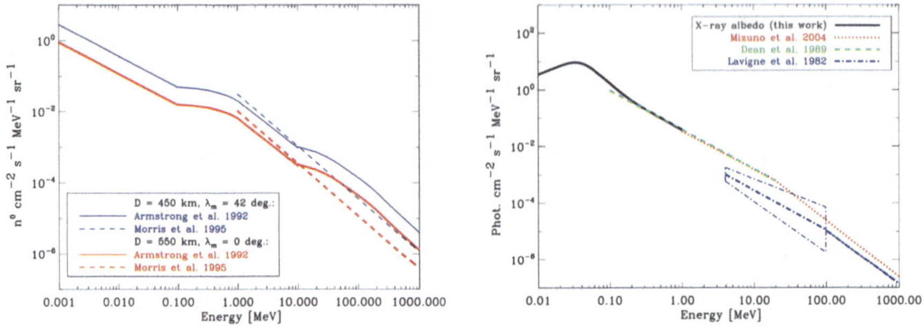

Fig. 3.36 *Left panel*: X-ray and gamma-ray albedo differential flux for an equatorial LEO orbit obtained from both models and in-flight measurements. *Right panel*: the albedo neutron differential model for two LEO configurations (see [105] and references therein)

of the mission, i.e. achieving its science goals. Before the advent of particle transport codes and the increase of computing performance, semi-empirical methods, scaling laws, on-ground measurements and balloon flights were used to predict the instrumental background detected once in orbit [110]. The different designs of the instruments and the complexity of particle interactions, together with poor models of the space radiation environment, would lead to large uncertainties in the models, far from the accuracy needed in the reconstruction of the true source signal.

Particle transport codes can be used to track particles and their secondaries along their path through matter. The occurrence of a physics interaction along their track is described in most cases by probability distribution functions and its realization requires a Monte Carlo method, a stochastic computational algorithm based on the use of randomly generated numbers to solve a problem once the probability distribution is known [111].

Monte Carlo particle tracking simulations, coupled with the advent of high power and low cost CPU technologies, are able to propagate the radiation through the telescope as a real observation in space, with energy and spatial resolutions of the order of few tens of eV and nanometers, respectively. Geant4 [112–114] is the most widely used toolkit library for the development of Monte Carlo simulations for space applications. Created by CERN for particle physics experiments in accelerators, it has been extended to lower energies (tens of eV) and it is now supported by an international community. The European Space Agency (ESA) is currently developing, under the AREMBES[2] (ATHENA Radiation Environment Models and X-Ray Background Effects Simulators) project, a Geant4-based software simulator of the radiation background effects on the L-Class science mission ATHENA [5], an activity that is comprised of the study of the ATHENA orbit radiation environment and an update of Geant4 physics models in the mission ranges of interest.

Geant4 is a C++ library to achieve a correct physical realization of the interactions, for building a virtual model of the telescope and handling both the input particles

[2]http://space-env.esa.int/index.php/news-reader/items/AREMBES.html.

(e.g. a parallel photon beam entering the field of view) and the desirable output (e.g. the energy deposits in the detector). The user has the task of building the software program using the Geant4 library, hence the responsibility in selecting the correct physics models and the general accuracy of the simulation. Several Geant4-based software frameworks (e.g. ESA/GRAS [115], BoGEMMS [116], MEGALib [117]) are available covering different areas of applications to speed-up the development process and ensure the scientific validity of the results.

Monte Carlo simulations have fundamental benefits along the entire mission development: the characterization and minimization of the background level, the shielding optimization, the calibration of its response towards X-ray and gamma-ray photons (e.g. detection efficiency, PSF, energy dispersion), the reconstruction, and scientific analysis. While the systematic uncertainty of the results is strongly linked to the correct modelling of the space radiation environment, the physics processes, and the geometry, the statistical error is proportional to the Poisson fluctuation of the total number N of simulated events ($1\sigma = \sqrt{N}$).

3.7.4 Background Evaluation

Once the radiation environment for the mission under study is modelled and the spectral distribution for each relevant population is available, the Geant4 mass model for the spacecraft instruments and payload is built and then exposed to both space radiation and particles (Fig. 3.37). The mass model includes dedicated shielding solutions, ranging from high-density materials (e.g. lead) to efficiently attenuate both primary high energy photons and secondary particles generated by hadronic cascades, to scintillation panels that trigger in coincidence with the detector if particles reach the focal plane outside the field of view, to time-of-flight systems able to discriminate background events on the basis of their direction, to magnetic deflectors that deviate the charged particles entering the field of view.

The user can create the spacecraft building blocks using the Geant4 C++ libraries or by importing a GDML (Geometry Description Markup Language) geometry description in XML format. Converters from CAD (Computer-Aided Design) to GDML have been made available recently, which allow for directly linking the engineering study to the scientific feasibility analysis. The shielding design and its optimization is one of the major goals of background simulations.

Given N simulated particles, randomly distributed over a certain energy range, they interact with the various volumes encountered along their path and some of them, or their secondaries, will eventually reach the detector and generate an energy deposit. The collection of these deposits is selected, summed, and filtered according to the specific instrument read-out system (e.g. all deposits within the same pixel generated by the same primary particles are summed). The conversion from a total number of counts C to an actual count rate R in counts s^{-1} requires the evaluation of the simulated equivalent exposure to the input flux. The algorithms used in this process depend on the specific configuration of the Geant4 source. In the case of

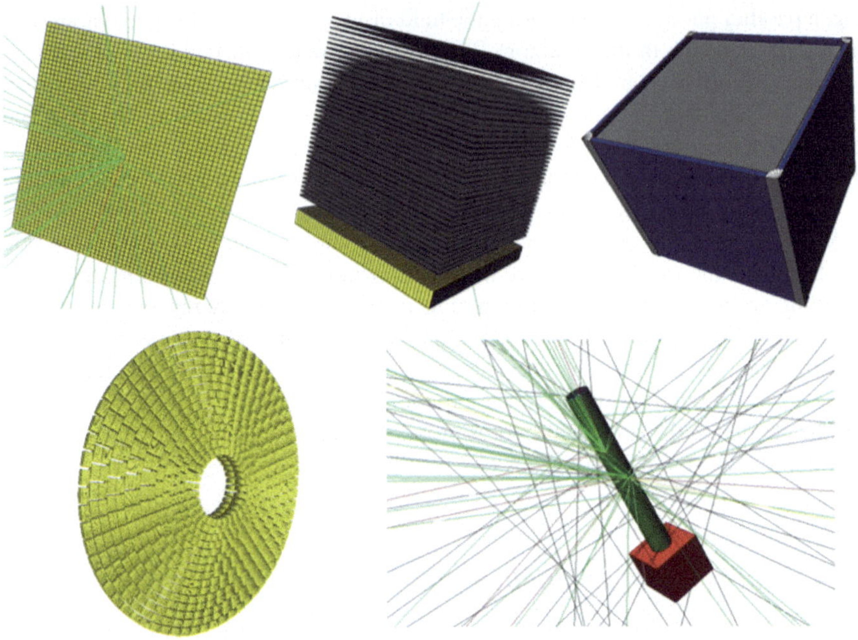

Fig. 3.37 Examples of Geant4 mass models for pair production tracking telescopes (top panels), ESA/ATHENA X-ray Silicon Pore Optics (bottom-left panel), and shielding systems for X-ray focusing telescopes (bottom-right panel). The coloured lines in the latter image are the tracks of the particles hitting the geometry, with different colours labeling different particles types

an isotropic source of background, a common case in the production of not-focused NXB, particles and photons are usually generated from a sphere surrounding the spacecraft within a cone subtending the model. For the input to be isotropic, i.e. to get the same intensity regardless of the direction of measurement, the angular distribution in the cone must follow a Lambert's or cosine-law (Fig. 3.38).

From the energy integrated particle intensity ϕ, in particles cm^{-2} s^{-1} sr^{-1}, the particle rate P, in units of particles s^{-1}, is given by:

$$\mathrm{P} = \phi \times A_{\mathrm{source}} \times \Omega \ , \tag{3.42}$$

where A_{source} and Ω are the surface emitting the particles and the cone solid angle, respectively. We will consider here a sphere as the particle source but the algorithm can be applied to any input surface. The intensity for each direction is proportional to the cosine of the angle θ between the source direction θ and the local normal to the emission sphere surface. The solid angle Ω of the emission cone of semi-aperture θ', can be expressed as:

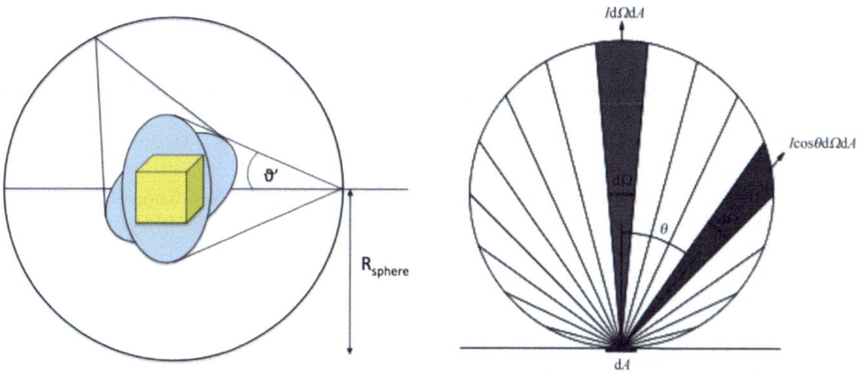

Fig. 3.38 Particle generation scheme for the simulation of an isotropic source in space (left panel) and cosine distribution (right panel) of the simulated intensity I in cm^{-2} s^{-1} sr^{-1} from a surface area dA as a function of the angle θ from the surface normal [118]. Each wedge represents equal solid angles $d\Omega$

$$\Omega = \int_0^{2\pi}\int_0^{\theta'} cos\theta sin\theta d\theta d\phi = \pi sin^2\theta' \text{ sr.} \tag{3.43}$$

The particle rate is then given by:

$$\mathrm{P} = \phi \times 4\pi \mathrm{R}^2_{\mathrm{sphere}} \times \pi \sin^2\theta' = \phi \times 4\pi^2 \mathrm{R}^2_{\mathrm{sphere}} \sin^2\theta' \text{ particles s}^{-1}. \tag{3.44}$$

The exposure time T for the simulated input particle flux is required to convert the number of detected counts, or any measurable quantity produced by the simulation, to a count rate. For an isotropic input flux it is obtained as follows:

$$\mathrm{T} = \frac{\mathrm{N}}{\mathrm{P}} = \frac{\mathrm{N}}{\phi \times 4\pi^2 \mathrm{R}^2_{\mathrm{sphere}} \sin^2\theta'} \text{ seconds.} \tag{3.45}$$

3.7.5 *Verification and Validation Activity*

Many sources of uncertainty lie behind the simple logic chain of background simulations: (i) the availability of data coverage for the input particles fluxes, (ii) the approximations in the mass model representation, (iii) the correctness of the physics models in the particle transport code, (iv) the minimum sampling of the particles tracks, and (v) the minimum production energy of their secondaries. For this reason, Monte Carlo simulations should always be combined with analytical computations based on tabulated data (e.g. the attenuation efficiency for X-ray photons), laboratory measurements to test the correctness of the physics models, and in-flight or balloon-borne background measurements.

References

1. F. Fuschino et al., Nucl. Instrum. Methods Phys. Res. A **588**(1–2), 17 (2008). https://doi.org/10.1016/j.nima.2008.01.004
2. V.Y. Young, G.B. Hoflund, *Handbook of Surface and Interface Analysis Methods for Problem-Solving*, 2nd edn. (CRC Press, Boca Raton, 2009)
3. G. Knoll, *Radiation Detection and Measurement*, 4th edn. (Wiley, New York, 2010)
4. J.H. Hubbell, Radiat. Phys. Chem. **75**(6), 614 (2006). https://doi.org/10.1016/j.radphyschem.2005.10.008
5. K. Nandra et al., ATHENA mission proposal (2013)
6. M. Barbera et al., in *Proceedings of the SPIE*, **9601** (2015). https://doi.org/10.1117/12.2189326
7. M. Tanabashi, Particle Data Group. Phys. Rev. D. **98**(3), 030001 (2018). https://doi.org/10.1103/PhysRevD.98.030001
8. W.B. Atwood et al., ApJ **697**(2), 1071 (2009). https://doi.org/10.1088/0004-637X/697/2/1071
9. A.C. Zoglauer, Ph.D. thesis, (2006)
10. A. Mirzaei et al., Electron. Mater. Lett. **14**(3), 261 (2018). https://doi.org/10.1007/s13391-018-0033-2
11. Hitomi Collaboration, F. Aharonian et al., Nature **551**(7681), 478 (2017). https://doi.org/10.1038/nature24301
12. A.M. Read et al., A&A **534**, A34 (2011). https://doi.org/10.1051/0004-6361/201117525
13. V. Fioretti et al., ApJ 896, 61 (2020). https://doi.org/10.3847/1538-4357/ab929a
14. P.W. Cattaneo et al., ApJ **861**(2), 125 (2018). https://doi.org/10.3847/1538-4357/aac888
15. F. Pajot et al., J. Low Temp. Phys. **193**(5), 901 (2018). https://doi.org/10.1007/s10909-018-1904-5
16. S.S. Wilks, Ann. Math. Stat. **9**, 60 (1938). https://doi.org/10.1214/aoms/1177732360
17. T.P. Li, Y.Q. Ma, ApJ **272**, 317 (1983). https://doi.org/10.1086/161295
18. T. Takahashi et al., **8443**, 84431Z (2012). https://doi.org/10.1117/12.926190
19. P. Murdin, *Encyclopedia of Astronomy and Astrophysics* (Nature Publishing Group, Philadelphia, 2001)
20. P. Gorenstein, Opt. Eng. **51**(1), 011010 (2012). https://doi.org/10.1117/1.OE.51.1.011010
21. D. Spiga, A&A **468**(2), 775 (2007). https://doi.org/10.1051/0004-6361:20077228
22. F.E. Christensen et al., Exp. Astron. **6**(4), 33 (1995). https://doi.org/10.1007/BF00419256
23. G. Pareschi, V. Cotroneo, in *Proceedings of the SPIE*, **5168** (2004). https://doi.org/10.1117/12.509470
24. P. O'Brien et al., Memorie della Societa Astronomica Italiana **89**, 130 (2018)
25. R. Giacconi, B. Rossi, J. Geophys. Res. **65**, 773 (1960). https://doi.org/10.1029/JZ065i002p00773
26. H. Wolter, Annalen der Physik **445**(4), 286 (1952). https://doi.org/10.1002/andp.19524450410
27. R.C. Chase, L.P. Van Speybroeck, Appl. Opt. **12**(5), 1042 (1973). 10.1364/AO.12.001042, http://ao.osa.org/abstract.cfm?URI=ao-12-5-1042
28. P. Friedrich et al., in *Proceedings of the SPIE*, **7011** (2008). https://doi.org/10.1117/12.788948
29. R. Willingale et al., in *Proceedings of the SPIE*, **9905** (2016). https://doi.org/10.1117/12.2232946
30. R. Hudec et al., Contributions of the Astronomical Observatory Skalnate Pleso **48**(3), 456 (2018)
31. M. Urban et al., Acta Astronaut. **140**, 96 (2017). https://doi.org/10.1016/j.actaastro.2017.08.004
32. J.R.P. Angel, ApJ **233**, 364 (1979). https://doi.org/10.1086/157397
33. W.K.H. Schmidt, Nucl. Instrum. Methods **127**, 285 (1975). https://doi.org/10.1016/0029-554X(75)90501-7

34. P. Gorenstein, in *Proceedings of the SPIE*, **8861** (2013). https://doi.org/10.1117/12.2024905
35. P. Gondoin et al., in *Proceedings of the SPIE*, **4140** (2000). https://doi.org/10.1117/12.409103
36. D.N. Burrows et al., in *Proceedings of the SPIE*, **4140** (2000). https://doi.org/10.1117/12.409158
37. M.C. Weisskopf, Opt. Eng. **51**(1), 011013 (2012). https://doi.org/10.1117/1.OE.51.1.011013
38. P.J. Serlemitsos et al., PASJ **59**, S9 (2007). https://doi.org/10.1093/pasj/59.sp1.S9
39. R. Iizuka et al., J. Astron. Telesc. Instrum. Syst. **4**, 011213 (2018). https://doi.org/10.1117/1.JATIS.4.1.011213
40. W.W. Craig et al., in *Proceedings of the SPIE*, **8147** (2011). https://doi.org/10.1117/12.895278
41. D. Della Monica Ferreira, et al., in *Proceedings of the SPIE*, **10699** (2018). https://doi.org/10.1117/12.2313275
42. D. Gotz et al., in *Proceedings of Swift: 10 Years of Discovery (SWIFT 10*, p. 74 (2014)
43. J.G. Timothy, *Microchannel Plates for Photon Detection and Imaging in Space* (Springer, New York, 2013), pp. 391–421. https://doi.org/10.1007/978-1-4614-7804-1_22
44. A. Busboom, H. Elders-Boll, H.D. Schotten, Exp. Astron. **8**(2), 97 (1998)
45. A. Goldwurm et al., in *Exploring the Gamma-Ray Universe*, ESA Special Publication, 459, p. 497 - 500, ed. by A. Gimenez, V. Reglero, C. Winkler (2001)
46. T. Takahashi, et al., PASJ, 59, 35 (2007). https://doi.org/10.1093/pasj/59.sp1.S35
47. V. Schönfelder et al., Data Anal. Astron. **185–200**, (1992)
48. S. Boggs et al., in *Proceedings of the SPIE*, **6266** (2006). https://doi.org/10.1117/12.670605
49. G. Pareschi et al., in *Proceedings of the SPIE*, **3445** (1998). https://doi.org/10.1117/12.330278
50. R.K. Smither, Rev. Sci. Instrum. **53**(2), 131 (1982). https://doi.org/10.1063/1.1136941
51. F. Frontera, P. von Ballmoos, X-Ray Optics and Instrumentation **2010**, 215375 (2010). https://doi.org/10.1155/2010/215375
52. R. Camattari, A&A **587**, A21 (2016). https://doi.org/10.1051/0004-6361/201526745
53. N. Barriére, et al., in *Proceedings of the SPIE*, **10566** (2017). https://doi.org/10.1117/12.2308289
54. A. Meuris, J. Instrum. **9**(5), C05019 (2014). https://doi.org/10.1088/1748-0221/9/05/C05019
55. F. Hartmann, Nucl. Instrum. Methods Phys. Res. Sect. A **666**, 25 (2012). https://doi.org/10.1016/j.nima.2011.11.005. Advanced Instrumentation
56. T. Takahashi, S. Watanabe, I.E.E.E. Trans, Nucl. Sci. **48**(4), 950 (2001). https://doi.org/10.1109/23.958705
57. M. Lesser, Publ. Astron. Soc. Pac. **127**(957), 1097 (2015). https://doi.org/10.1086/684054
58. J. Müller-Seidlitz et al., in *Proceedings of the SPIE*, **9905** (2016). https://doi.org/10.1117/12.2235408
59. E. Gatti, P. Rehak, Nucl. Instrum. Methods Phys. Res. **225**(3), 608 (1984). https://doi.org/10.1016/0167-5087(84)90113-3
60. A.D. Butt, Ph.D. thesis, Politecnico di Milano (2016)
61. P. Lechner et al., Nucl. Instrum. Methods Phys. Res. A **458**(1–2), 281 (2001). https://doi.org/10.1016/S0168-9002(00)00872-X
62. A. Vacchi et al., Nucl. Instrum. Methods Phys. Res. A **326**(1–2), 267 (1993). https://doi.org/10.1016/0168-9002(93)90362-L
63. A. De Angelis et al., Experim. Astron. **44**(1), 25 (2017). https://doi.org/10.1007/s10686-017-9533-6
64. Y. Evangelista et al., J. Instrum. **13**(9), P09011 (2018). https://doi.org/10.1088/1748-0221/13/09/P09011
65. R. Campana et al., Memorie della Societa Astronomica Italiana **89**, 137 (2018)
66. J.W. den Herder et al., A&A **365**, L7 (2001). https://doi.org/10.1051/0004-6361:20000058
67. C.R. Canizares et al., PASP **117**(836), 1144 (2005). https://doi.org/10.1086/432898
68. F.S. Porter, I.S.S.I. Sci, Rep. Ser. **9**, 459 (2010)
69. H.M. Günther, R.K. Heilmann, J. Astron. Telesc. Instrum. Syst. **5**, 021003 (2019). https://doi.org/10.1117/1.JATIS.5.2.021003
70. D. Barret et al., in *Proceedings of the SPIE*, vol. 10699 (2018), p. 106991G. https://doi.org/10.1117/12.2312409

71. B. Aschenbach et al., Space Sci. Rev. **30**(1–4), 569 (1981). https://doi.org/10.1007/BF01246075
72. B.W. Grefenstette et al., in *Proceedings of the SPIE*, **10709** (2018). https://doi.org/10.1117/12.2312844
73. J.A. Van Allen, Sci. Amer. **200**, 39 (1959)
74. R. Campana et al., Exp. Astron. **37**, 599 (2014). https://doi.org/10.1007/s10686-014-9394-1
75. S.N. Virani et al., in *Proceedings of the SPIE*, **4012**. (2000). https://doi.org/10.1117/12.391606
76. Y.I. Logachev, K. Kecskeméty, M.A. Zeldovich, Solar Phys. **208**, 141 (2002). https://doi.org/10.1023/A:1019689101515
77. J.M. Ryan, J.A. Lockwood, H. Debrunner, Space Sci. Rev. **93**, 35 (2000). https://doi.org/10.1023/A:1026580008909
78. R.E. Gold et al., Space Sci. Rev. **86**, 541 (1998). https://doi.org/10.1023/A:1005088115759
79. E. Perinati et al., Exp. Astron. **33**, 39 (2012). https://doi.org/10.1007/s10686-011-9269-7
80. K. Hagino et al., J. Astron. Telesc. Instrum. Syst. **4**(2), 021409 (2018). https://doi.org/10.1117/1.JATIS.4.2.021409
81. V. Fioretti et al., ApJ **867**, 9 (2018). https://doi.org/10.3847/1538-4357/aade99
82. R. Willingale, XRT-LUX-RE-011/1 Technical Report (2000)
83. S.L. O'Dell, et al., in *Proceedings of the SPIE*, **4140** (2000)
84. S.L. O'Dell et al., in *Proceedings of the SPIE*, **6686** (2007). https://doi.org/10.1117/12.734594
85. M. Marelli et al., Exp. Astron. **44**, 297 (2017). https://doi.org/10.1007/s10686-017-9542-5
86. J.A. Gaskin et al., JATIS 5, 021001 (2019). https://doi.org/10.1117/1.JATIS.5.2.021001
87. E.G. Berezhko, Nucl. Phys. B Proc. Suppl. **256**, 23 (2014). https://doi.org/10.1016/j.nuclphysbps.2014.10.003
88. IceCube Collaboration, M.G. Aartsen et al., Science **361**, 147 (2018). https://doi.org/10.1126/science.aat2890
89. J.G. Simpson, Ann. Rev. Nucl. Part. Sci. **33**, 323 (1983)
90. O. Adriani et al., Bull. Russ. Acad. Sci. Phys. **73**, 568 (2009). https://doi.org/10.3103/S1062873809050104
91. J. Beringer et al., Phys. Rev. D. **86**(1), 010001 (2012). https://doi.org/10.1103/PhysRevD.86.010001
92. E.G. Stassinopoulos et al., IEEE Trans. Nucl. Sci. **43**, 369 (1996). https://doi.org/10.1109/23.490756
93. B. Heber et al., ApJ **699**, 1956 (2009). https://doi.org/10.1088/0004-637X/699/2/1956
94. L.J. Gleeson, W.I. Axford, ApJ **154**, 1011 (1968). https://doi.org/10.1086/149822
95. M.E. Wiedenbeck et al., in *International Cosmic Ray Conference*, vol. 2 (2005)
96. C. Störmer, Q.J.R. Meteorolog, Soc. **82**, 115 (1955). https://doi.org/10.1002/qj.49708235123
97. D.F. Smart, M.A. Shea, Adv. Space Res. **36**, 2012 (2005). https://doi.org/10.1016/j.asr.2004.09.015
98. T. Mizuno et al., ApJ **614**, 1113 (2004). https://doi.org/10.1086/423801
99. D.H. Lumb et al., A&A **389**, 93 (2002). https://doi.org/10.1051/0004-6361:20020531
100. K. Dennerl, Space Sci. Rev. **157**(1–4), 57 (2010). https://doi.org/10.1007/s11214-010-9720-5
101. R. Gilli, Memorie della Societa Astronomica Italiana **84**, 647 (2013)
102. M. Ackermann et al., ApJ **799**(1), 86 (2015). https://doi.org/10.1088/0004-637X/799/1/86
103. A.A. Abdo et al., Phys. Rev. Lett. **104**(10), 101101 (2010). https://doi.org/10.1103/PhysRevLett.104.101101
104. D.E. Gruber et al., ApJ **520**(1), 124 (1999). https://doi.org/10.1086/307450
105. V. Fioretti et al., in *Proceedings of the SPIE*, **8453** (2012). https://doi.org/10.1117/12.926248
106. J. Alcaraz et al., Phys. Lett. B **472**(1–2), 215 (2000). https://doi.org/10.1016/S0370-2693(99)01427-6
107. J. Alcaraz et al., Phys. Lett. B **484**(1–2), 10 (2000). https://doi.org/10.1016/S0370-2693(00)00588-8
108. O. Adriani et al., J. Geophys. Res. (Space Phys.) **114**(A12), A12218 (2009). https://doi.org/10.1029/2009JA014660

109. P. Zuccon et al., Astropart. Phys. **20**(2), 221 (2003). https://doi.org/10.1016/S0927-6505(03)00160-9
110. N. Gehrels, NIM A **239**, 324 (1985). https://doi.org/10.1016/0168-9002(85)90732-6
111. N. Metropolis, S. Ulam, J. Amer. Stat. Assoc. **44**, 335 (1949)
112. S. Agostinelli et al., Nucl. Instrum. Methods Phys. Res. A **506**, 250 (2003). https://doi.org/10.1016/S0168-9002(03)01368-8
113. J. Allison et al., IEEE Trans. Nucl. Sci. **53**(1), 270 (2006). https://doi.org/10.1109/TNS.2006.869826
114. J. Allison et al., NIM A **835**, 186 (2016). https://doi.org/10.1016/j.nima.2016.06.125
115. G. Santin et al., IEEE Trans. Nucl. Sci. **52**, 2294 (2005). https://doi.org/10.1109/TNS.2005.860749
116. A. Bulgarelli et al., in *Proceedings of the SPIE*, **8453** (2012). https://doi.org/10.1117/12.926065
117. A. Zoglauer, R. Andritschke, F. Schopper, NewAR **50**, 629 (2006). https://doi.org/10.1016/j.newar.2006.06.049
118. X.Y. Zhao et al., Chin. Phys. C **37**(12), 126201 (2013). 10.1088/1674-1137/37/12/126201. https://doi.org/10.1088

Chapter 4
Past, Present, and Future X-Ray and Gamma-Ray Missions

Andrea Bulgarelli and Matteo Guainazzi

4.1 The Era of Sounding Rockets and Proportional Counters

4.1.1 The Initial Steps

X-ray astronomy is a comparatively young branch of scientific investigation. The discovery of the Sun as an X-ray emitting source dates back to 1949 [77]. The first celestial X-ray source outside the solar system was discovered by a sounding rocket experiment in 1962 [81]—bestowing on his discoverer the Nobel Prize for Physics 50 years later. During the first decade after Giacconi's pioneer discovery, sounding rockets sent to an altitude of at least 100 km carrying proportional counters with modest spectral ($\sim$20% at 6 keV) and angular ($\sim$1') resolution with flight duration of a few minutes. They allowed a first set of discoveries. Those most relevant to accreting black hole systems (the main focus of this chapter) are the detection [81] and spectral characterization of a source of diffuse background in X-rays ("Cosmic X-ray Background", CXB; [29, 86]), and the detection of X-rays from the galaxy M 87 in the Virgo cluster [34]. Two missions were specifically designed to study the CXB, the *Orbiting Solar Observatories* (OSO) carrying a small (area $\sim 10^{-3}$ m^2) NaI crystal with CsI shield sensitive in the 8–20 keV energy range and 23° FWHM (Full Width Half Maximum) collimation. They produced an all sky map of the 10–40 keV CXB emission showing it to be isotropic to within $\sim$4% [212].

For more than two decades the main advancements in the field were due to missions carrying proportional counters. Following the classification in the historical review paper by [35], we can identify a "small satellite era" until 1977, and a "large satellite era" extending through the decade of the 80s' with the successful operations of the first non-US observatories.

A. Bulgarelli (✉)
INAF/OAS Bologna, Via Piero Gobetti 93/3, 40129 Bologna, Italy
e-mail: andrea.bulgarelli@inaf.it

M. Guainazzi
European Space Agency, ESTEC, Keplerlaan 1, 2201AZ Noordwijk, The Netherlands
e-mail: Matteo.Guainazzi@sciops.esa.int

C. Bambi (ed.), *Tutorial Guide to X-ray and Gamma-ray Astronomy*,
https://doi.org/10.1007/978-981-15-6337-9_4

4.1.2 The "Small Satellites' Era"

The first mission entirely dedicated to X-ray astronomy was the *Uhuru* satellite launched in 1970 [82]. It carried a large area (0.084 m^2) proportional counter sensitive in the 2–20 keV energy band with a $1° \times 10°$ collimation (FWHM). It surveyed the entire sky down to an intensity of the order of 10^{-3} the Crab Nebula (by that time already established, rather misleadingly, as an X-ray "standard candle"). The final catalogue included 339 sources. While many of them clustered along the Galactic plane, an unexpectedly large number of Seyfert galaxies and cluster of galaxies were also detected, distributed homogeneously in the sky. The distribution of the X-ray emission in the latter class of objects was diffuse, suggesting the presence of a dilute hot gas pervading the space among the constituent galaxies [91, 112]. [We know now that the hot gas responsible for this emission constitutes $\simeq$85% of the visible baryons in the largest gravitationally-bound structures in the Universe.] *Uhuru* detected rapid aperiodic variability in the flux intensity of the brightest X-ray source in the Cygnus constellation (Cyg X-1) [167]. Follow-up ground based observations identified the optical counterpart. The measurement of the optical mass function led to Cyg X-1 being established as the first binary system containing a (stellar-mass) black hole. Rockets and balloon experiments discovered flares of millisecond duration [203]. A decade later a 300-d periodicity was discovered [189] by the *Vela* series US classified satellites, carrying X-ray scintillation detectors with a $6° \times 6°$ FWHM collimation [48]. Cyg X-1 was also the subject of investigation by the initial X-ray astronomy program in the Soviet Union [10].

As far as super-massive black holes are concerned, the first astrophysical characterization of their X-ray properties was achieved on one of the brightest AGN of the X-ray sky, Centaurus A. The OSO-7 satellite, launched in 1971, carried a system of multiple proportional counters with a wide energy response (1–60 keV; [45]). It detected a variability of the X-ray emission of Centaurus A by 60% in 6 days [255]. A variation of a factor of 4 in less than 2 years was reported later by the *Copernicus* satellite, carrying a suite of small proportional counters in the collimation plane of grazing incidence reflecting collectors [33, 205]. However, only with the advent of the *Ariel-5* mission ([215]; 1974–1980) it was recognized that Seyfert 1 galaxies are universal X-ray emitters [59]. With its combination of a scanning slat-collimated proportional counters of $\sim$0.03 m^2 area [247] and a wide-field pinhole X-ray camera [102], *Ariel-5* produced a catalogue of 251 sources [150, 250]. The simultaneously operated *SAS-3* mission (1975–1979), also carrying proportional counter arrays with slat and tubular collimators [38, 134] led to the discovery of the first quasar through its X-ray emission [200].

4.1.3 *The "Large Satellites' Era"*

4.1.3.1 Collimated Payload: HEAO-1, EXOSAT, *Ginga*

In the late 70s, NASA launched a series of scientific missions with large scientific payload: the High Energy Astronomy Observatories. The first in this series (HEAO-1; 1977–1979) was a spinning satellite scanning a great circle on the sky in a direction perpendicular to the Earth-Sun line every 30 min. In this way a deep survey of the sky achieved over a six-month period. The sky was surveyed almost three times during the mission science operations. The payload allowed to cover the whole energy range from 0.2 keV to 10 MeV with a highly complementary suite of instruments:

- the Large Area Sky Survey Experiment (LASS), a 1 m^2 proportional counter, sensitive in the 1–20 keV energy band [76]. It unveiled aperiodic variability in Cyg X-1 down to a time-scale of 3 ms [153]
- the *Cosmic X-ray Experiment* (CXE; [202]), a smaller ($\simeq$0.4 m^2) experiment designed primarily to study the CXB in the 0.2–60 keV energy range. It led to the first complete flux-limited sample of high-latitude (i.e., primarily extragalactic) sources. The "Piccinotti AGN sample" [180], extracted from the survey, has been a reference catalogue for decades, leading to the first characterization of the spectral properties of different classes of AGN such as BL Lac [259]—established for the first time as another class of X-ray emitters—and Seyfert 1 galaxies [163]
- a Modulation Collimator (MC) experiment [90] for the accurate determination ($\sim$1') of celestial positions, yielding to a catalogue of >600 1–20 keV X-ray sources (in combination with the LASS; [198])
- an high-energy experiment extending the energy range until $\sim$10 MeV, with a specific small aperture ($\simeq$0.020 m^2) devoted to the 15–100 keV energy range [178]. It yielded a catalogue of >40 high-energy sources [133].

While the subsequent HEAO satellites moved away from the standard proportional-counter based payload, other international players entered in the field of X-ray astronomy in the decade of the 80s. The first X-ray observatory of the European Space Agency, EXOSAT [228] carried a 0.16 m^2 proportional counter (Medium Energy, ME) sensitive in the 1–50 keV energy range with a collecting area comparable to that of *Uhuru* ($\simeq$0.08 m^2) and a small field-of-view (0.75°) reducing the risk of source confusion [244]. A unique feature of EXOSAT was its highly eccentric orbit (as opposed to the low-Earth orbit in which all prior X-ray satellites had operated), with an orbital period of 90 h, allowing several days of uninterrupted observations of the same position in the sky at the modest price of a higher background dependent on the solar activity. The scientific success of EXOSAT is undeniably tied to the discovery of Quasi Period Oscillations in the spectrum of several binary system with neutron stars such as GX5-1 [246], Sco X-1 [190] or Cyg X-2 [99]. On the other hand, discoveries on black-hole systems were equally remarkable. A spectral survey of almost 50 Seyfert galaxies unveiled an ubiquitous "soft excess" component whose origin is still debated today [243]. At that time it was suggested to be a signature of

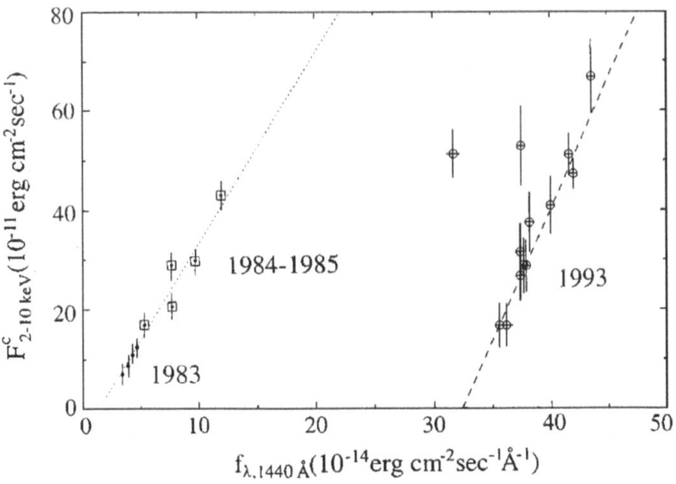

Fig. 4.1 Comparison between the 2–10 keV absorption-corrected flux and the UV flux (IUE, 1440Å) for the Seyfert galaxy NGC 4151. The 1993 X-ray data are from ASCA observations. The *dotted line* on the left represents the linear correlation derived by [174] using EXOSAT and IUE data. Figure extracted from [251]

the accretion disk, an hypothesis that still holds today as reprocessing of the AGN primary emission by the its hot atmosphere [50], or Comptonization of the accretion disk thermal photons by a relativistic corona [57]. The good correlation with UV (Fig. 4.1) led to the start of another fruitful stream of investigation. Whether X-rays lead (reprocessing) or trail (accretion rate fluctuations) the variability is still a matter of debate. Short-term variations were discovered to be common in AGN [13]. They exhibit a "red noise" pattern without evidence for a characteristic time-scale in EXOSAT data [132, 149][1] EXOSAT pioneered also the era of the study of the Fe line emission in black-hole accreting systems through the detection of this feature in many AGN [243], as well as the detection of Doppler variations of the thermal (centroid energy 6.7 keV) line in SS 443, supposed to arise from the near side of its jet [252]. However, it was the Japanese X-ray mission program that put the Fe-line spectroscopy science on a solid foothold.

"Cold" (i.e., corresponding to fluorescence by neutral or mildly ionized ions) Fe lines were detected in AGN [144] by the second X-ray Japanese satellite, *Tenma* (1983–1984) [222], that carried a large area ($\simeq$0.06 m^2) gas scintillation proportional counter with excellent spectral resolution ($\simeq$9.5% at 6 keV). The third Japanese X-ray mission, *Ginga*, carried a very large area ($\simeq$0.4 m^2) proportional counter sensitive in the 1.5–37 keV energy range with a 1° × 2° collimation. Thanks to the excellent sensitivity, *Ginga* could obtain unprecedentedly accurate spectra of

[1]Two decades later, a black hole mass and accretion rate-dependent time-scale was discovered by long-term monitoring campaign with RXTE (Sect. 4.3.1).

many X-ray bright AGN. They demonstrated that emission and absorption features in the Fe regime were common in Seyfert 1 spectra [145, 165, 187], as well as in quasars [113, 242], pointing to the ubiquitous presence of optically-thick matter surrounding the accreting super-massive black hole. Moreover, *Ginga* could prove for the first time that Seyfert 2 spectra are systematically more heavily absorber than Seyfert 1 spectra [9], one of the fundamental pieces of evidence supporting the unified AGN scheme for radio-quiet AGN [6]. This will turn out to be one of the key information to solve the AGN-background "spectral paradox", i.e., the observational fact that the spectrum of the CXB is significantly flatter than that of the quasars of whose unresolved integrated light it should be the resultant, as confirmed by *Ginga* observations of quasars at redshift up to $\simeq$1.4 [169, 254]. New evidence of dramatic X-ray variability in the X-ray light curves of the quasars 3C 279 [140] and PKS0558-504 [197] suggested strongly beamed X-ray emission. A wide-field sky monitor [240] led to the discovery of two new bright transients—GS 2023+338 [118] and GS 2000+25 [241]. Their similarity with A 0620-00 suggested that they are all binary systems hosting a stellar-mass black-hole. Finally, a relative delay in the hard X-ray versus soft X-ray variability pattern was discovered in Cyg X-1 [161].

4.1.3.2 The Dawn of Focusing X-Ray Astronomy: *Einstein*

Einstein (the re-christening of HEAO-2) opened the field of grazing incidence focusing optics for X-ray astronomy [80]. *Einstein* could rotate one of the following instruments in the focal plane of its double-reflection X-ray mirrors, with an effective area of 0.04 m^2 at 0.25 keV and 0.02 m^2 at 2 keV: (a) an Imaging Proportional Counter (IPC) with a modest spatial resolution ($\simeq$1'); (b) a High-Resolution Imager (HRI) with an effective FWHM resolution of 4"; (c) a Solid State Spectrometer (SSS) with a spectral resolution largely exceeding that of proportional counters; and d) a Bragg Focal Plane Crystal Spectrometer (FPCS) with unprecedented spectral resolution ($\lambda/\Delta\lambda \simeq 300$ at 1 keV). Besides the largely complementary payload, the in-focusing capabilities allowed *Einstein* to obtain an enormous gain in sensitivity for diffuse emission in extended objects as well as point-like sources, down to 10^{-7} the intensity of the Crab Nebula. This allowed to confirm that the overwhelming majority of quasars are X-ray emitters [253], and that the integral of the unresolved emission of many of them could account for a large fraction of the CXB [138]—modulo the "spectral paradox" mentioned earlier. X-ray jets, well aligned with the radio jets, were discovered in the cores of the AGN Centaurus A [68] and M 87 [211]. Samples of almost one thousands X-ray sources in the deepest observations allowed the first determination of the logN-logS relation [84].

4.2 Serendipitous Source Catalogues in the Last 3 Decades

4.2.1 The ROSAT PSPC and the ROSAT All Sky Survey

ROSAT, the ROentgen SATellite, was the first German-led X-ray observatory, with participation by the US and the United Kingdom. The satellite was launched in June 1990, and was turned off in February 1999. At the focal plane of a 2.40 m focal length mirror assembly of 4 nested Wolter-I mirrors, two sets of instruments were placed:

- a set of two Position Sensitive Proportional Counters (PSPC), providing modest energy resolution (ΔE/E = 0.43 at about 1 keV), high spatial resolution ($\simeq$25 arc-seconds) over a 2 degree diameter field-of-view, with a relative time resolution $\geq$130 μs
- a High Resolution Imager (HRI), very similar to the Einstein Observatory HRI. The ROSAT HRI provides a 38 arc-min side square field-of-view, with $\simeq$2 arc-seconds spatial resolution (FWHM), and time resolution $\geq$61 μs

While ROSAT performed a very large number of pointed observations, arguably the main and most enduring contribution of this mission to the development of the X-ray astronomy has been the ROSAT All Sky Survey (RASS) and its serendipitous source catalogue [248]. The survey was carried out with a specific operational mode of the PSPC between June 1990 and August 1991. With the latest data processing [32], 135,000 (71,000) X-ray sources were detected in the 0.1–2.4 keV energy band down to a likelihood threshold of 6.5 (9), yielding a spurious detection fraction of $\simeq$30% (5%). The lowest measured 0.1–2.4 keV flux (somewhat dependent on the assumed spectral model) is $\sim$$10^{-13}$ erg cm^{-2} s^{-1}. In the "Bright" RASS catalogue (count rate $\geq$ 0.05 s^{-1}; [248]), 90% of the original 18,811 detections were identified, 7117 thereof associated with a unique source. Out of these, 2966 were extragalactic sources, split between galaxies (2553) and galaxy clusters (413).

4.2.2 Later Serendipitous Source Catalogs

Sizable catalogues of serendipitous sources have been collected by several missions after ROSAT, even if none has the unique combination of large-area and sensitivity of the RASS (yet). The main X-ray source catalogues recently published are:

- the XMM-Newton/EPIC serendipitous source catalogue, currently in its 4th release (4XMM; [201]) (see Table 4.1). This is based on pointed observations with the XMM-Newton satellite [106]. The latest release of the 4XMM catalogue, 4XMM-DR9 covering about 1150 square degrees in the sky includes 810,795 X-ray detections comprising 550,124 unique X-ray sources. Spectra and light curves are provided for the $\simeq$289,000 brightest sources. The astrometric accuracy is $\leq$3" at the 99% confidence level. The median flux in the total photon energy band

(0.2–12 keV) is $\sim$2.3 $\times$ 10^{-14} erg cm^{-2} s^{-1}; in the soft energy band (0.2–2 keV) the median flux is $\sim$5.3 $\times$ 10^{-15} erg cm^{-2} s^{-1}, and in the hard band (2–12 keV) it is $\sim$1.2 $\times$ 10^{-14} erg cm^{-2} s^{-1}. About 23% of the sources have total fluxes below 10^{-14} erg cm^{-2} s^{-1}

- the XMM-Newton Slew Catalogue (XMMSL2, [206]), currently in its release 2.0. It covers now about 84% of the sky. In the coverage corresponding to the deepest exposures, almost 30,000 sources were detected in the full 0.2–12 keV energy band. The median flux is $\sim$3.0 $\times$ 10^{-12} erg cm^{-2} s^{-1} (1.2 and 9.3 in the same units in the E $\leq$ 2 keV and $\geq$2 keV energy range, respectively). The catalogue positions have a 1σ astrometric error of 8" which for point sources (much larger for extended sources). 4924 sources were observed at least two times, with some sources having been observed up to 15 times.
- the Chandra Source Catalogue (CSC; [60]). The last version (2.0) is the largest X-ray source catalogue ever produced in terms of sheer detections (almost one million), out of them almost 320,000 are unique sources (about 1,300 correspond to extended sources). Thanks to its low-background and sub-arc-second angular resolution of the *Chandra* payload, CSC sources reach an unprecedented low flux level of $\sim$ 10^{-18} erg cm^{-2} s^{-1} (in the 0.7–7 keV energy band). However, most of them have fluxes in the $\sim$ 10^{-13} to 10^{-15} erg cm^{-2} s^{-1}. The sky coverage is about 320 square degrees. The excellent angular resolution of the *Chandra telescope* implies also that the astrometric accuracy is excellent. The mean offset against SDSS QSO is $\simeq$0.22" for CSC sources within 3' of the aim-point, $\simeq$0.62" for sources within 10', and 1.2" overall.
- the *Swift*/BAT All-Sky hard X-ray Survey [168]. The Burst Alert Telescope (BAT, [15]) on board the *Swift*-Gehrels observatory [79] is coded mask imaging detector sensitive in the 14–195 keV energy band. It scans regularly the whole sky. Its sensitivity is $\simeq$8.4 $\times$ 10^{-12} erg cm^{-2} s^{-1} over 90% of the sky, and $\simeq$7,2 $\times$ 10^{-12} erg cm^{-2} s^{-1} over 50% of the sky. The corresponding published catalogue, now corresponding to 105 months of observations, includes 1632 hard X-ray sources at a significance level 4.8σ or higher. An extensive identification program allowed to unveil that 34% of the 422 new detections are Seyfert AGN in nearby galaxies (z $\leq$ 0.2), and $\simeq$10% are blazar or BL Lac objects. In this context, it is worth mentioning that NuSTAR (cf. Sect. 4.3.3) has recently produced the first source catalogue based on a focusing instrument above 10 keV [130], with an areal coverage of $\simeq$13 squared degrees, and 497 sources detected over the 3–24 keV energy range.

4.2.3 The Future: eROSITA

The future of X-ray survey is linked to eROSITA (extended ROentgen Survey with an Imaging Telescope Array), the primary instrument on the Russian Spektrum-Roentgen-Gamma (SRG) mission [156]. Successfully launched in July 2019, eROSITA will perform a deep survey of the entire X-ray sky over the initial 4 years of the operational mission. In the soft X-ray band (0.5–2 keV), this will be about 20 times more sensitive than the RASS, while in the hard band (2–10 keV) it will provide the first ever true imaging survey of the full sky at those energies. The eROSITA payload consists of 7 identical mirror modules with 54 mirror shells each, with a an angular resolution $<15''$ on-axis, degrading to an average of $\leq 30''$ over the 30" diameter field-of-view. At the focal plane, X-ray CCD cameras will yield an energy resolution of about 130 eV at 6 keV (cf. Table 4.1). eROSITA is supposed to detect $\sim 10^5$ galaxy clusters out to redshifts $z>1$, allowing to investigate the large scale structure in the Universe, and test and characterize cosmological models including Dark Energy. eROSITA is also expected to detect around 3 millions AGN.

4.3 The Broad-Band Perspective

4.3.1 RXTE

The era of accurate X-ray timing measurements was ushered by the Rossi X-ray Timing Explorer (RXTE). RXTE could cover time scales from microseconds to years over a spectral spectral range from 2 to 250 keV. Launched in 1995, RXTE was operated during 16 years before being decommissioned at the beginning of 2012. The mission carried two collimated instruments (about 1 degree FWHM field-of-view); the Proportional Counter Array (PCA) with an the energy range until about 50 keV; and the High Energy X-ray Timing Experiment (HEXTE) covering the upper energy range. In addition, an All-Sky Monitor (ASM) scanned about 80% of the sky every orbit, allowing monitoring over time scales of 90 min or longer down to a sensitivity limit of tenths of a Crab.

Thanks to its high throughout and fast readout, the timing properties of bright black hole accreting binary systems could be followed through extensive studies of their Power Spectrum Density (PSD) over different states. The micro-quasar GRS 1915+105 is a typical example. Its complex spectral-timing behaviour can be broadly classified in 3 classes ('A', 'B', and 'C'), based on the spectral hardness [19], with transitions occurring on time-scales of seconds. The different spectral states correspond to different PSD shapes, with broad continuum components (due to a varying combination of red noise power-law and Lorentzian components) dominating at all states while Quasi Period Oscillations (QPOs) are present only in the hardest state (Fig. 4.2).

Complex relations between the timing and spectral states were discovered in other accreting black hole systems [196]. Particularly noteworthy was the discovery of High Frequency QPOs at time frequencies ≥ 100 Hz in stellar-mass black hole systems, analogue to those already known in neutron star systems [191]. Their stability suggests that they may trace fundamental properties of the black hole such as the mass and/or spin.

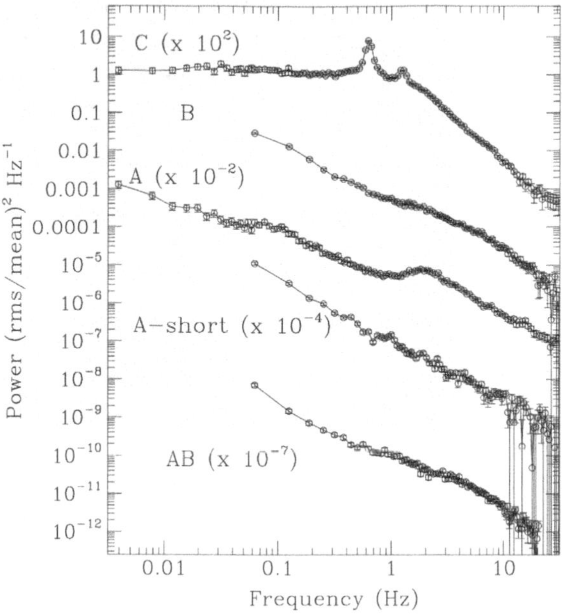

Fig. 4.2 *Left panel*: representative examples of the power spectrum density of the micro-quasar GRS 1915 + 105 in three characteristic source states (labeled "A", "B", and "C"). Figure extracted from [194]

4.3.2 BeppoSAX

More focused on X-ray broadband spectroscopy, BeppoSAX was a major program of the Italian Space Agency with participation of the Netherlands Agency for Aerospace Programs and ESA. It was launched on April 1996 and was operated for 6 years. The scientific payload covered more than three decades of energy—from 0.1 to 300 keV—with a relatively large effective area, medium energy resolution and imaging capabilities in the range of 0.1–10 keV through four Narrow Field Instruments (NFIs):

- the Low Energy Concentrator Spectrometer (LECS), covering the 0.1–1 keV energy range, with an effective area of 22 cm^2 at 0.28 keV, a field-of-view of 37' diameter, and an angular resolution of 9.7' FWHM at 0.28 keV
- the Medium Energy Concentrator Spectrometer (MECS), covering the 1.3–10 keV energy range, with an effective area of 150 cm^2 at 6 keV, a 56' diameter field-of-view, and an angular resolution 75" HEW.
- the High pressure Gas Scintillator Proportional Counter (HPGSPC) covering the 4–120 keV energy range, with an effective area of 240 cm^2 at 30 keV
- the Phoswitch Detection System (PDS) covering the 15–300 keV energy range, with an effective area of 600 cm^2 at 80 keV

The NFIs were complemented by a Wide Field Camera covering the 2–30 keV energy range with a field of view 20° × 20°.

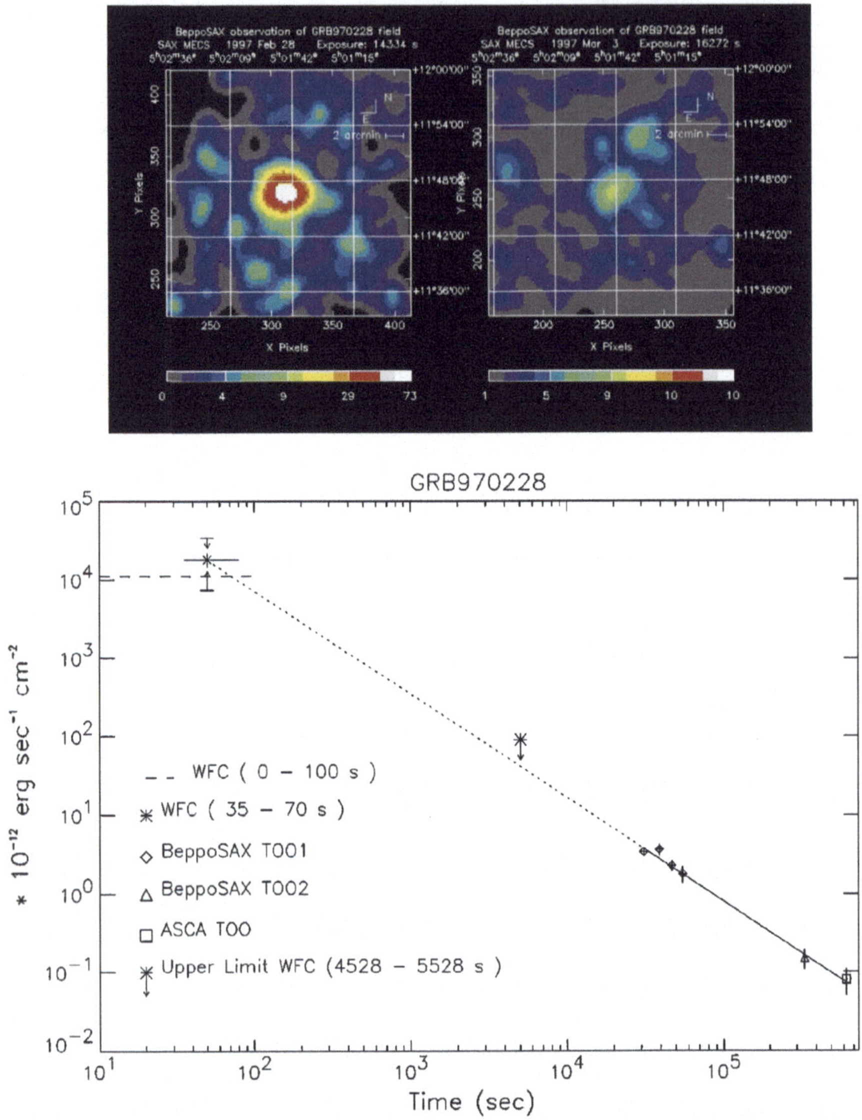

Fig. 4.3 *Top panel:* Discovery images of the X-ray afterglow of GRB 970228 with the BeppoSAX NFIs. *Bottom panel*: The X-ray afterglow light curve of GRB 970228 taken with the BeppoSAX WFI and NFI, as well as with the ASCA instrument. Figures extracted from [49]

BeppoSAX is remembered for the discovery of Gamma Ray Burst X-ray afterglows (Fig. 4.3), detected independently on-board by the lateral shields of the PDS in the range of 60–600 keV, and later pointed by the NFIs

However, its scientific payload gave contribution in several fields of astronomy such as obscured AGN (the PDS instrument remaining one of the most sensitive non-focusing instruments above 10 keV ever flown) or relativistic spectroscopy of accreting black holes (see Sect. 4.4.5) over the whole mass range. In this respect, the capability of BeppoSAX of accurately modelling the continuum underling the key Fe K_{α} line was key to reliably assess the shape, and therefore the physical drivers of the line broadening in XRB (Fig. 4.4) as well as in AGN [88].

The unprecedented sensitivity of the PDS permitted also detailed spectroscopic studies of the hard states of XRBs (see, e.g., [261]).

4.3.3 NuSTAR: The Focusing Revolution

In this context, the crucial role played by NuSTAR shall be emphasized. This mission [95] carries the first orbiting telescopes capable of focusing X-rays above 10 keV (up to 79 keV). Its payload consists of two co-aligned grazing incidence telescopes. Each optical module contain 133 nested multilayer-coated grazing incidence shells in a conical approximation to a Wolter-I geometry. Solid state CdZnTe pixel detector [94] surrounded by a CsI anti-coincidence shield are located in the focal plane at the end of an extendable optical bench to achieve a 10-meter focal length. The unique payload combination enables an improved sensitivity by two orders of magnitude over previous missions that had operated at these X-ray energies, thanks to the combination of effective area and low background enabled by the imaging capabilities. This has contributed greatly to several areas of research involving accreting black holes, such as resolving a $\simeq$80% of the CXB at the energies of its Spectral Energy Distribution (SED) peak [96], the measurements of the temperature of the electron distributions in AGN coronae [236], suggesting that AGN coronae are hot and compact, probably intrinsically limited and regulated by runaway pair production [65]. The latter discovery, emerging from the location of AGN coronae in the compactness-temperature diagram, leads to the conclusion that pair production and annihilation are essential ingredients in the coronae of AGN and, probably, BHB.

4.3.4 ASTROSAT

ASTROSAT is a multi-wavelength Indian astronomy mission, as well as the first Indian X-ray observatory. It was launched on September 28, 2015. The expected operating life time of the satellite will be more than five years, and at the time this book was being written it was operating nominally. ASTROSAT is another example of an X-ray mission devoted to broad-band spectroscopy, thanks to a unique suite of scientific instruments:

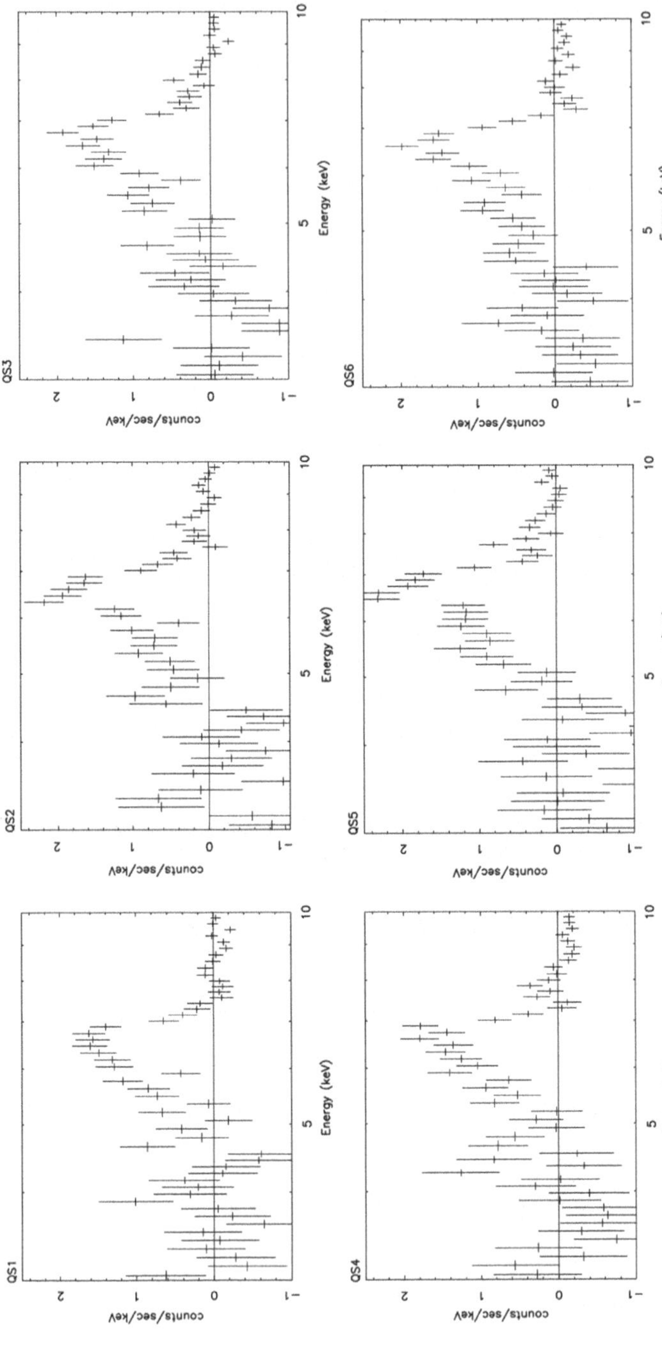

Fig. 4.4 Residual spectra extracted from a *BeppoSAX*/MECS observation of GRS1915+105 during 6 time intervals. They correspond to fits with a continuum function excluding the 3–7.5 keV energy range. The show a profile of an emission line consistent with K_{α} fluorescence from neutral or mildly ionized iron broadened by general relativistic effects close to an accreting stellar-mass black hole. Figure extracted from [142]

- A Soft X-ray Telescope (SXT) focusing X-ray into a CCD detector, covering the energy range 0.3–8 keV, and an effective area will be about 120 cm^2 at 1 keV.
- Three units of Large Area Xenon Proportional Counters (LAXPC) covering an energy range from 3 to 80 keV with an effective area of 8000 cm^2 at 10 keV.
- A Cadmium-Zinc-Telluride coded-mask imager (CZTI), covering hard X-rays from 10 to 150 keV, a $\simeq$6 degrees field of view and 480 cm^2 effective area.

This pointed X-ray payload is complemented by a Scanning Sky Monitor (SSM) consisting of three one-dimensional position-sensitive proportional counters with coded masks. The assembly is placed on a rotating platform capable to scan the available sky once every six hours. To extend the sensitive band-pass of ASTROSAT to longer wavelength, a 38-cm aperture Ultraviolet Imaging Telescopes (UVIT) covers from the Far-UV to optical bands.

The main scientific themes of ASTROSAT are: simultaneous multi-wavelength monitoring of intensity variations in a broad range of cosmic sources; monitoring the X-ray sky for new transients; sky surveys in the hard X-ray and UV bands; broadband spectroscopic studies of XRBs and AGN, among other classes of objects; and studies of periodic and non-periodic variability of X-ray sources.

4.3.5 HXMT/Insight

A Phoswitch scintillation detector is at core of the Chinese Hard X-ray Modulation Telescope, rechristened *Insight* after launch. It is an array of 18 NaI(Tl)/CsI(na) scintillation detectors, collimated to 5.7° × 1° overlapping fields of view, with an effective area of 286 cm^2 each, and covering the 20–200 keV energy range. They are complemented by a medium energy X-ray telescope (5–30 keV), and a low energy X-ray telescope (1–15 keV).

4.4 X-Ray CCD Astrophysics

Pioneered by the Japanese mission ASCA, the era of the X-ray CCD astrophysics started with the end of the last century. CCD detectors ensures imaging capability, low background, fast readout, and energy resolution >10 over the two decades in energy between 0.1 and 10 keV with good quantum efficiency. A comparison of the main performance of the currently operational CCD detector systems is shown in Table 4.1.

Table 4.1 Comparison of the main properties of past and operational X-ray CCDs. The energy resolution refers to the boresight/aim-point. In all cases the standard grade/pattern selection for spectral analysis is assumed. The value is provided beginning of life

Name	Satellite	Operations	Energy resolution (eV @6 keV)	Area (cm^2 @6 keV)
SIS	ASCA	1993–2001	150	130
ACIS	*Chandra*	1999–	150	200[a]
EPIC-pn	XMM-Newton	2000–	180	900
EPIC-MOS	XMM-Newton	2000–	160	300
XIS	*Suzaku*	2005–2015	150	250/220[b]
XRT	*Swift*	2004–	140	70
SXI	*Hitomi*	2016	170	300
eROSITA	Spektrum X-γ	2019–	140	250

[a] ACIS-S
[b] front/back-illuminated

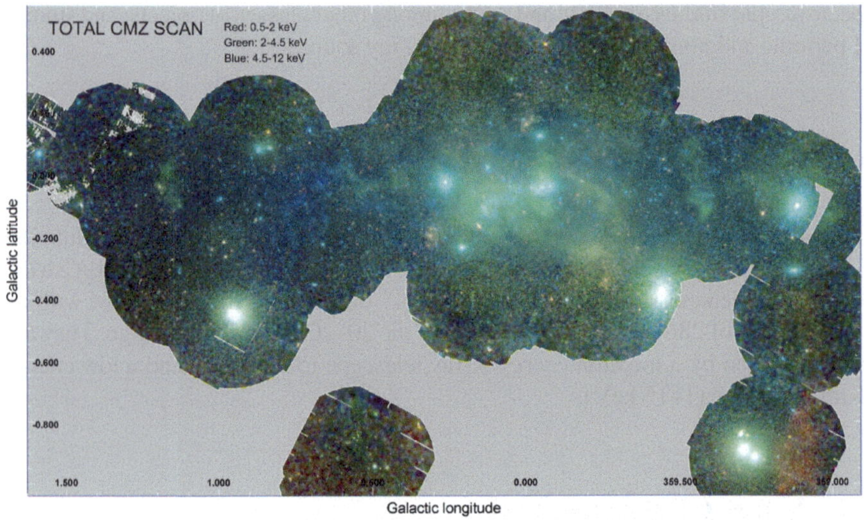

Fig. 4.5 Broad energy band RGB mosaic image of all EPIC XMM-Newton observations within 1° of Sgr A*. Figure extracted from [182]

4.4.1 *Sgr A**

An example of the power of the spectral imaging capability opened by the era of X-ray CCD astrophysics is Fig. 4.5, showing the XMM-Newton view of the 1° degree surrounding the location of the super-massive black hole at the center of our own galaxy, Sgr A*. The image is a combination of individual transient and persistent sources, bright stray light arcs due to bright sources outside the field-of-view, and diffuse emission.

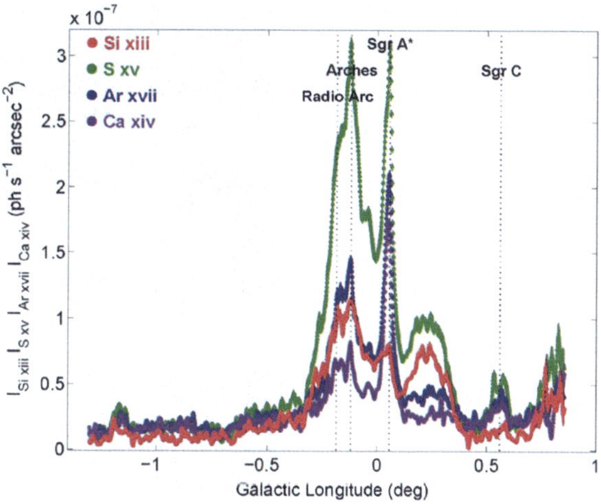

Fig. 4.6 Intensity profiles of the SiXIII (*red*), SXV (*green*), ArXVII (*blue*) and CaXIX (*violet*) emission lines as a function of Galactic longitude, integrated in latitude. Figure extracted from [182]

The spectroscopic capabilities of the CCD cameras allows to derive the spatial profiles in different emission lines (Fig. 4.6).

4.4.2 *AGN Feedback*

4.4.2.1 Radio Mode

Even more powerful in terms of imaging spectroscopy is the combination of a CCD detector (the ACIS) with the sub-arcseconds spatial resolution of the *Chandra* telescope. An example is shown in Fig. 4.7, a multi-color X-ray image of the core of the Perseus galaxy cluster. The X-ray emission in the InterCluster Medium (ICM) is due to thermal bremsstrahlung and line radiation at million degrees temperature. In the core of the cluster, a radio-loud AGN (NGC 1275) is present. Jets from the nucleus of that galaxy inflated bubbles (see as dark "voids" in the image), displacing the ICM [31, 62]. The Perseus Cluster is one of the most spectacular examples of "radio-mode" feedback, relativistic particles in the form of a highly collimated jet ejected from an active nucleus and capable of injecting sufficient energy in the ICM to prevent it from collapsing, cooling, and generating new stars.

Similarly, trace of the shocks induced in the atmosphere surrounding the a SMBH were detected by the *Chandra* ACIS as a nearly circular ring of outer radius 13 kpc around the radio galaxy M 87 (Fig. 4.8).

Fig. 4.7 *Chandra* ACIS multi-color X-ray image of the Perseus cluster core [63]. The *red*, *green*, and *blue* channels correspond to the 0.3–1.2 keV, 1.2–2 keV, and 2–7 keV energy bands, respectively

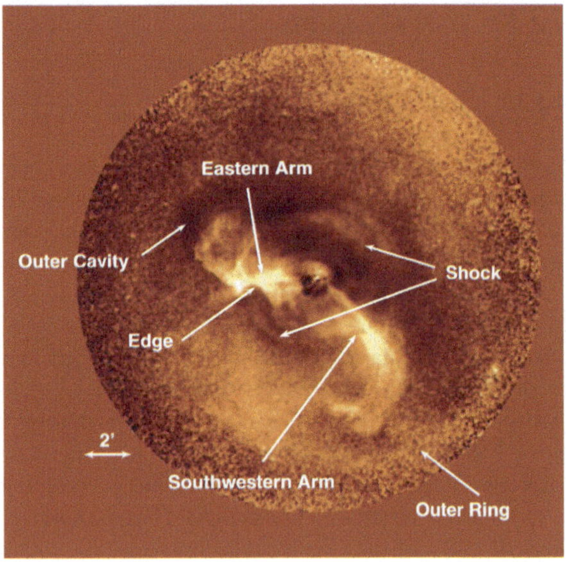

Fig. 4.8 *Chandra*/ACIS 0.5–2.5 keV surface brightness from a region surrounding the nearby radio galaxy M 87. A radially averaged surface brightness model and point sources were subtracted. The main features of the interaction between the jet and the intergalactic medium are shown. Figure extracted from [74]

4.4.2.2 Quasar Mode

X-ray CCD spectroscopy with XMM-Newton takes instead advantage of the unprecedented effective area of the telescope to perform detailed spectroscopy particularly in the energy range above 2 keV where the current grating systems on-board *Chandra* and XMM-Newton have insufficient effective area or energy coverage. One of the key results of the XMM-Newton EPIC cameras in this area is the discovery of highly ionized, high-column density outflows in the spectra of a large fraction (30–50%) of the bright AGN in the nearby universe [233]. This discovery was later subsequently confirmed by the XIS on board the Japanese satellite *Suzaku* [85], and intriguingly extended also to a smaller sample of radio-loud AGN [232]. The experimental evidence are deep absorption features at energies larger then the resonant transitions of He- and H-like Fe ionized species (Fig. 4.9).

While in a few cases the exact identification of these lines is difficult, and detailed modelling with photoionisation codes are required to solve potential degeneracy [115]. it is now uncontroversial that at least some of these features correspond to outflows with relativistic velocities, up to 0.3–0.4 c.

Some of them shows indication of a P Cygni profile (Fig. 4.10; see also [166]), indicative of a large covering factor.

This is one of the key results of XMM-Newton in the field of accreting supermassive black holes. Sub-relativistic outflows could carry sufficient mass and kinetic energy that—if efficiently coupled with the InterStellar Medium—could provide the long-sought source of either negative and positive feedback, stopping star formation

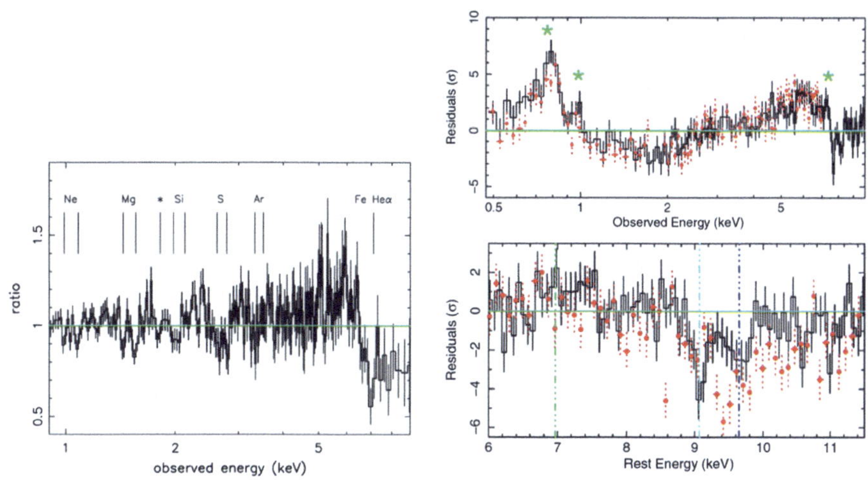

Fig. 4.9 Ratio of EPIC-pn spectrum to a simple power-law continuum for an XMM-Newton observation of PG1211+143 (*left panel*; after [184]) and PDS 456 (*right panel*; after [193]. The deep absorption features at 7 keV are the imprinting of highly ionized outflowing gas along the line-of-sight to the AGN

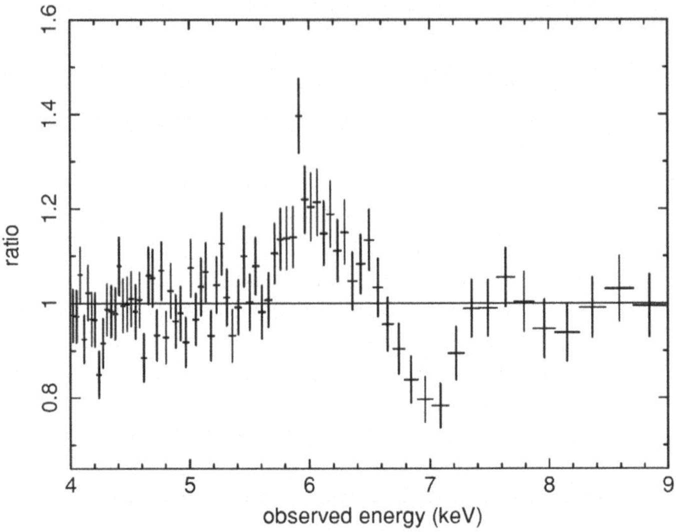

Fig. 4.10 The profile of the Fe K_α fluorescent emission line in the EPIC-pn spectrum of the XMM-Newton observations of PG 1211+143. The P Cygni like profile shows with a comparable equivalent width of the emission and blue-shifted absorption components indicate the highly ionized outflow has a large covering factor. Figure extracted from [185]

via evacuation or over-heating of cold molecular gas [55, 103] or, on the contrary, stimulating star formation via shock-induced gas compression. The so-called *Ultra-Fast Outflows*, discovered by XMM-Newton, are in principle sufficiently powerful to be good candidates [136, 235] for the former mechanism. They could be the ultimate driver of fundamental relations such as the tight correlation between the star velocity dispersion in the galactic bulge and the super-massive black hole mass hosted at the core of most galaxies [114, 147].

4.4.3 X-Ray Jets in AGN

The sub-arcseconds resolution of the *Chandra* telescope has been also the key performance capable of unveiling the ubiquitous nature of X-ray relativistic jets in radio-loud quasar. A very large fraction of detections has been demonstrated by dedicated surveys [141]. The spatial *Chandra* resolution matches well that of HST in the optical/UV and of the VLA in radio (Fig. 4.11).

This has allowed the identification of the mechanism responsible for the X-ray emission of individual knots in structured jets, separating those due to synchrotron emission in a jet well approximated by a single flow from those where the different spectral shapes indicates dominance by Inverse Compton of the Cosmic Microwave Background. The former is more common in radio galaxies, tracing in-situ accelera-

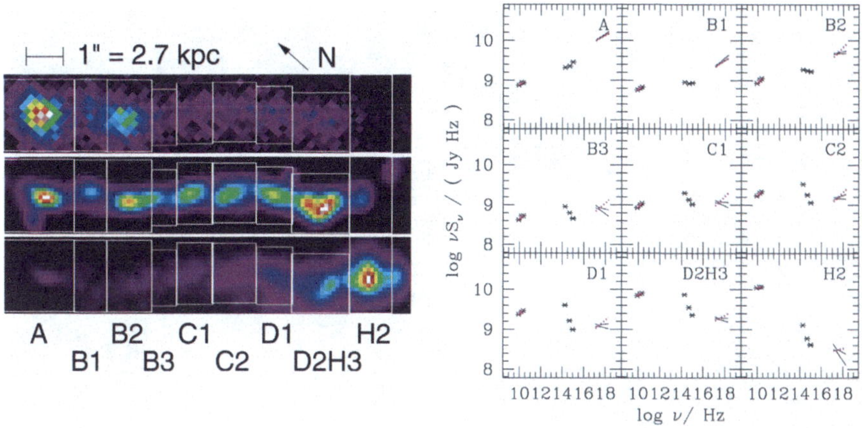

Fig. 4.11 *Left panel*: Jet in 3 C 273 observed with Chandra (*top*), HST ($\lambda = 620$ nm, *middle*), and VLA ($\lambda = 3.6$ cm, bottom). The Spectral Energy Distributions of the boxy regions defined in the *left panel* are shown in the *right panel*. Figure extracted from [107]

tion and variable AGN input; the latter are more common in quasars, tracing highly relativistic bulk flow and variable particle acceleration [258]. Relativistic jets have been discovered by *Chandra* also in quasars at high redshift (Fig. 4.12).

4.4.4 Surveys and the XRB

The energy density of the X-ray sky is dominated by a diffuse radiation which is mostly of cosmic origin: the X-ray background (XRB). While in the soft X-rays it is dominated by components of Galactic origin, at energies $\geq$1 keV it is dominated by sources of extragalactic origin, as shown by the shadow that the Moon profile shed on it in a famous image taken by ROSAT [213]. Deep surveys of the X-ray sky with *Chandra* and XMM-Newton have resolved $\simeq$80% of the X-ray sky up to 10 keV in discrete sources, primarily obscured and unobscured AGN.

The way to achieve these results is trough painstaking "X-ray surveys". Surveys of different coverage and depths have been performed by all major modern X-ray observatories. They can be broadly classified in "shallow/large area" and "deep/pencil beam" [36] with a gradual transition from the two classes.

The RASS is the most extreme example of the first class, reaching a 0.5–2 keV flux limit of a few 10^{-13} erg cm^{-2} s^{-1} over the whole sky. On the other hand of the scale, the *Chandra* surveys of the Hubble Deep Fields reaches sensitivities of a few 10^{-17} erg cm^{-2} s^{-1} on about 0.1 degrees^2. At these faint level, galaxies start competing with AGN as the dominant X-ray source population (Fig. 4.13).

Due to its broader PSF, the XMM-Newton Lockman Hole survey reaches fluxes $\sim 10^{-16}$ erg cm^{-2} s^{-1} at its confusion limit (reached with exposures of about 2 Ms),

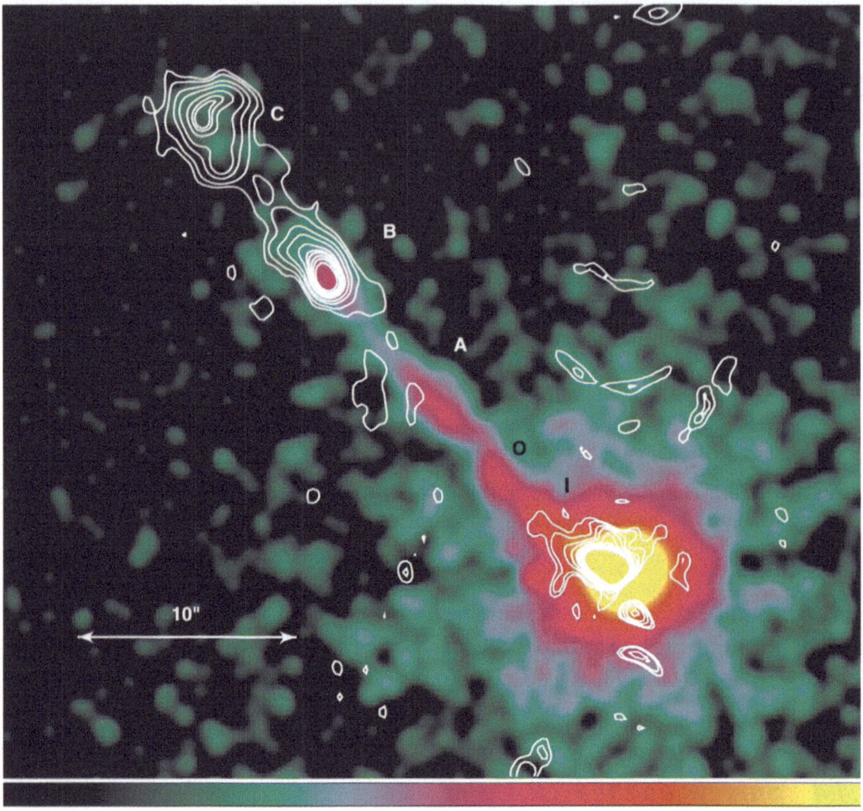

Fig. 4.12 X-ray image of the z = 1.18 quasar PKS1127-145, smoothed with a Gaussian kernel ($1\sigma = 0.615$"), overlaid with the 8.5 GHz radio contour map (radio core subtracted). The look-up-table of the X-ray contours is shown at the bottom of the image, ranging between 10 and 10^5 photons cm^{-2} s^{-1} $pixel^{-2}$. Figure extracted from [214]

allowing, however, to collect many more photons and therefore a better characterization of the spectral properties of each individual source. Despite the fact that samples of thousands of photometrically and spectroscopically identified AGN have been unveiled in these surveys, fundamental question such as the level of agreement between the predictions of the Unified AGN scenario, the fraction of obscured AGN and the dependence of this fraction on luminosity and redshift are still far from being ultimately answered [157].

The availability of large sample of AGN at different redshift allows us also to study their evolution. More luminous AGN and quasars peak earlier in the history of the universe, while the low-luminosity ones arise later (a phenomenon referred to as "cosmic downsizing"). Moreover, for objects with an X-ray luminosity $\geq 10^{44}$ the space density decreases significantly. As the comparison in Fig. 4.14 shows, models

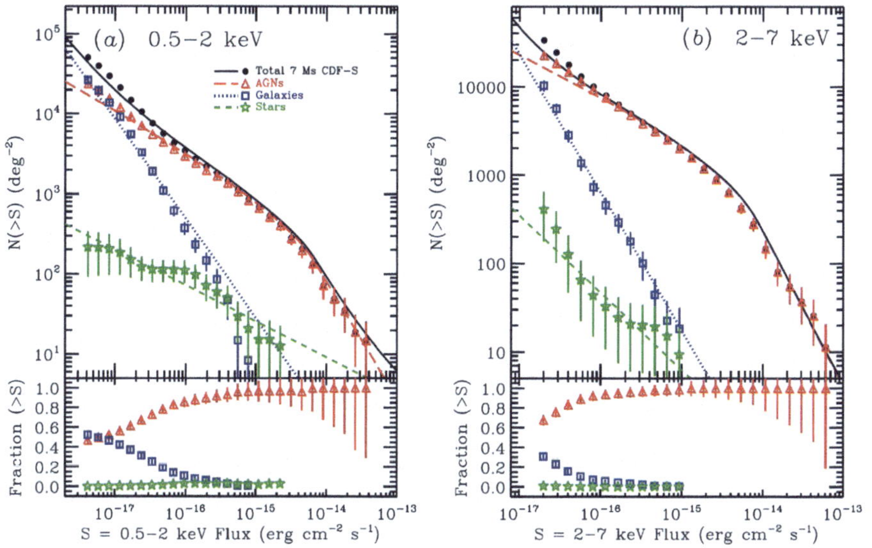

Fig. 4.13 Cumulative number counts (number of sources brighter than a given flux) for the source catalog extracted from the *Chandra*/ACIS 7 Ms Chandra Deep Field North observation in the soft X-ray band (0.5–2 keV; *left panel*), and (2–7 keV; *right panel*) energy ranges. The number counts are further broken down into AGN (*open red triangles*), normal galaxies (*open blue squares*), and Galactic stars (*open green stars*). Figure extracted from [137]

Fig. 4.14 The space density of AGN as a function of redshift in different luminosity classes. From *top* to *bottom*: $\log(L_X)$ in the ranges 42.0–43.0, 43.0–44.0, 44.0–045.0, 45.0–46.0, ≥46.0. The *lines* constitute the best-fit PLE and LDDE models. Figure extracted from [100]

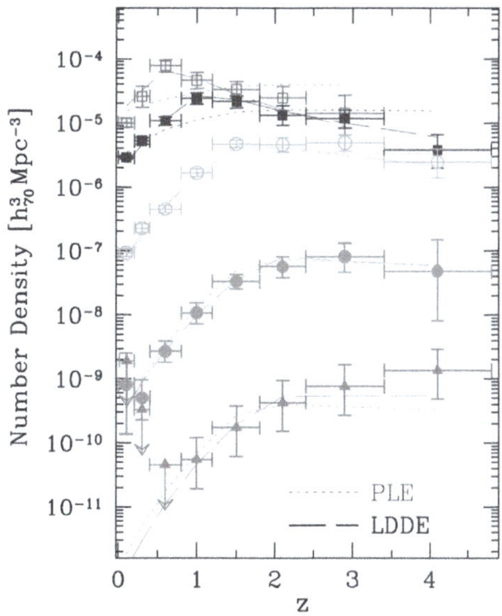

of "Luminosity-Dependent Density Evolution" (LDDE) fit these results better than simple Pure Luminosity Evolution (PLE) ones.

4.4.5 Relativistic Spectroscopy

Studying the profiles of emission lines emitted in X-ray illuminated relativistic accreting disk a few gravitational radii from the accreting black hole event horizon down to the disk innermost stable circular orbit may allow us to determine the geometry and the emissivity profile of the accretion flow, as well as the black hole spin [61, 131]. This stream of research was inaugurated with the X-Ray CCD on-board ASCA (Fig. 4.15), even if the complementary scintillator proportional counter on board the same mission (the Gas Imaging Spectrometer, GIS) could also give an important contribution in the study of these spectral features in XRBs [158].

Later observations with the CCD on-board *Suzaku* and XMM-Newton confirmed in details the shape of the line profiles detected by ASCA (Fig. 4.16).

Nowadays, about 30 AGN exhibit broad and skewed iron line profiles that are consistent with being produced in the innermost regions of an accretion disk. Most of them are in the local Universe, even if at least an example at $z = 0.66$ is known [195].

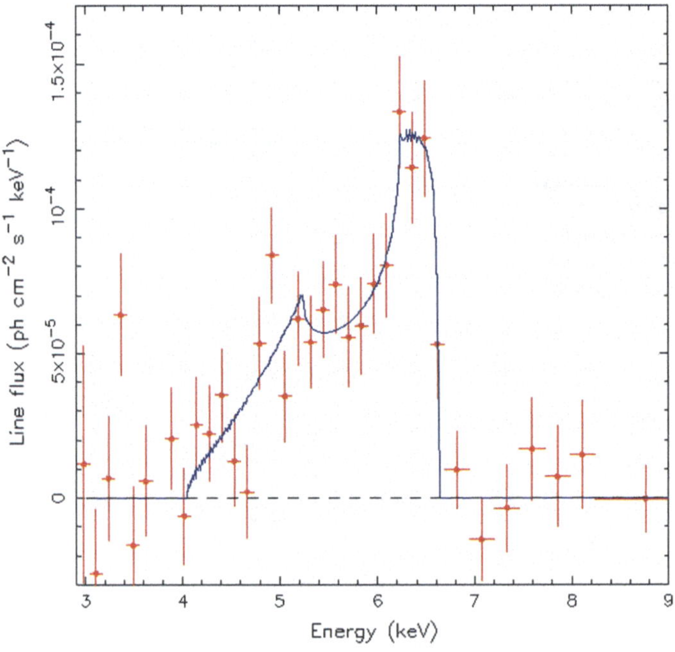

Fig. 4.15 Ratio between the best-fit continuum and the ASCA/SIS spectrum of the Seyfert 1 Galaxy MCG-6-30-15 (*left panel*). Figure extracted from [223]

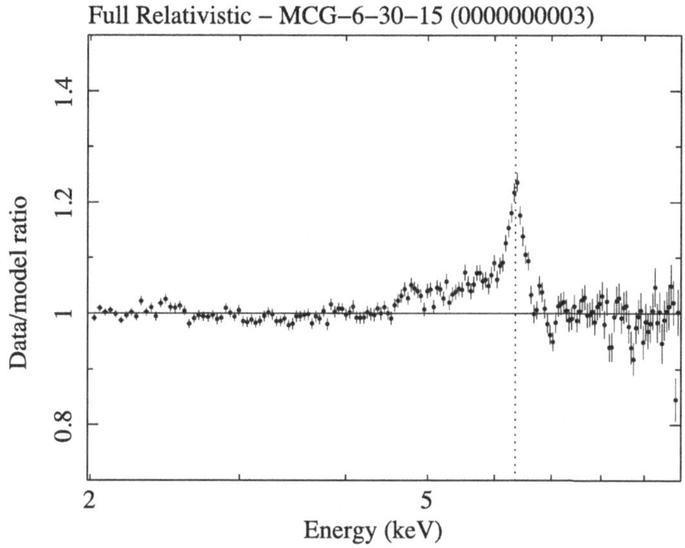

Fig. 4.16 Ratio between data and continuum for the best-fit of the 3–8 keV spectrum of the nearby Seyfert 1 Galaxy MCG-6-30-15 with the XMM-Newton/EPIC-pn. The vertical line represents the position of a fluorescent neutral K_α line in the source rest frame. Figure extracted from [53]

The census of black hole spin amounts to about 20 objects [199]. The distribution is largely skewed towards maximally spinning black holes. While it would be tempting to interpret is as an evolutionary effect of the host galaxy via merging and coherent accretion [22], the current results are likely to be biased because highly spinning accreting black holes are likely to be radiatively more efficient [64].

4.5 High-Resolution Spectroscopy

High-resolution X-ray spectroscopy has the potential to reveal the physics of the plasma state that accounts for much of the material in the observable universe. The X-ray band contains the inner-shell transitions of all abundant elements from C through N, O, Ne, Mg, Si, S, Ca, Fe and Ni. X-ray spectra of cosmic sources typically consist of a mixture of ionic or atomic emission or absorption lines and electron continuum components. Observed spectra can be modelled in order to identify the underlying physics and dynamics of the plasma concerned. They often depart from the ideal conditions of equilibrium and spatial uniformity. It is not coincidental that the availability of high-resolution data in the X-ray domain at the beginning of the current millennium has triggered important development in the modeling of X-ray emitting plasma [18, 47], in the collisionally ionized as well as in the photo-ionized regime.

Table 4.2 Comparison of the main properties of operational X-ray gratings

Name	Satellite	Sensitive bandpass	Resolving power	Area (cm^2 @ 1 keV)
LETG/HRC-S	*Chandra*	0.07–10 keV	>1000[a]	30
LETG/ACIS-S	*Chandra*	0.2–10 keV	$\sim 20 \times \lambda$	20
HETG	*Chandra*	0.4–10 keV	200[b]	60[c]
RGS	XMM-Newton	0.3–2 keV	300[d]	60[d]

[a] 50–160 Å
[b] @6.5 keV for a point source up to 14 counts per second; 1000–65 in the full band
[c] with ACIS-S
[d] @15 Å, first order

4.5.1 Comparison of Grating Properties

A giant leap in this field was achieved with the launch of *Chandra* and *XMM-Newton* carrying high-sensitivity, high-resolution grating systems. Their properties are listed in Table 4.2.

The *Chandra* grating ensures a better energy resolution, at the price of a smaller effective area. The higher collecting area of the RGS is shown by the apparent emission "bump" between 12 and 20Å, due to the blending of several hundreds of unresolved emission lines. Moreover the HETG extends the sensitive band pass of the *Chandra* grating to 7 keV, even with a very limited effective area.

In the field of AGN, high-resolution spectroscopy allows us to perform detailed plasma diagnostics of AGN outflows (conceptually similar to the UFOs discussed in Sect. 4.4.2.2, even if the relation between relativistic and non-relativistic outflow is still a matter of debate [128, 235]. Known since more than 30 years [92], the somewhat misleadingly christened "warm absorbers" are a fundamental ingredient of any AGN structure model. Phenomenologically, they appear as a forest of absorption lines due to resonant transitions of He- and H-like ions of elements from Carbon to Iron (Fig. 4.17).

They are present in $\geq$75% of nearby AGN in flux-limited, complete samples [127], with outflow velocity ranging from a few 1000s km s^{-1} to a fraction of the speed of light, a broad distribution of ionization parameters ($\log(\xi)$ ranging from 0 to 10^4 erg s^{-1} cm^{-1}) and column densities (N_H ranging from 10^{20} to 10^{24} cm^{-2}) [17]. Plasma diagnostics of the outflowing gas indicate that the gas is photoionised, most likely by the radiation field of the AGN. However, it is still unclear which is the ultimate mechanism driving the absorbing clouds. The main contenders are radiation pressure by the AGN radiation field, magnetic forces, or "thermal winds".

Observationally, from high-resolution spectra one can measure directly (by identifying lines): ionization state, multi-ion components, column density for each ionic species, outflow velocity. Comparison of the data with the predictions of grids of photoionisation models enables to derive the global column density, the ionization parameters and temperature distribution, once energy and ionization balance, and a

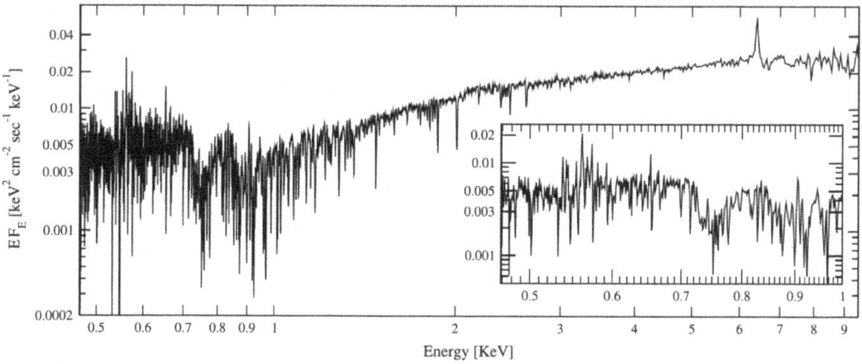

Fig. 4.17 Combined MEG and HEG first-order 900 ks of the Seyfert 1 Galaxy NGC 3783, binned to 0.01 Å. A forest of absorption features due to intervening ionized gas to the active nucleus is apparent. Figure extracted from [110]

given Spectral Energy Distribution (SED) of the ionizing continuum (typically that of the AGN) are assumed. A more elusive quantity is the covering fraction, being X-ray absorption lines typically not saturated). A possibility is estimating it from the faction of AGN showing absorbers in well-defined, flux-limited samples [127].

There is an intrinsic degeneracy between the volume density of the outflow and the distance between the absorber and the ionizing source, due to the definition of the ionization parameter:

$$\xi \equiv \frac{L}{n_e R^2}$$

where L is the integrated ionizing luminosity, n_e is the volume electron density, and R is the distance between the ionizing source and the innermost face of the ionised cloud. The product at the denominator can be disentangled in its factors only if an independent estimate of one of the two is available. The most reliable method is based on the response time-scale of the ionising clouds to changes of the ionising continuum (as traced by the AGN X-ray flux, when seen to be variable). This is proportional to n_e^{-1}. Knowing how fast the outflow clouds respond (do not respond) to changes of the ionising continuum allows to determine an upper limit (lower limit) to the density, translating into a lower limit (upper limit) on the launch radius (assuming it is commensurable to the location of the innermost clouds). While large uncertainties exists on their launch radius and location, the most relativistic outflows are believed to be emitted as disk winds at a few tens-hundreds of gravitational radii from the black hole event horizon, while the slower ones are probably due to evaporation of clouds from the outer surface of the torus [30, 234]. From the estimate of the launch radius, the mass and kinetic energy outflow rates can be calculated [125], and the potential feedback effect estimated [235]. Equatorial disk winds have been shown to be ubiquitous also in the soft states of XRB [183], and in one case demonstrated to be driven by magnetic fields [159].

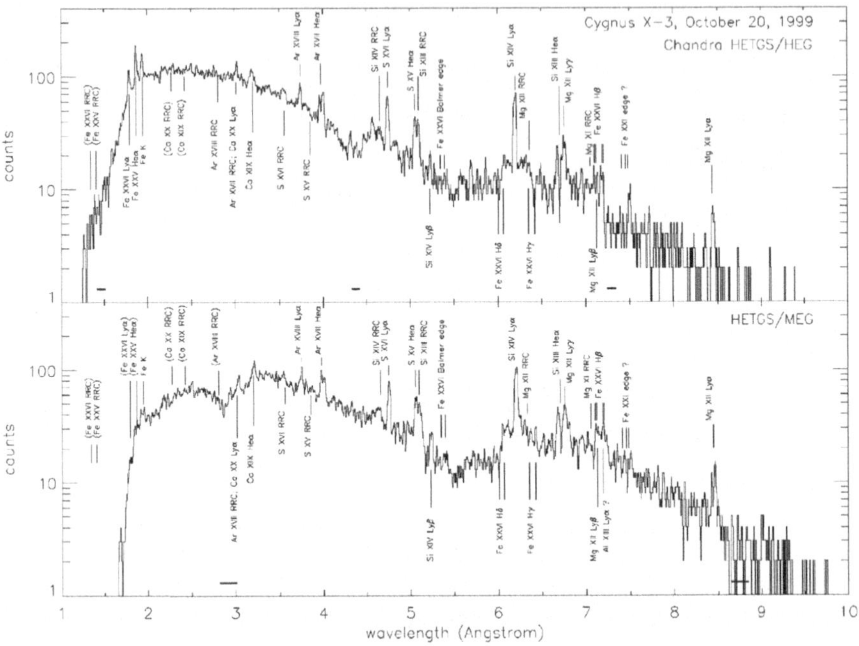

Fig. 4.18 The 1–10 Åspectrum of Cyg X-3 as observed by the *Chandra* HETG: HEG (*top panel*) and the MEG (*bottom panel*) spectra are binned in 0.005 Åbins, with positive and negative first orders added. The labels indicate the positions of various discrete spectral features. Figure extracted from [173]

X-ray photoionised winds can be studies in XRB also through high-resolution spectroscopy of emission lines (Fig. 4.18).

Emission line spectroscopy of highly obscured AGN allows instead one to investigate the properties of the X-Ray Extended Narrow Line Regions [89, 108, 117].

4.5.2 The Era of Micro-Calorimeters: Hitomi *and XRISM*

A huge step forward was expected with the JAXA-led X-ray observatory *Hitomi*, whose successful launch on 2016 February 17 led to hope that a new era in observational X-ray spectroscopy had started. *Hitomi* carried the *Soft X-ray Spectrometer* (SXS), micro-calorimeter detector with an unprecedented $\leq$5 eV energy resolution over the 0.3–12 keV energy band [111]. Such a performance corresponds to an improvement in resolving power larger then one order of magnitude at 6 keV with respect to the *Chandra* high-energy transmission gratings, together with an one-order-of magnitude larger effective area at the same energies. Regrettably, the spacecraft was lost after only six weeks of operations due to a chain of anomalies

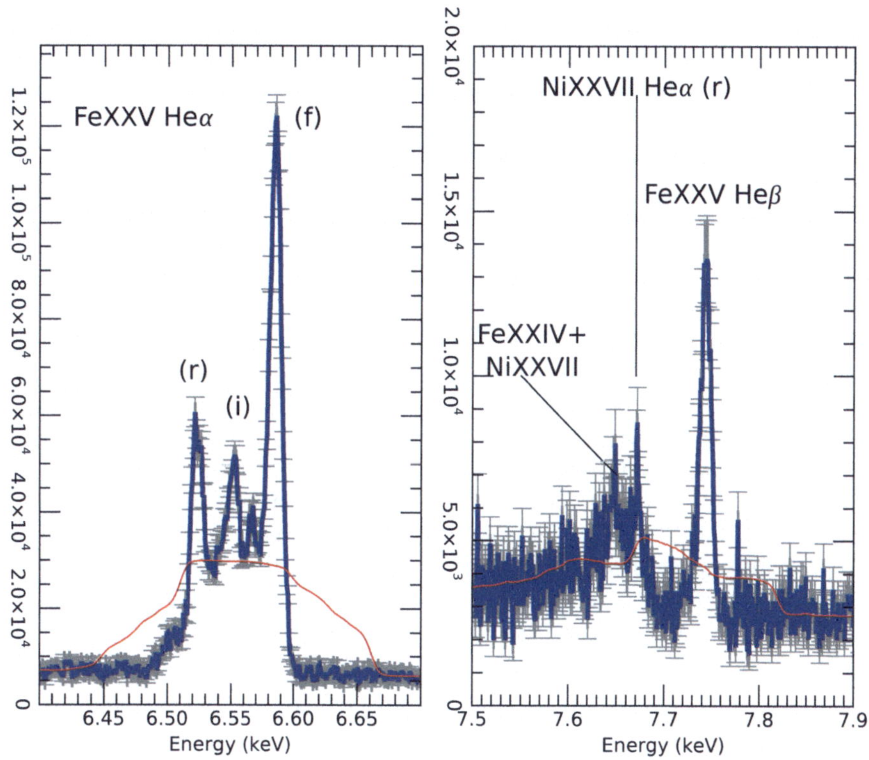

Fig. 4.19 *Hitomi*/SXS spectrum of the Perseus Cluster in the 6.4–6.7 keV (*left panel*) and 7.5–7.9 keV energy range (*right panel*), respectively [87]. The main emission lines discussed in the manuscript are labeled. The error bars correspond to 1σ Poissonian uncertainties. The *red lines* represent the same spectrum at CCD-resolution. Technical details: the spectrum was extracted from the combined calibrated event lists of Obs.# 100040020, –30, –40, and –50 ($\simeq$2.43 $\times$ 10^5 s net exposure time), downloaded from the *Hitomi* science archive. It corresponds to a circular extraction region around the cluster core with a 150" radius

of the attitude control system coupled to human errors. However, careful planning during the commissioning phase led to *Hitomi* observing for almost one week the core of the Perseus galaxy cluster, one of the X-ray brightest in the local Universe ($z = 0.0179$). While the SXS was still not formally commissioned, it was operated with the closed "gate valve" with a $\sim$300 μm beryllium window along the optical path as contamination prevention (leading to the total suppression of photons at energies $\leq$2 keV), and its data were largely self-calibrated [239], its spectra provided a transformational view of the properties of the ICM in Perseus (Fig. 4.19).

The first surprise came from the analysis of the dynamical state of the ICM. Based primarily on the Fe XXV multiplet (see the *left panel* of Fig. 4.19), the SXS measured a line-of-sight velocity dispersion of the turbulent gas of 164 $\pm$ 10 km s^{-1}, in a region 30–60 kpc from the central nucleus, with only a marginally larger velocity in the core (187 ± 13 km s^{-1}) [46]). This level of precision is unprecedented in X-ray astronomy. The measured velocities correspond to a turbulent pressure $\simeq$4% (<8%) of the thermal pressure. This is puzzling, because a powerful radio-loud Active Galactic Nucleus (AGN; NGC1275) is present at the cluster core, injecting an amount of energy in the ICM sufficient to evacuate bubbles filled with relativistic plasma [31, 62]: one of the most spectacular examples of "radio-mode" AGN feedback, postulated to prevent radiative cooling of the ICM core gas. The *Hitomi* result is important not only for its implications on the astrophysics of the ICM but also because, if confirmed on a larger sample of clusters at higher redshift, would imply that the deviations from the hydrostatic equilibrium in the ICM are small, and therefore that X-ray derived galaxy cluster masses using this assumption are reliable probes of cosmological parameters [5].

In the light of the extraordinary results obtained by *Hitomi* during its, alas, too short!, operational life, JAXA and NASA decided to propose a mission to recover one of its fundamental scientific objectives: "Resolving astrophysical problems by precise high-resolution X-ray spectroscopy". This is the **X-Ray Imaging and Spectroscopy Mission** (XRISM) [224]. The European and Canadian Space Agencies, as well as European institutes participate in the mission development together with a wide range of scientific institutions in Japan and the United States. The XRISM payload closely replicates the soft X-ray telescopes and instruments on *Hitomi*:

- a micro-calorimeter (*Resolve*) with a requirement energy resolution $\leq$7 eV in the 0.3–12 keV energy range over a 3' × 3' field-of-view covered by an array of 35 sky sensitive pixels;
- an array of CCD detector (*Xtend*) with a large field-of-view (larger than 30' × 30'), and an energy resolution $\leq$200 eV at 6 keV at the beginning of the operational life;
- a large-area, light weight soft X-ray telescope with a $\simeq$1.7' Half Energy Width (HEW) or better, and an area comparable to that of the soft X-ray telescopes on-board *Hitomi*.

4.5.3 *Athena*

Due to launch in early 2030s, *Athena* [164] will embody the next step in the decade-long effort of the European space industry and scientific community to investigate the X-ray sky. Selected by the ESA Science Program Committee in 2014, *Athena* is a "Large" (L-)class mission designed to address the science theme of "The Hot and Energetic Universe". More specifically, *Athena* will address two fundamental questions in modern astrophysics:

- How does baryonic matter assemble in the large-scale structures we observe today? How do the baryons locked in the cosmic web evolve from the structure formation epoch to the present day?
- How do black holes grow and shape galaxies?

While the aforementioned core science themes drive the design of the mission, *Athena* is posed to produce breakthrough discoveries across all corners of astrophysics, through a community-driven observational program where a dominant fraction of the observing time during the nominal operational phase will be allocated through a competitive, peer-reviewed process.

In the context of multi-messenger astrophysics, a key feature of the *Athena* spacecraft is the fast response time of $\leq$4 h to observe any Target of Opportunity (ToO) in a random position of the sky for at least 50 ks with a 50% efficiency.

The *Athena* science goals will be achieved through an innovative payload, constituted by an active pixel sensor Silicon detector, the **Wide Field Imager** [154] with a wide field of view of 40' $\times$ 40' and spectral-imaging capability with an energy resolution $\leq$140 eV at 6 keV; and the **X-ray Integral Field Unit** [14], a cryogenic imaging spectrometer with $\leq$2.5 eV energy resolution at 7 keV (corresponding to a resolving power >2000 at 5 keV) over a 5' diameter effective field-of-view and a $\leq$5" pixel size. These two instruments are placed on the focal plane of a modular X-ray telescope based on the innovative, European-led Silicon Pore Optics technology [16], with 12 m focal length and an effective area $\geq$1.4 m^2 at 1 keV and $\geq$0.25 m^2 at 6 keV.

This payload ensures that the prospective *Athena* science performance will exceed any operational or planned X-ray missions by at least one order of magnitude over several parameter spaces simultaneously. The effective area at 1 keV will exceed that of instruments with comparable energy resolution in *Chandra*, XRISM, or XMM-Newton by a factor $\geq$70 and $\geq$10 for the X-IFU and the WFI, respectively. This implies an X-IFU sensitivity to the detection of weak lines (as measured by the square root of the product of the effective area divided by the energy resolution) higher with respect to the *Resolve* instrument in XRISM by a factor of $\simeq$10 and $\simeq$5 at 1 keV and 6 keV, respectively [87]. Furthermore, the X-IFU will be able to perform spatially-resolved spectroscopy on scales at least 30 times smaller (in area) than *Resolve* due to the smaller pixels ($\leq$5" against $\simeq$30"), well matching the *Athena* optics Point Spread Function (5" Half-Energy Width requirement). As far as the ToO performance are concerned, observations with *Athena* will accumulate one order-of-magnitude more counts on a typical GRB X-ray afterglow (assuming a total 50 ks exposure time) than the *Swift/XRT* or the XMM-Newton/EPIC-pn thanks to its optimal combination of large effective area and rapid response. This implies a sensitivity to the detection of weak lines in the afterglow spectra by the X-IFU a factor $\geq$15 larger than the XMM-Newton/RGS, and $\geq$20 with respect to the *Chandra*/LETG.

4.6 The γ-Ray Missions

γ-rays are electromagnetic radiation above some hundreds of keV. There are several classifications of energy ranges used in γ-ray astronomy. We can adopt the following conventions: low-energy domain (from 100 keV up to 1 MeV), medium-energy domain (between 1 MeV until few MeV), high-energy domain (from few MeV until tens of GeV), and very-high-energy (VHE) domain (above tens of GeV). γ-ray astrophysics deals with only the non-thermal nature of the underlying emission processes as compared with X-ray space astrophysics, as well as particle acceleration processes that can be very extreme.

Details on detection techniques and instruments are described in Chap. 3. Here we report only a summary of the main interaction processes and the main differences between X- and γ-ray instruments.

In the energy band from a few hundred keV up to several MeV, Compton scattering is the dominant interaction mechanism. Compton telescopes must be developed to explore the Universe in this energy band. The detector materials which provide the best possible energy resolution and telescope topology are fundamental to determine the direction of motion of the scattered γ-ray. No Compton telescopes are currently in operation: the last Compton telescope was CGRO/COMPTEL.

Photons with energy higher than some tens of MeV interact almost exclusively through the process of pair production. Reflection (mirrors) and refraction (lenses) are not applicable to γ-ray detectors. A material to act as a pair converter is needed and pair-tracking telescopes must be developed to track the $e^+ - e^-$ pair; this tracking is a mandatory requirement because the properties of the incident γ-ray can only be derived from measurements of the two charged particles. Technological development in event classification for background rejection and tracking techniques for photon energy and arrival direction reconstruction are a crucial aspect of γ-ray astrophysics.

These basic principles explain why all medium- and high-energy γ-ray detectors use instrumentation developed for nuclear and particle physics. Detection techniques of X- and γ-ray detectors are entirely different: current X-ray detectors are based on focusing optics and CCD detectors, whereas the current generation of γ-ray instruments reveal individual photons and mirrors can no longer be used because γ-rays are highly penetrating particles. The number of photons that can be collected by γ-ray instruments is less than those obtainable by X-ray detectors, and the background is also completely different from X-ray astronomy: γ-ray instrumental background has a rate 10^4 times more significant compared to the X-ray background. For both X- and γ-ray astronomy telescopes are carried by space satellites to circumvent the Earth's atmospheric opacity (except for VHE domain, where ground-based telescopes are used). The choice of satellite orbit, especially its inclination and altitude, has a huge effect on the instrumental background.

The two main instruments in the high-energy domain from the last decade and still in operation (2020) are the AGILE telescope, launched on April 2007, and the Fermi Gamma-ray Space Telescope launched in June 2008. The Fermi/LAT (Large Area Telescope) yields observations in the energy range between 20 MeV to 300 GeV and AGILE/GRID (Gamma-Ray Imaging Detector) yields observations in the energy range 30 MeV to 50 GeV.

In recent years, a large number of sources have been detected, revealing the existence of a very heterogeneous population of astronomical objects emitting in γ-ray high-energy domain. Nowadays the inventory counts over 5,000 steady γ-ray sources as reported in the 4FGL catalogue [70], where more than 3,130 of the identified or associated sources are blazar, a class of active galactic nucleus (AGN).

Progress has also been significant in the very-high-energy domain (above tens of GeV), with well over 100 confirmed sources.[2] The γ-ray low-energy domain and the γ-ray medium-energy domain have not experienced a comparable development so far, and the number of MeV steady sources is of the order of several tens in the COMPTEL catalogue [210].

In addition to localised emitters (see Sect. 4.6.1), the γ-ray sky shows also the presence of (1) an isotropic γ-ray background of extra-galactic origin (see Sect. 4.6.2), and (2) a diffuse background of Galactic origin (see Sect. 4.6.3). For a review on recent development of γ-ray astrophysics see [52, 122, 226].

4.6.1 γ-Ray Sources

Since its advent about 50 years ago, γ-ray astronomy has revealed an unexpected variety of objects that release a significant, sometimes even dominant, fraction of their energy through non-thermal processes. In our Galaxy, several pulsars (the most important fraction of Galactic objects) [41], pulsar wind nebulae (PWNe), supernova remnants (SNRs) and microquasars have been detected to emit such energetic radiation. It is important to mention the most studied Galactic source of γ-ray: the Crab Nebula, a typical PWN, the leftover of the supernova explosion that occurred in 1054 A.D., powered by the pulsar PSR B0531+21 at its centre. Still, about 80% of the γ-ray sources are associated with extra-galactic objects. Outside the Milky Way, γ-rays have mostly been observed from galaxies with an exceptional rate of star formation and from ultra-relativistic jets of particles, either due to accreting supermassive black holes (SMBHs) exceeding millions of solar masses or to collapsing stellar systems (GRBs).

Active Galactic Nuclei (AGNs) are active galaxies hosting a SMBH at their center. Electromagnetic radiation from AGNs spans over the entire observable wavelengths, from radio to sometimes VHE γ-ray. About 10% of AGN population shows high activity in the radio band with highly collimated jets launched from the vicinities of the galaxy center. These jets are believed to be driven by the enormous amount of power released when gas and dust accrete onto the deep gravitational potential of SMBHs. The energy of the accreting matter is converted very efficiently into the kinetic energy of relativistic, highly magnetized outflows. A jet component can dominate the observed spectrum of a system due to the beaming effect, i.e. as it happens in Blazars. Blazars are a class of AGNs whose relativistic jets are aligned with our line-of-sight. They emit γ-rays with high intensities and show variability that indicates the existence of complex processes capable of radiating a significant fraction of accretion luminosity into high-energy γ-rays. The spectral energy distribution (SED) of blazars is characterized by two broad-band non-thermal components due to: (i) synchrotron radiation of relativistic electrons, which peaks in optical/UV/X-rays; (ii) inverse Compton radiation, which peaks in γ-ray.

[2] http://tevcat.uchicago.edu/.

Table 4.3 A summary of the classification of AGNs in the 4FGL catalogue. The first number is the number of associated sources, the second number is firmly identified sources

Description	Number of sources (4FGL)
BL Lac type of blazar	1094 + 22
FSRQ type of blazar	644 + 42
Radio galaxy	36 + 6
Non-blazar active galaxy	17 + 1
Radio galaxy	36 + 6
Blazar candidate of uncertain type	1327 + 3
Narrow line Seyfert 1	5 + 4
Seyfert galaxy	1 + 0

Since the first discovery of the γ-ray emission from the blazar 3 C 273 by COS-B, AGNs have been observed by all γ-ray space missions. 3C 273 was identified in the COS-B data from the position coincidence of the quasar [25, 220]. Before the launch of the CGRO, the quasar 3C 273 was the only extra-galactic source of high-energy γ-ray emission known.

The First CGRO/COMPTEL source catalogue [210] reported ten more Active Galactic Nuclei. The CGRO/EGRET instrument performed the first all-sky survey above 100 MeV. A third EGRET catalogue [98] of γ-ray sources was released with a hundred blazars, among which 3 C 279 was one of the most active to be observed. For the first time, campaigns involving observes from radio telescopes to high-energy space satellites were organised, marking the birth of multifrequency γ-ray astrophysics (see Sect. 4.10.2.3 and Fig. 4.28 for more details). Unfortunately, EGRET small field of view did not allow systematic monitoring of blazars flaring in γ-rays, and only on a few occasions these multifrequency campaigns could be arranged.

With the advent of AGILE and Fermi (see Sects. 4.10.3.1 and 4.10.3.2 respectively), AGN monitoring and multifrequency campaigns finally became effective thanks to large field of views of about 60° in radius. Since the beginning of the AGILE observations in 2007, the high-energy extra-galactic sky was found to be dominated by blazars (with FSRQs and BL Lacs providing major contributions to the total γ-ray emission) and mostly time variable above 100 MeV. In the second AGILE catalogue of γ-ray sources [39], about sixty blazars were identified within the first two years of observations, with equivalent exposure for each source of a few weeks or months. After more than ten years of activity, thousands of blazars have been detected by Fermi/LAT. In the 4FGL [70] catalogue, about 60% of sources is associated with extra-galactic objects. The largest source population in 4FGL is that of AGN, with 3131 blazars, 42 radio galaxies and 35 other AGN. The blazar sample comprises of 686 FSRQ, 1116 BLL and 1330 BCU. Table 4.3 reports a summary of the classification of AGNs in the 4FGL catalogue.

Many questions arise in connection with the blazar properties observed at other wavelengths. What are the physical properties of their relativistic jets? How and when are the γ-rays produced? What is that triggers the γ-ray emission? What can be learned by simultaneous observations in the radio, optical, X-ray and γ-ray bands? The γ-ray astrophysics of blazars is extremely rich (a review of γ-ray observations of AGNs can be found in [139], a overview of AGN multi-wavelength properties can be found in [171]) and will continue to be the focus of observations and theoretical investigations by γ-ray telescopes.

Gamma-ray bursts (GRBs) are yet another class of extra-galactic objects that contribute to the γ-ray sky. GRBs are daily recorded transient events characterised by a two-phased emission: a prompt emission (highly energetic, brief and strongly variable in time) produced by the central engine, and an afterglow emission (less energetic, more lasting and regularly decaying in time) due to the subsequent interaction of outflows with the external medium. GRBs can be classified in "short" and "long" based on the duration of their prompt emission [21], linked to the astrophysical nature of the progenitor. Short GRBs are identified by a prompt emission lasting less than two seconds and are associated with neutron star binaries mergers [1]. Long GRBs have a prompt emission of more than two seconds and are associated with the extremely energetic collapse of massive stars [78] such as hypernovae [105]. The energy spectrum is non-thermal, peaking at around a few hundred keV and extending up to several GeV or TeV.

4.6.1.1 Unidentified γ-Ray Sources: The Geminga Case

About one-third of the γ-ray sources above 100 MeV are still unidentified, i.e. sources not associated with any objects known at other wavelengths. The history of Geminga is an example of the difficulty of the association of a γ-ray emission with known objects at different wavelengths. For this reason, in this section, special attention is devoted to its discovery. To provide a firm identification, some characteristic of the γ-ray source, such as a correlation in the variability or the periodicity with candidate counterparts at other wavelengths, must be established; the positional association is not sufficient, due to the PSF extension of γ-ray telescopes (e.g. some degrees at 100 MeV for AGILE/GRID and Fermi/LAT). In the 4FGL catalogue, only 7% of the sources are identified, 67% are positionally associated, and the remaining 26% are unidentified.

Geminga, a pulsar believed to be the remnant of a supernova exploded 300,000 years ago [121], and located in the Galactic Anticenter region, was the first example of an unidentified γ-ray source. In particular, Geminga was the first example of a radio-quiet pulsar.

SAS-II first detected Geminga (named $\gamma 195 + 5$) as a significant excess of γ-rays over the expected background of diffuse Galactic emission [72]; a pulsation in the γ-ray band was reported (with a period of approximately 59 s), but with a limited confidence level. In addition, the exact location of the source was uncertain, constrained only to be within a relatively large "error region" of some known sources (see Fig. 4.20). None of these known sources were associated, and the SAS-II team suggested that an undiscovered radio-pulsar was the most likely progenitor [230].

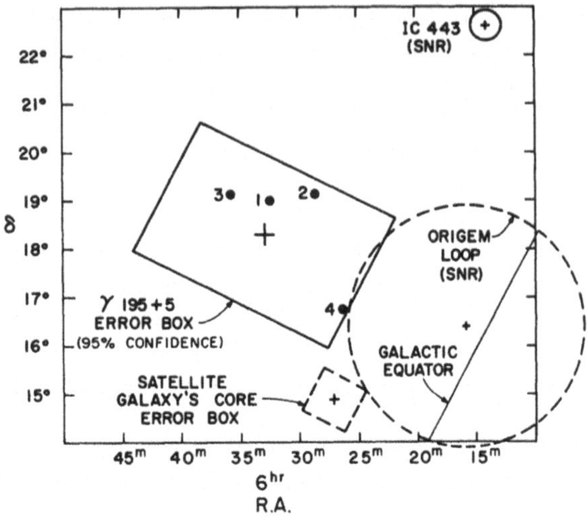

Fig. 4.20 The region of the sky containing $\gamma 195 + 5$, shown in celestial coordinates. The points marked 1, 2, 3, and 4 are the four strongest radio sources within the $\gamma 195 + 5$ error box: 4C+19.22, 4C+19.21, 4C+19.23, and 4C+16.17 [230]

With the COS-B satellite the source $\gamma 195 + 5$, classified as 2CG 195+04, remained unidentified, even though the 59 s pulsation was confirmed [20, 143]. The Einstein Observatory (HEAO-2) detected the X-ray source 1E 0630+178 [27] in the COS-B error box with an error circle of 3", with unusual properties. In [28] was subsequently applied the name "Geminga" to this mysterious object, meaning "it's not there" in the Milanese dialect.

The detection of periodic X-ray emission from 1E 0630+178 [93] using ROSAT data with a period of 0.237097 sec allowed a successful search for the same periodicity in the γ-ray data of CGRO/EGRET [23]. The mystery of Geminga was solved. Retrospectively, the same periodicity was also found in the COS-B [26] and SAS-II data [146].

4.6.2 Isotropic γ-Ray Background

An extra-galactic isotropic γ-ray background was detected for the first time by SAS-II and confirmed by subsequent missions (COS-B, EGRET, AGILE/GRID, Fermi/LAT). In general, this isotropic γ-ray background is composed of unresolved extra-galactic emissions and residual Galactic foregrounds. The Fermi/LAT has provided a measurement of a faint isotropic diffuse γ-ray background (IGRB) from 100 MeV to 820 GeV [4]. At an energy of 100 GeV, half of the total IGRB intensity has been resolved into individual sources by the LAT, predominantly blazars of the BL Lac type (see Sect. 4.6.1). The IGRB emission could contain the signature of some of the most interesting phenomena of γ-ray astrophysics. Intergalactic shocks produced by the assembly of large scale structures, γ-ray emission from galaxy clusters, and emission from starburst and normal galaxies, are candidates for the generation of diffuse GeV emission. Also, a signal from dark matter annihilation could be imprinted in the IGRB.

4.6.3 Galactic Diffuse Emission

In the high-energy domain (above tens of MeV), the dominant contribution (about 80% in the GeV range) to the γ-ray flux is due to Galactic diffuse emission, dominated by emission from interstellar processes in the Milky Way. A large fraction of this emission above 30 MeV is attributable to (1) cosmic ray (CR) protons interacting with Galactic interstellar gas nuclei, producing γ-rays through neutral pion decay, and contributing to the soft component of the gamma emission; (2) CR electrons scattering off interstellar radiation fields which contribute to the hard component of the γ-ray emission via Bremsstrahlung and Inverse Compton (IC) scattering of CR electrons with low-energy (infrared and ultraviolet) photons of the interstellar radiation field (ISRF) also produces γ-rays. The column densities of interstellar gas and intensity of the ISRF are low enough that once produced, the γ-rays with energies below several hundred GeV typically undergo no further interaction in crossing out of the galaxy.

This emission is brightest at low Galactic latitudes, and is highly structured, primarily reflecting the spatial structure of the interstellar gas. This was observed for the first time by OSO-3 (see Sect. 4.7), and then confirmed also by SAS-II and COS-B (see Sect. 4.10.2). On larger scales, variations in the density of CRs and the ISRF also affect the observed intensity. This interstellar γ-ray emission is considered background with respect to discrete γ-ray sources. Considering the low counting statistics of γ-ray observations and the limited angular resolution of γ-ray telescopes, an accurate model of the Galactic interstellar emission is usually developed and used as a background template against which discrete γ-ray sources must be detected and characterised.

In addition to the diffuse emission, two huge bubble-like structures, called Fermi Bubbles, extending $\sim 50°$ for about 25 thousand light-years distant above and below the galactic centre, have been observed [56, 218, 219]. This structure may be the remnant of an eruption from a supersized black hole at the centre of our galaxy. The γ-ray emission from these structures exhibits a power-law spectrum with spectral index 1.9 ± 0.2, significantly harder than the spectrum of the diffuse emission from the Galactic disk, with cut-off energy of 110 ± 50 GeV [3].

4.7 Early Phase: Scintillators for γ-Ray Astronomy

In the 1950s, based on studies of cosmic rays, the understanding of nucleosynthetic production of elements, and the development of radio astronomy, it became evident that in the Universe there are processes able to produce γ-ray s. These energy bands are detectable only above the Earth's atmosphere with payloads on balloons or spacecraft. Many balloon-born experiments were launched in the second half of the last century. In the meantime, NASA planned several series of missions designed to explore the electromagnetic spectrum at high energies.

Direct detection of the first celestial γ-ray sources was achieved in 1958 in the γ-ray low-energy domain (below 1 MeV) [179]. The authors detected a burst of γ-rays coincident with a Class II solar flare via a balloon-borne experiment. These authors were the first to make use of the term "gamma-ray burst", which would be associated with a different phenomenon 15 years later. In the same year, Morrison [162] summarized predictions regarding the emission of γ-rays from celestial sources. These calculations, even if optimistic, were the drivers in the field of observational γ-ray astronomy for the next several years.

The first satellite carrying an instrument sensitive to high-energy γ-rays was the First Orbiting Solar Observatory (OSO-1), launched on March 7, 1961 to observe the Sun, but no such radiation was detected. The first satellite performing a detection of γ-rays from space in the high-energy domains was Explorer XI (1961), carrying an instrument designed to detect γ-rays of energy above 50 MeV, and launched on April 27, 1961. This was the last effort in a program initiated in 1957 whose purpose was to search for γ-rays produced in interactions of cosmic rays with interstellar matter. The main results were the detection of 22 events reconstructed as cosmic γ-rays which came from a variety of directions in space [124] and approximately 22,000 events from cosmic radiation. The instrument (see Fig. 4.21) consisted of a sandwich crystal scintillator (CsI and NaI) and a Lucite Cherenkov counter (see

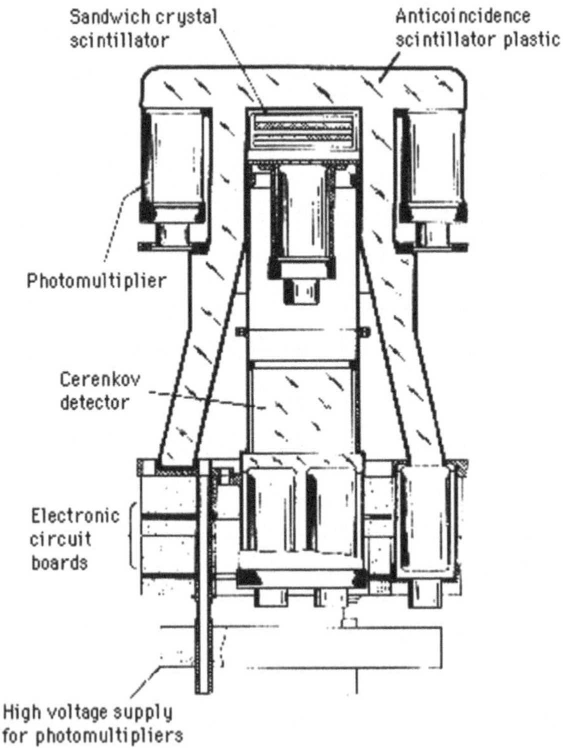

Fig. 4.21 Diagram of the high energy γ-ray detector flown on Explorer XI. [Credits: NASA]

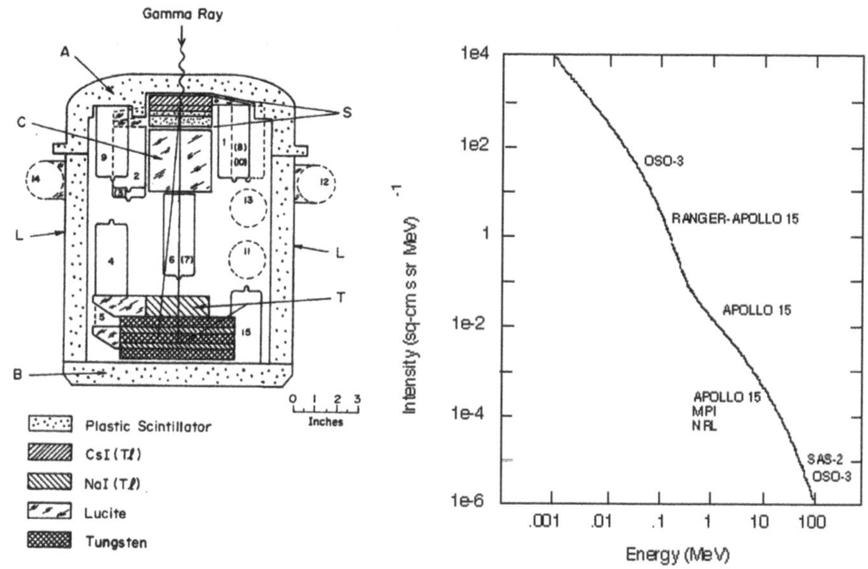

Fig. 4.22 Diagram of the high energy γ-ray detector flown on OSO-3 (photomultipliers are labelled 1-15, left panel) and the spectrum of the diffuse γ-ray background in the 0.001–100 MeV energy range as measured by OSO-3, Rangers 3 and 5, Apollo 15, and SAS II (*right panel*). [Credits: NASA]

Chap. 3), surrounded by a plastic anticoincidence scintillator. The two detectors in coincidence served to define the solid angle of the instrument to about 17° half-angle.

In the following year, a serendipitous discovery of a flux of diffuse γ-ray background at energies from 0.1 to 3 MeV was made via an isotropic counter aboard Ranger 3 [7], which flew to the Moon. The total count rate between 0.5 and 2.1 MeV was 0.27 ± 0.01 counts/cm^{2}s; above 2.1 MeV, it was 0.67 ± 0.02 counts/cm^2 see (with a detector area of about 58 cm^2) (see Fig. 4.22, right panel).

OSO-3, the Third Orbiting Solar Observatory, was launched on March 8, 1967, into a nearly circular orbit inclined at 33° with a mean altitude of 550 km [44, 123]. It was the first space mission to detect a broad peak in intensity towards the Galactic Center derived from cosmic ray interactions with Galactic gas [66]. A complete sky survey showed that the celestial distribution of γ-rays is anisotropic and concentrated along the galactic equator. It recorded the observation of a single flare episode from Sco X-1. At the end of the mission, 621 cosmic γ-ray events above 50 MeV were recorded. Like other OSO experiments, OSO-3 was compounded of two main segments. The so-called "Sail" segment was oriented towards the Sun and carried Sun-pointing experiments for solar physics. The "Wheel" segment was a rotating section with a spin-period of roughly 1.7 s, carrying sky scanning instruments; between them, the High Energy Gamma-Ray experiment, an improved version of the same instrument flown on Explorer XI, operating at energies above 50 MeV was part of the payload (see Fig. 4.22). The detector was a multilayer scintillation counter telescope,

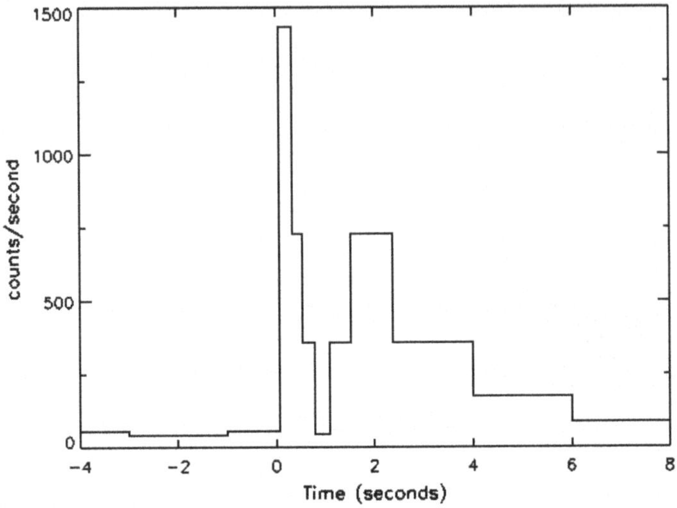

Fig. 4.23 The event recorded by Vela 4a,b on July 2, 1967 which also triggered the still operational Vela 3 satellites. The event appeared to be a cosmic γ-ray burst but at the time the constellation of satellites did not have sufficient timing resolution at the trigger to make a good determination of direction to the burst source. [Credits: NASA]

made of layers of CsI and plastic. The directional Cherenkov counter was a 3×3 in a cylinder of lucite. The energy detector contained several layers of NaI, sandwiched by layers of tungsten, enclosed by plastic scintillator anti-coincidence system [176, 177]. This detector obtained only a preliminary view of the γ-ray sky with a clear indication of enhancement of emission in the Galactic plane. No individual sources were detected. A review of the main results is reported in [11].

Another mission playing a key role in γ-ray astrophysics was the Vela constellation, that detected the first cosmic Gamma-Ray Burst (GRB) ever observed, on July 2, 1967, via the Vela 4a and b satellites (see Fig. 4.23). Due to a military classification this discovery would not be made public for several years. In 1969, while looking back over Vela 4 data just before the launch of Vela 5, Klebesadel and Olsen found an event recorded by Vela 4a,b which also triggered the still operational Vela 3 satellites. The event appeared to be a cosmic γ-ray burst. Yet, at the time, the constellation of satellites did not have sufficient timing resolution at the trigger to make a good determination of the direction to the burst source. In [120] are reported a list of GRBs detected by the Vela constellation.

With these missions the path toward the observation of the γ-ray sky was opened; to move a step forward, technological improvements were necessary to gain information on photon energies and directions and to improve sensitivity and particle background rejection in the high-energy regime.

The OSO-7 satellite, launched from Cape Canaveral on September 29, 1971, detected γ-ray emission lines from solar flares in August 1972 via an experiment aboard OSO-7, the UNH Solar Gamma-Ray Monitor designed to observe solar flares in the energy range 0.3–10 MeV and made with a NaI(Tl) scintillation spectrometer in a CsI(Na) active anti-coincidence shield. This includes the 511 keV positron annihilation line, the 2.223 MeV neutron-capture line, and the weak detection of C and O de-excitation lines at 4.438 and 6.129 MeV. Continuum photons are also detected up to 10 MeV.

The HEAO-3 satellite (the 3rd High Energy Astrophysical Observatory) was launched on September 20, 1979, into an orbit of 500 km altitude, 43.6° inclination. It carried three instruments, a high-resolution γ-ray spectrometer (the largest germanium spectrometer placed in orbit at that time), and two instruments measuring the composition of cosmic rays. The first instrument, the High Resolution Gamma-Ray Spectrometer (HRGRS) had an energy range of 0.05–10 MeV, a FoV of 30°, and an effective area of 75 cm^2 at 100 keV. A large CsI shield surrounded the detectors in electronic anti-coincidence, which was segmented in order to provide crude directionality.

The GRANAT, a Russian/European satellite, was launched on December 1, 1989, and operated until November 27, 1998. It carried several instruments that detected X-rays and γ-rays. The satellite was initially placed in a highly eccentric 96-h orbit with an apogee of 200,000 km and a perigee of 2,000 km. The orbit circularized so that by 1991 the perigee had increased to 20,000 km. After an initial period of pointed observations, GRANAT was placed in survey mode in September 1994, when the attitude control gas was exhausted. The main instruments on-board were a Coded-mask X-ray telescope (SIGMA), with an energy range of 0.03–1.3 MeV, an effective area of 800 cm^2, and a FoV of $5° \times 5°$; an additional Coded-mask X-ray telescope (ART-P) with an energy range of 4–60 keV, an effective area of 1250 cm^2, and a FoV of $1.8° \times 1.8°$; an X-ray proportional counter spectrometer (ART-S) with an energy range of 3–100 keV, an effective area of 2400 cm^2 at 10 keV, and a FoV of $2° \times 2°$; an All-sky monitor (WATCH) with a full FoV, an energy range of 6–120 keV, an effective area of 45 cm^2; a Gamma-ray burst experiment (PHEBUS) with an energy range of 0.1–100 MeV, six units of 100 cm^2 each, and an all-sky FoV; a Gamma-ray burst experiment (KONUS-B), with an energy range of 0.02–8 MeV, seven units of 315 cm^2 each, and an all-sky FoV; the Gamma-ray burst experiment (TOURNESOL) with an energy range of 0.002–20 MeV, a FoV of $5° \times 5°$. Particularly, the PHEBUS experiment was designed to record high energy transient events and was made of two independent detectors; each detector consisted of a BGO crystal 78 mm in diameter by 120 mm thick, surrounded by a plastic anti-coincidence. There were 116 energy channels. With all these instruments, GRANAT was designed to observe the Universe in $2keV$ to 100 MeV energy range. Specific instruments were meant to study GRBs and other X-ray transient sources while others were intended to image X-ray sources near the Galactic Center. For example, the WATCH instrument was designed to monitor the sky continuously and alert the other instruments of new or interesting X-ray sources. The main GRANAT results included very deep imaging of the galactic centre region, the discovery of electron-positron annihilation lines from the Galactic micro-quasar 1E1740-294 and the X-ray Nova Muscae, and the study of spectra and time variability of black hole candidates.

The next generation of high-energy telescopes, instead of just recording the passage of electron-positron pairs in scintillators, developed detectors to image the interactions of incoming photons and their secondary particles in greater detail. We can divide the next steps of this history based on the main interacting principles of photons into matter or with particular imaging techniques. In Sect. 4.9 we report a description of the Compton telescopes, in Sect. 4.10 a description of the pair-tracking telescopes in the γ-ray high-energy regime, and in Sect. 4.11 coded masks for γ-ray astronomy are described. Before this, in Sect. 4.8 the Compton Gamma Ray Observatory, that carried both Compton and pair-tracking telescopes, is introduced.

Fig. 4.24 The Compton Gamma-Ray Observatory (CGRO) as it was being deployed from the space shuttle Atlantis (STS-37). COMPTEL is the instrument located in the center of this S/C, between the solar panel mounts. [Credits: NASA]

4.8 The CGRO Era

The Compton Gamma Ray Observatory (CGRO) was a space observatory launched from Space Shuttle Atlantis during STS-37 flight on April 5, 1991 (see Fig. 4.24), deployed on April 7, and operated until its deorbit on June 4, 2000. The CGRO was one of NASA's four "Great Space Observatories". The satellite was launched in low earth orbit at 450 km to avoid the Van Allen radiation belt. It was the heaviest (17,000 Kg) astrophysical payload ever flown at that time. The large size was necessary because the number of γ-ray interactions that can be detected is directly related to the mass of the detector. In addition, the number of celestial γ-ray photons is very small; large instruments are needed to detect a significant number of γ-ray in a reasonable amount of time. CGRO was composed of four main instruments in one spacecraft, with an energy range from 20 keV to 30 GeV, an energy band of six orders of magnitude in photon energy. The main instruments were:

1. **BATSE**. The Burst and Transient Source Experiment (BATSE) searched for GRBs in the energy band 20–600 keV, conducting a full-sky survey. It consisted of eight identical detector modules, one at each of the satellite's corners. Each of them consisted of both a NaI(Tl) Large Area Detector covering the 20 keV–2 MeV energy range, and a NaI Spectroscopy Detector, which extended the upper energy range up to 8 MeV, all surrounded by a plastic scintillator for an active anti-coincidence to veto the large background rates due to cosmic rays and trapped radiation.
2. **OSSE**. The Oriented Scintillation Spectrometer Experiment (OSSE) was composed of four detector modules, which could be pointed individually to detect γ-rays entering the field of view of any module in the 0.05–10 MeV range. Each detector module had a central scintillation spectrometer crystal of NaI(Tl) optically coupled with a CsI(Na) crystal viewed by seven photomultiplier tubes. The CsI backing crystal acted as an active anti-coincidence shield, vetoing events from the rear. A further CsI shield, also in electronic anti-coincidence, surrounded the central detector on the sides and provided coarse collimation, rejecting γ-rays and charged particles. The four detectors typically operated in pairs of two: during an observation, one detector took observations of the source, while the other slew slightly off source to measure the background levels. The two detectors routinely switched roles, allowing for more accurate measurements of both the source and background.
3. **COMPTEL**. The Imaging Compton Telescope, (COMPTEL) is described in more details in Sect. 4.9.1
4. **EGRET**. The Energetic Gamma Ray Experiment Telescope, (EGRET), operating in the energy range 20 MeV–30 GeV, is described in Sect. 4.10.2.3.

The main goals of CGRO were to perform the first ever all-sky survey in γ-ray astronomy, followed by deep observations of individual sources, thanks to the large fields of view of EGRET and COMPTEL (co-aligned in their viewing directions). BATSE detected nearly 3,000 γ-ray bursts, which were distributed isotropically over the sky.

4.9 Compton Telescopes

In the energy band from a few hundred keV up to several MeV (depending on the scatter material), dominated by Compton scattering, the atmosphere is opaque to MeV photons. For this reason, high-altitude balloon or satellite missions are needed for its exploration. Around 2 MeV (this depends on the material) the interaction probability shows a minimum, and for this reason a significant amount of material is needed to stop γ-rays in the active detector material. From a few MeV the most demanding challenge is the high internal instrumental background induced by the space radiation environment. This background generally leads to very low signal-to-background ratios; it is essential to reduce background with passive and active background rejection techniques (see Chap. 3), coupled with data analysis techniques for γ-ray background suppression. Compton telescopes are surrounded by an anti-coincidence shield made of a thin plastic scintillator to discard charged particles or of a significantly heavier material (CsI, BGO) to also reject photon components.

The key objective for a Compton telescope is to determine the direction of motion of the scattered γ-ray. The reconstruction of the origin and energy of photons becomes feasible at energies above a few hundred keV with Compton scattering, which produces scattered or secondary events that must be tracked in the detector. Recording energies and directions of the secondary and scattered particles allows for the retrieval of the origin, energy, and sometimes polarisation of the primary photon. If one measures the position of the initial Compton interaction, energy, and direction of the recoil electron as well as the direction and energy of the scattered γ-ray, the origin of the photon can be identified. In Chap. 3 there is a representative topology for a Compton event in a Compton telescope.

Several instruments were developed to explore this energy regime, including GRE on SMM [75], TGRS on WIND [170], and the instruments aboard the CGRO/COMPTEL [208]. Currently, the instruments aboard INTEGRAL [256] and RHESSI [135] measure photons up to several MeV.

4.9.1 CGRO/COMPTEL

The most important Compton telescope was COMPTEL aboard the Compton Gamma-ray Observatory, which pioneered the observation of the γ-ray sky in the energy range between 0.75 and 30 MeV.

COMPTEL was the first double-scattering Compton telescope designed for γ-ray astrophysics to operate on a satellite platform. A detailed description of the COMPTEL instrument can be found in [208].

The instrument consisted of two planes of detector arrays, D1 and D2, separated by 1.58 m (see Fig. 4.25). The D1 detector consisted of seven cylindrical modules filled with NE 213A organic liquid scintillator, with a low average atomic number to optimize the occurrence of a single Compton scatter. The D2 detector consisted

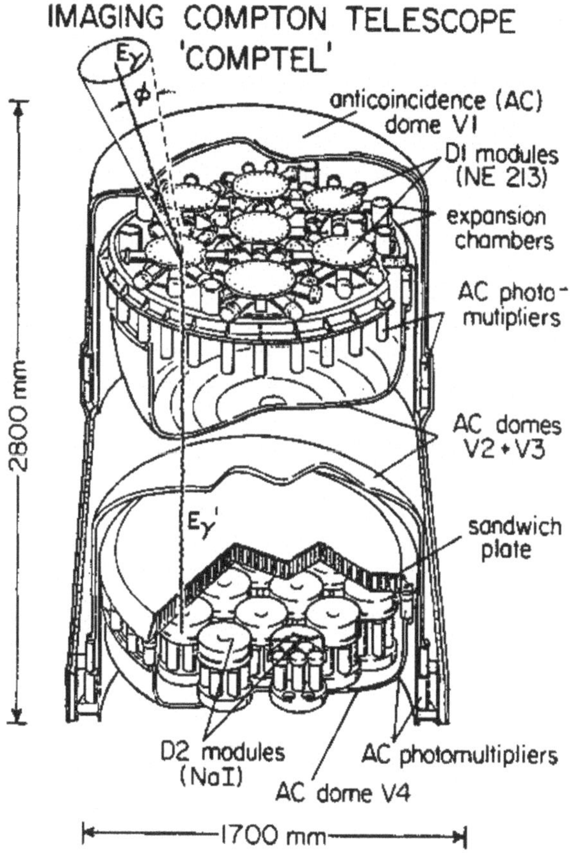

Fig. 4.25 An incoming γ-ray Compton scatters in the first detector layer—the electron from that interaction is detected through scintillation interactions. The resulting photon passes through the detector to the second layer, where it is absorbed, causing a pulse of light from the detecting material [Credits: NASA]

of 14 cylindrical NaI(Tl) crystals with a high density and average atomic number to maximize their photon absorption properties. The locations of the interactions and energy losses in both detectors are measured to determine the overall energy and angular resolution of the telescope, similar to an optical camera; the upper detector directs the light onto the second detector in which the scattered photon is absorbed. Although the photons were not focused, data obtained by COMPTEL was used to reconstruct sky images over a wide field of view with a resolution of a few degrees.

Each detector array was surrounded by an anti-coincidence dome manufactured of NE 110 plastic scintillator to reject charged particle triggers of the telescope.

On the sides of the telescope's structure, two small plastic scintillator detectors were placed containing weak ^{60}Co sources that provided tagged photons for in-flight energy calibration.

The COMPTEL instrument accepted and recorded coincident triggers in a single D1-D2 module pair within the coincidence time window of $\sim$40 ns in the absence of a veto signal from the four charged particle shields as valid events. The origin of this interaction could be a single photon or multiple photons and/or particles. Among other parameters, the time-of-flight (ToF) value was recorded for each event. The ToF measures the time difference between the triggers in the D1 and D2 detectors and is used to discriminate forward scattered (from D1 to D2) events, due to celestial photons, from backward scattered (from D2 to D1) background events.

The summed energy deposits in the two detectors are a measure of the total energy of the incident photon while the photon scatter angle is determined using the Compton kinematics.

The main results of COMPTEL can thus be summarised:

1. first all-sky map of Galactic ^{26}Al line, tracing sites of recent stellar nucleosynthesis;
2. detection of ^{57}Co line emission from SN 1987A;
3. limits on ^{22}Na line emission from ONeMg novae, tightly constraining nova models;
4. fundamental insight into the high-energy physics of solar flares. For example, temporal variations as fast as 0.1 s were observed at energies above 10 MeV constraining acceleration models;
5. results on the extragalactic diffuse γ-ray emission compatible with power-law extrapolations and showed no evidence for a MeV excess.

The first COMPTEL source catalogue [210] contains firm as well as marginal detections of continuum and line emitting sources. The numbers of the most significant detections are 32 for steady sources and 31 GRBs. Among the continuum, sources are spin-down pulsars, stellar black-hole candidates, supernova remnants, interstellar clouds, nuclei of active galaxies, and the Sun during solar flares. Line detections have been made of the 1.809 MeV ^{26}Al line, the 1.157 MeV ^{44}Ti line, the 847 and 1.238 MeV ^{56}Co lines, and the neutron capture line at 2.223 MeV.

4.10 Pair-Tracking Telescopes

4.10.1 General Principles

Pair tracking telescopes are used for the measurement of γ-ray cosmic photons above about 10 MeV, where pair production is the dominant photon interaction in most materials. The core of this type of telescope is a pair-tracking device, coupled with a γ-ray absorber (a calorimeter that could be a scintillator detector), an anti-coincidence system, and on-board trigger logic electronics. An optional time-of-flight measurement system could be present.

The standard design of the pair-tracking device is an instrument made of a vertical series of layers, with converter layers interleaved with tracking material layers. The converter is typically a heavy metal (such as tungsten or lead) which provides the target for creating the initial electron-positron pair while the tracking material detects the electron and positron tracks that is recorded by an electronic readout system. The pair-tracking device is also the trigger telescope.

The absorber device is used to measure the energy of particles produced within the tracking device. The calorimeter is mounted below the pair-tracking device and could be composed of inorganic crystals (e.g. NaI(Tl)). The particles enter the calorimeter and initiate a particle shower, and the energy of the particle is deposited, collected, and measured. The energy may be measured in its entirety, requiring total containment of the particle shower, or partially.

The direction (to determine its origin on the sky) and the total energy of the original γ-ray are evaluated through the reconstruction of the charged particle tracks as it passes through the vertical layers. In particular, the calculation of the total energy of the initial γ-ray is determined through analysis of the energy-dependent scattering of the pair into the tracking device or through absorption of the pair by the absorber device after they exit the tracking device.

The pair-tracking and absorber devices are surrounded by an anti-coincidence system, a charged particle detector with the shape of a box or a dome, closed on top and the sides, that could be made of thin foils of plastic scintillator for the rejection of charged particles.

A time-of-flight system could also be present. This is a detector which determines the relative times at which the pair travel through the telescope. In this way, it can be determined whether the pair came from the expected direction.

The on-board trigger logic electronics decide on the acquisition of γ-ray cosmic photons and background rejection. They are usually structured in many levels and are based on the identification of event topology.

4.10.2 Early Phase: Spark Chambers

The general principles of spark chambers are described in Chap. 3.

4.10.2.1 SAS II

SAS-II (Small Astronomy Satellite II), also known SAS B, SAS 2 or Explorer 48, was a NASA γ-ray telescope launched on November 15, 1972, from the San Marco platform off the coast of Malindi, Kenya, into a low Earth equatorial quasi-circular orbit. It completed its observations on June 8, 1973, when the low-voltage power supply failed. Eight thousand photons were recorded during the lifetime of the experiment, where approximately 55% of the celestial sphere, including most of the galactic plane, was surveyed.

The primary objective of the SAS-II was to measure the spatial and energy distribution of primary galactic and extragalactic γ-ray emission in the energy range 30–200 MeV, with an angular resolution of the telescope varied as a function of energy and arrival direction from 1.5 to 5°.

The telescope was made of two spark-chambers, four plastic scintillation counters, four Cherenkov counters, and an anticoincidence scintillation counter dome. The spark chamber assembly consisted of 16-wire spark-chamber modules with a magnetic core readout system. A plane of plastic scintillator formed by the four scintillation counters was present between the two spark-chambers for the γ-ray conversion into a $e^+ - e^-$ pair. It provided a means of determining the energy and direction of the particles and, from this information, the energy and direction of the γ-ray were determined. The scintillation counters and the four directional Cherenkov counters that were placed below the second spark chamber assembly constituted four independent counter coincidence systems. The anticoincidence plastic scintillator dome surrounded the whole assembly except at the bottom to discriminate against charged particles. Figure 4.26 reports a schematic view of the instruments.

SAS-II provided the first detailed map of γ-ray emission due to the Milky Way. This satellite provided a profile of the diffused γ-ray emission along the galactic plane and sparked discussions on the cosmic origin of this emission [73].

SAS-II first detected Geminga (more details and the history of this source is reported in Sect. 4.6.1.1) as an intense and unassociated localized source near galactic coordinates $(l, b) = (195, 5)°$. Intensity peaks and the pulsed emission corresponding to the position of two known pulsars, Crab and Vela [229], were discovered. Also, the detection of the well known Cygnus X-3 source, identified by its 4.8 h periodicity, was achieved [129].

4.10.2.2 COS-B

COS-B was the first European mission to study γ-ray sources. It was launched August 9, 1975, from Western Test Range, California, U.S.A. The scientific purpose was to study in detail sources of cosmic γ-ray photons above 30 MeV. An eccentric orbit, with the orbital plane inclined at 90° to the Earth's equator, an apogee of 100,000 Km, and a period of 37 h, was chosen to ensure that for most of the time the satellite was outside the Earth's radiation belts providing an efficient viewing program at the price of a high cosmic-ray background.

The payload was a single instrument, the "Gamma-ray telescope" [24]. The experiment features a 16-gap wire-matrix spark chamber for the identification and determination of the arrival direction of the γ-rays, triggered by a three-element scintillation counter telescope; a CsI calorimeter for energy determination; a plastic-scintillator anticoincidence shield for the rejection of charged particles. Also included was a proportional counter sensitive to X-rays in the 2–12 keV range, to synchronize on the possible short period pulsations of γ-ray emission from sources detected to pulsate at X-ray wavelengths.

The effective area reached a maximum value of about 50 cm^2 at 400 MeV (20 cm^2 at 100 MeV). The angular resolution, given as the FWHM, was about 3.6° at 100 MeV. The energy resolution had its best value (40% FWHM) at about 150 MeV.

COS-B increased the amount of γ-ray data by a factor of 25. Scientific results included the 2CG Catalogue, listing around 25 γ-ray sources [221], and the first

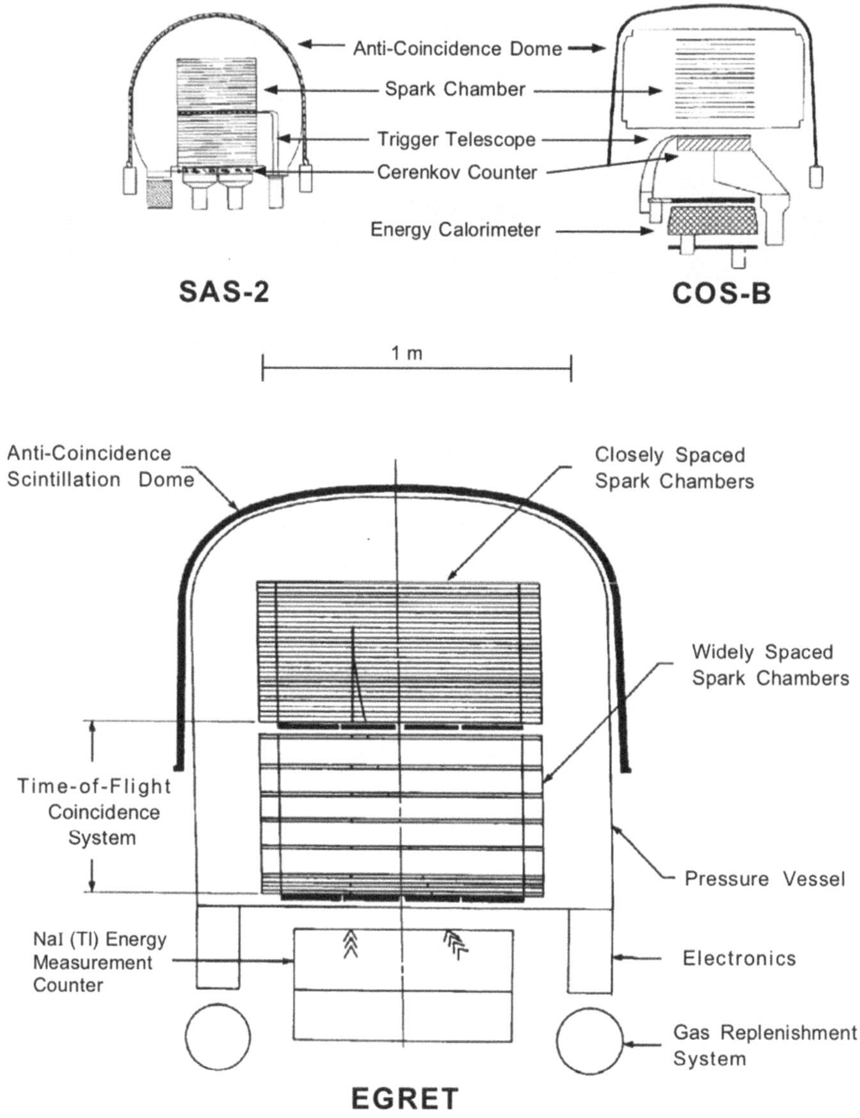

Fig. 4.26 Schematic view of the EGRET detector as compared to the earlier γ-ray instruments SAS-II and COS-B [109]

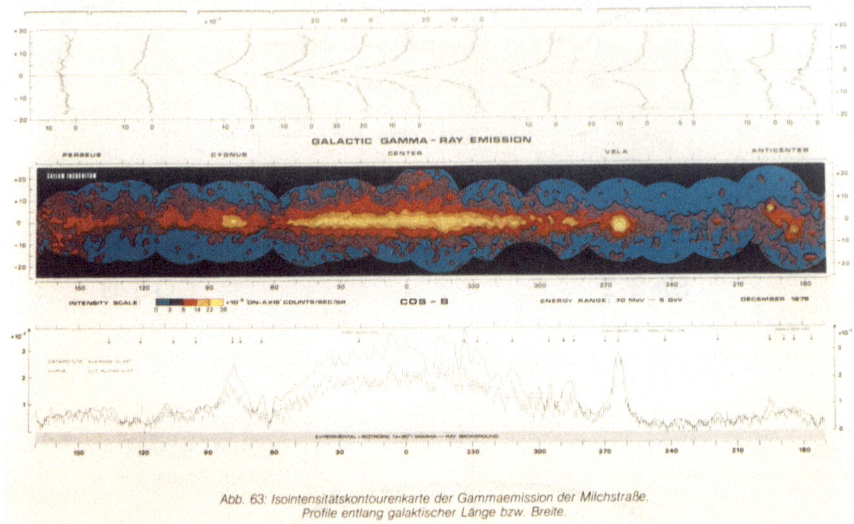

Fig. 4.27 An extensive survey of the Galaxy in the energy range 50 MeV–5 GeV performed by ESA COS-B satellite [Credits: ESA]

full γ-ray map of the Milky Way (Fig. 4.27). The satellite also observed the X-ray binary Cygnus X-3 [101] and the first γ-ray AGN, 3C 273 [25, 181]. The spectrum of Geminga was derived from locating it to within 0.25° permitting counterpart searches.

The spark chamber was still performing credibly at the end of the mission in spring 1982.

4.10.2.3 The CGRO/EGRET Telescope

The CGRO/Energetic Gamma Ray Experiment Telescope (CGRO/EGRET) [231] covered the energy range 20 MeV-30 GeV using a γ-ray telescope with more than an order of magnitude higher sensitivity and better angular and energy resolution than instruments previously flown, with a weight of 1,830 Kg and a power consumption of 190 W. The instrument was composed of a multilevel thin-plate spark chamber system to detect γ-rays by the $e^+ - e^-$ pair production process. A calorimeter made of NaI(Tl) was placed beneath the instrument to provide a better energy measurement of the primary photon. The instrument was covered by a plastic scintillator anti-coincidence dome to avoid event triggers not associated with γ-rays. Figure 4.26 reports a schematic view of the EGRET detector as compared to the earlier γ-ray instruments SAS-II and COS-B. A γ-ray entering the telescope has a probability (about 35% above 200 MeV) of converting into a $e^+ - e^-$ pair in one of the thin plates between the spark chambers in the upper portion of the telescope. If the time-of-flight coincidence system detects at least one particle of the pair as a downward moving particle, and if there is no signal in the anti-coincidence dome, the track imaging system is triggered, providing a digital picture of the γ-ray event. Incident charged particles are rejected by the anti-coincidence dome, but low energy backwards-moving charged particles which do not trigger the anti-coincidence are rejected by the time-of-flight measurement.

It had a very large field of view, approximately 80° in diameter, although the instrument point-spread function and the effective area degraded significantly beyond 30° off-axis. The effective area on-axis was more than $1,000\ \mathrm{cm}^2$ between 100 MeV and 3 GeV. The angular resolution was strongly energy-dependent, with a 68% containment radius of 5.5° at 100 MeV, falling to 0.5° at 5 GeV on axis; bright γ-ray sources could be localised with approximately 10' accuracy. The energy resolution of EGRET was 20–25% over most of its range of sensitivity. Absolute arrival times for photons were recorded with approximately 50 μs accuracy.

Each γ-ray was detected individually, where less than half of the events triggered inside the spark chamber satisfied the rigid rules for finally accepted celestial γ-ray events. The most critical part of the ground analysis was event analysis. Each event contained spark coordinate information in two orthogonal views. The reconstruction of the event (i.e. the arrival direction and the energy of the γ-ray) involved the correlation of the multiple track projections in the two views of the instrument. The usual signature of a γ-ray interaction is an inverted "V" formed by the electron and positron originating from the point where the γ-ray conversion occurs.

If the tracks are appropriately structured, a multiple-scattering analysis is then performed for each track to estimate the energy of each secondary particle. A least-squares fit is made to the points of each track in each view. A weighted bisector of the angle between the tracks of each orthogonal view is constructed. It is used, together with the pointing direction of the instrument, to determine the arrival direction of the γ-ray. In summary, the event processing provides arrival time, energy, and direction for each γ-ray. This automated analysis was also monitored by trained personnel who studied the two orthogonal views of analysed events; the most time-consuming task was the individual γ-ray event studies on the graphics units for those events which had been flagged by the automatic analysis as needing more study for final classification.

The determination of the exposure factors was very complicated for EGRET. Typically, the determination of exposure involves integrating the product of live-time, detection efficiency, and solid angle for each bin on the sky, taking into account the time-varying mode changes.

The main scientific objectives of the EGRET instrument were to perform an all-sky survey of high-energy γ-ray emission from diffuse emission and γ-ray sources.

Among the significant EGRET discoveries was the identification of blazars as γ-ray emitters, detecting dozens of blazars and finding them to be quite variable in flux with flares with time scales from days to hours (see Fig. 4.28). Before the launch of the CGRO, the quasar 3 C 273 was the only extragalactic source of high-energy γ-rays emission known. In the 3EG Catalogue of γ-ray sources [98], the EGRET instrument identified 271 high-energy γ-ray sources above 100 MeV. The most significant part of the sources, 170, are unidentified. The recognised sources

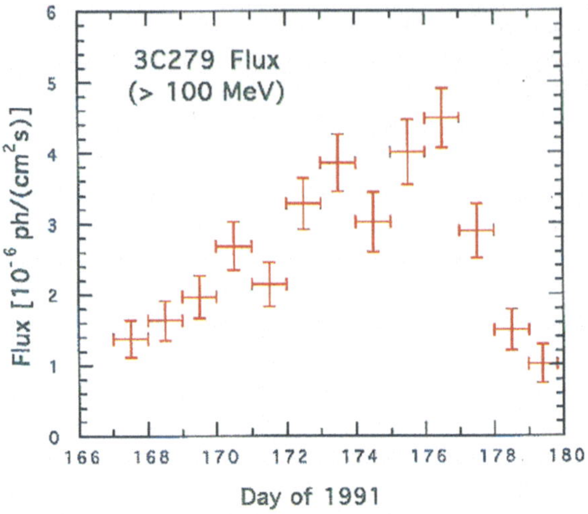

Fig. 4.28 This EGRET light curve shows the emission from the blazar 3 C 279 over a two weeks period in June of 1991 above 100 MeV [97]. The flux is plotted for each day during this period. The vertical size of the symbols represents the uncertainty in each measurement. The flux levels are seen to change dramatically in just a few days. Prior to the launch of the CGRO the quasar 3 C 273 was the only known extragalactic source of high-energy γ-rays. [Credits: NASA]

include 93 blazars, five pulsars, a radio galaxy (Cen A), and a normal galaxy (LMC). The LMC, which is the first detection in γ-rays of a normal galaxy beyond the Milky Way, were used to confirm that γ-rays produce a diffuse emission from the collisions of cosmic rays with interstellar gas and photons within that galaxy [217], also concluding that the cosmic-ray density in the LMC is the same as in the Milky Way. EGRET also obtained the most sensitive map of the diffuse γ-ray emission of the Milky Way and made a reliable measurement of the isotropic, presumably extragalactic, diffuse emission.

Several GRBs were also detected by EGRET, including a case with a 18-GeV photon more than one hour after the prompt emission of the burst [104].

As reported in Sect. 4.6.1.1, EGRET data confirmed Geminga as a radio-quiet pulsar, the first detected in γ-rays.

4.10.3 Silicon Tracker Telescopes: AGILE and Fermi

The single most important development from the EGRET era to the current generation of γ-ray telescopes is the advent of the silicon microstrip detectors (see Chap. 3) that offered significant advantages over the older gas detectors: higher spatial resolution, no pressure vessel or gas degradation, self-triggering, faster readout, and extremely high efficiency. Silicon microstrip detectors allowed for the development of silicon trackers as pair-tracking devices.

Silicon trackers are stacks of silicon microstrip detectors with the primary purpose of providing a compact imager for γ-ray photons. A silicon tracker plays two roles at the same time: it converts the γ-rays in heavy-Z material layers, where the photon interacts producing a e^+e^- pair (that are MIPs, Minimum Ionizing Particles) in the detector, and records the e^+e^- tracks by a sophisticated combination of silicon microstrip detectors and associated readout. An event is a collection of all the e^+e^- interactions into the microstrips of the silicon detector (each interaction generates a cluster that is a group of neighbouring strips collecting the charge deposited by the particle) together with the energy deposit in the calorimeter, and the information from the anti-coincidence plastic scintillators (used as veto logic). A complete representation of the event topology allows the reconstruction of the incoming direction and energy of the γ-ray.

The acquisition of this information by the on-board satellite electronics must be coupled with the dramatic increase in on-board computing power. Simulations at the individual detector readout level became possible, allowing not only optimization of the instrument design but also a parallel development of data analysis software to reject background and extract information from the pair production particles. On-board of satellites some steps of event reconstruction could be done, giving far more flexibility in data acquisition and in particular for a first background/photon discrimination.

In Sect. 4.10.3.1 the AGILE Observatory is described and in Sect. 4.10.3.2 the Fermi Observatory is described. A full review of the main results of AGILE and Fermi is reported in [226]. Table 4.4 reports a Comparison Between the AGILE/GRID and Fermi/LAT instruments.

An essential difference between the two instruments is the silicon tracker readout system, analogue for AGILE/GRID and digital for Fermi/LAT. Although the AGILE/GRID analogue readout is structured to read-only odd-numbered strips with no signal pick up at even-numbered strips (the floating strips), the capacitive coupling between adjacent strips allows for a complete sampling of the particle hit positions, discriminating between hits involving directly read and not-read microstrips.

4.10.3.1 AGILE

AGILE (Astrorivelatore Gamma ad Immagini LEggero (Light Imager for Gamma-Ray Astrophysics) is a scientific mission of the Italian Space Agency (ASI) launched on April 23, 2007 [227] (Fig. 4.29). The Gamma-Ray Imaging Detector (GRID) is used for observations in the 30 MeV-50 GeV energy range. The AGILE payload detector consists of the Silicon Tracker (ST) [12, 40, 42, 188], the SuperAGILE X-ray detector [71], the CsI(Tl) Mini-Calorimeter (MCAL) [126], and an anti-coincidence system (ACS) [175]. The AGILE event processing is operated by the Payload Data Handling Unit (PDHU) that provides on-board trigger logic algorithms.

Table 4.4 A Comparison Between the AGILE/GRID and Fermi/LAT instruments. The Instrument Response Functions Pass 8 Release 3 Version 2 for Fermi/LAT, and H0025 for AGILE/GRID have been used for the effective area and PSF. The first 12 planes of Fermi/LAT are the "front converter" and are interleaved with W absorber of 0.03 radiation length each; the Fermi/LAT "back converter" is made of 4 additional planes with W absorber of 0.18 radiation length each. The AGILE Silicon Tracker consists of a total of 12 trays, the first 10 with tungsten converter followed by a pair of silicon microstrip detectors plans, the last two trays consisting only of silicon detectors

Parameter	AGILE/GRID	Fermi/LAT
Total weight	100 kg	2,789 kg
Total power	60 W	650 W
Number of towers	1	16
Total number of Tracker planes	12	12 (front) + 4 (back)
Vertical spacing (s) between adjacent planes	1.8 cm	3.2 cm
Silicon tile size	9.5×9.5 cm^2	8.95×8.95 cm^2
Silicon detector array for each plane	4×4	4×4
Silicon-strip pitch (δ_P)	121 μm	228 μm
Readout pitch	242 μm	228 μm
Signal readout	Analog	Digital
Ratio (δ_P/s)	0.007	0.007
Tungsten converter thickness per plane	0.07 X_o	0.03 X_o (front) and 0.18 (back)
Number of planes with W converter	10	12 (front) and 4 (back)
On-axis total radiation length	0.9	0.5 (front) and 0.8 (back)
Total n. of readout channels	36,864	884,736
Power consumption/channels	400 μW	180 μW
Energy range	30 MeV–50 GeV	20 MeV–300 GeV
Field of view	2.5 sr	2.5 sr
Effective area on-axis at 100 MeV	400 cm^2	1900 (front) + 1400 (back) cm^2
Monocromatic angular resolution at 100 MeV (68% cont. radius) on-axis	3.5°	5.1° (total), 4.1° (front) + 7° (back)
Energy resolution at 100 MeV on-axis	40%	20%
Absolute time resolution	2 μs	1 μs
Launch date	April 23, 2007	June 11, 2008
Orbit inclination	2°	25°
Orbit altitude	550 km	550 km

Fig. 4.29 The AGILE satellite integrated with the fourth stage of the PSLV C-8 rocket in the ISRO Sriharikota launch base (April 15, 2007) [Credits: AGILE Team]

The combination of ST, MCAL, and ACS, forms the Gamma-Ray Imaging Detector (GRID) (Fig. 4.30). Accurate timing, positional, and attitude information are provided by the Precise Positioning System and the two Star Sensors units. Both AGILE/GRID PSF and effective area are characterised by very good off-axis performance and are well-calibrated up to almost 60°, showing very smooth variations with the angle relative to the instrument axis [43]. The scientific performances of the AGILE/GRID are summarised in Table 4.4.

The silicon tracker is the core of the AGILE/GRID, and it relies on the process of photon conversion into electron-positron pairs. It consists of a total of 12 trays, the first 10 of which consist of a tungsten converter followed by a pair of silicon microstrip detectors with strips orthogonal to each other, the last two consisting only of silicon detectors. Each plane is made of two layers of 16 single-sided, AC-coupled, 410 mm thick, $9.5 \times 9.5\,\mathrm{cm}^2$ silicon detectors. The 16 detectors of each plane side (view) are grouped in 4 ladders each one consisting of 4 detectors wire-bonded one after the other along the direction of the strips. The two views of the plane are organized in an x-y configuration. The γ-rays are converted in the tungsten or silicon layers and electron and positron interact in the silicon microstrips that are connected with readout electronics that acquire and process the data. The physical strip pitch is 121 mm while the readout one is 242 mm: this "floating strip" scheme allows for the reading of one strip every two thus decreasing the number of readout channels (and thus decreasing the power consumption of the instrument) while maintaining an excellent spatial resolution.

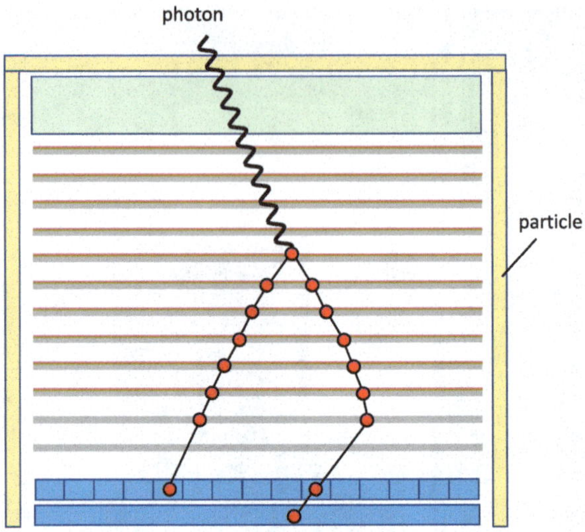

Fig. 4.30 Schematic view of the AGILE instrument [227]. The AGILE scientific instrument consists of the X-ray imager SuperAGILE (green), the γ-ray silicon tracker (orange lines represent tungsten layers, grey lines are silicon plans), and the Mini-Calorimeter (blue). The anticoincidence system is partially displayed (yellow) and is used for particle discrimination. No lateral electronic boards and harness are shown for simplicity. The AGILE instrument "core" is approximately a cube of about 60 cm size and weight approximately equal to 100 kg. A typical event topology is shown, where red circles represent the interaction of the electron and positron

The SuperAGILE instrument [71] is an X-ray detector part of the AGILE satellite that detects photons in the energy range 18–60 KeV.

The Mini-Calorimeter is composed of 30 CsI(Tl) scintillator bars each one $15 \times 23 \times 375$ mm^3 in size, arranged in two orthogonal layers, for a total thickness of 1.5 radiation lengths.

The AC system is aimed at a very efficient charged particle background rejection (99.99%). It surrounds all AGILE detectors (Super-AGILE, ST and MCAL).

Despite its relatively small size (about 100 kg, total power 60 W) the AGILE instrument represented the first experiment with no consumable elements, and therefore of potentially long duration in orbit. The AGILE FoV at GeV energies (2.5 sr) was unprecedented at the time of the launch; together with an excellent PSF (that turns out to be similar to Fermi/LAT, see Sect. 4.10.3.2), the mission is competitive for fast detection of bright transients and sources especially in the range 100 MeV–1 GeV with unique co-axial imaging capabilities in the 20–60 keV by SuperAGILE. An excellent timing and a special trigger support the operation of the Mini-Calorimeter for GRB detection, capable of independently detecting GRBs and other transients in the 350 keV–100 MeV energy range. Until October 15, 2009, the satellite operated in pointing mode. Since 2010 the satellite is in "spinning mode," rotating every 7 min around the satellite-Sun axis and sweeping about 80% of the entire sky for each revolution.

Fig. 4.31 The first half of the payload fairing is moved into place around NASA's Gamma-Ray Large Area Space Telescope within the mobile service tower on Launch Pad 17-B at Cape Canaveral Air Force Station. The fairing is a moulded structure that fits flush with the outside surface of the Delta II upper stage booster. [Credits: NASA]

4.10.3.2 Fermi

The NASA Fermi mission was launched by NASA on June 11, 2008, on a Delta II Heavy Launch vehicle in an orbit with 25° inclination (Fig. 4.31).

The LAT (Large Area Telescope) instrument [8] (Fig. 4.32) consists of a tracker to measure the tracks of the e^+e^- pair created in the range 30 MeV–300 GeV by pair production of the γ-rays, a calorimeter to determine the energy of the pair (and

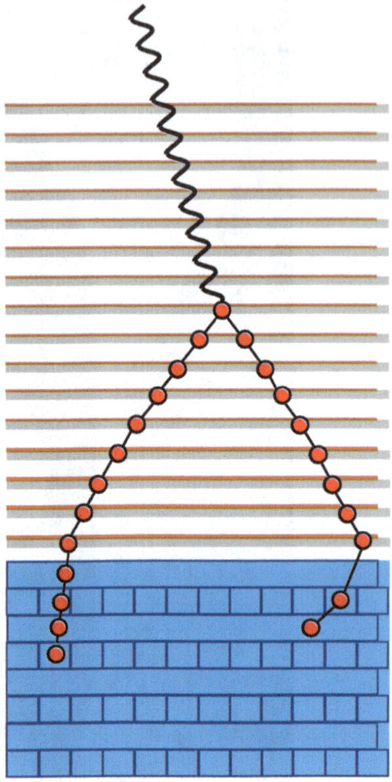

Fig. 4.32 Schematic diagram of one of the sixteen towers of the Fermi/LAT instrument [8]. The γ-ray silicon tracker (orange lines represent tungsten layers, grey lines are silicon planes), and the calorimeter (blue) are shown, with a typical γ-ray event topology, where red circles represent the interaction of the electron and positron. The thickness of the tungsten layer of the last four planes is greater than the first twelve (see text for more details). The anticoincidence system and readout electronics are not displayed

hence of the primary γ-ray), and an anti-coincidence detector to veto the charged particle background. The LAT γ-ray imager is a silicon tracker with 36 layers of silicon microstrip detectors interleaved with 16 layers of tungsten foil to facilitate the pair-creation (12 thin layers, 0.03 radiation length, at the top followed by four thick layers, 0.18 radiation length, in the back section, adding up to a total radiation length of 1 radiation lengths X_0). The total surface area of the LAT is 1.8 m $\times$ 1.8 m with about 80% of that being sensitive area. The calorimeter is composed of 96 CsI(Tl) crystals, arranged in an 8-layer hodoscopic configuration for a total depth of 8.6 total radiation lengths. The segmented anti-coincidence detector surrounding the tracker consists of plastic scintillators read out by photomultiplier tubes. The total LAT weight is 2,789 kg, and total power 650 W. More details are reported in Table 4.4.

The gamma-ray burst monitor (GBM) [152] instrument is devoted to GRB and other fast transients detection. The GBM includes two sets of detectors: twelve sodium iodide (NaI) scintillators sensitive from a few keV to about 1 MeV, and two cylindrical bismuth germanate (BGO) scintillators. Each NaI crystal is 12.7 cm in diameter by 1.27 cm thick, while the BGO is 12.7 cm in diameter and 12.7 cm in height. The NaI detectors are sensitive from a few keV to about 1 MeV and provide burst triggers and locations of a transient source. The BGO detectors cover the energy range 150 keV to about 30 MeV, providing a good overlapping with the NaI at the lower end and with the LAT instrument at the higher energies. The GBM has a very large field of view of about 9.5 sr (75% of the whole solid angle). Transient bursts of γ-rays are detected by an important change in the count rate in at least two NaI scintillators with a programmable triggering algorithm. After a trigger, the GBM on-board logic calculates a preliminary position and spectral information and send this to the ground and possible autonomous repointing of the satellite.

4.11 Coded Masks for γ-Ray Astronomy: INTEGRAL

One way of determining the origin of γ-rays is obscuring parts of the field-of-view of the detector and allowing only a fraction of the γ-rays to hit the detector. From the detected modulated signal and knowledge of the obscuration scheme, the original distribution of the photons can be recovered by employing a deconvolution. Different sources on the sky generate different shadows of the mask onto a spatially resolving detector system.

INTEGRAL is an ESA mission launched on October 17, 2002, for imaging of γ-ray sources in the energy range 15 keV–10 MeV with concurrent source monitoring in the X-ray (4–35 keV) and optical (V-band, 550 nm) energy ranges. Three coded mask instruments (SPI, IBIS and JEM-X) operate aboard INTEGRAL [256], from which two, the imager IBIS and the spectrometer SPI, extend into the MeV range. JEM-X (an X-ray monitor) and an optical camera (OMC) support these instruments as a monitor. All four instruments are co-aligned and observe the same region of the sky simultaneously. Figure 4.33 shows a schematic diagram of the INTEGRAL spacecraft.

The spectrometer SPI (Spectrometer on INTEGRAL) measures photons in the energy range 18 keV–8 MeV, with an effective area of 500 cm^2, a spectral resolution of 2.33 keV FWHM at 1.33 MeV, and a field of view of about 16°. This is accomplished by using a detector made up of an array of 19 hexagonal high purity germanium detectors cooled by a Stirling cooler system to an operating temperature of −188 °C. It makes the spectrometer heavy with a mass of 1,300 Kg. A hexagonal coded aperture mask, of 3 cm thick tungsten and consisting of 127 hexagonal elements of which 63 are opaque and 64 transparent, is located 1.7 m above the detection plane to image large regions of the sky with an angular resolution of 2°. BGO crystals shield the detectors and act as the anti-coincidence system.

The imager on-board INTEGRAL (IBIS) measures photons in the energy range 15 keV–10 MeV. The detector is made of two parallel planes of pixels located one on top of the other and separated by 90 mm. The top layer (ISGRI) is made of 16,384 CdTe pixels with a detector area of 2, 600 cm^2 to detect the low-energy γ-rays. The second layer (PICsIT) consists of 4,096 caesium iodide (CsI) pixels with a detector area of 3100 cm^2 to capture high-energy γ-rays. IBIS has a spectral resolution of ~10% at 1 MeV and imaging of 12' FWHM. The telescope includes a tungsten coded-aperture mask located at 3.2 m above the detector.

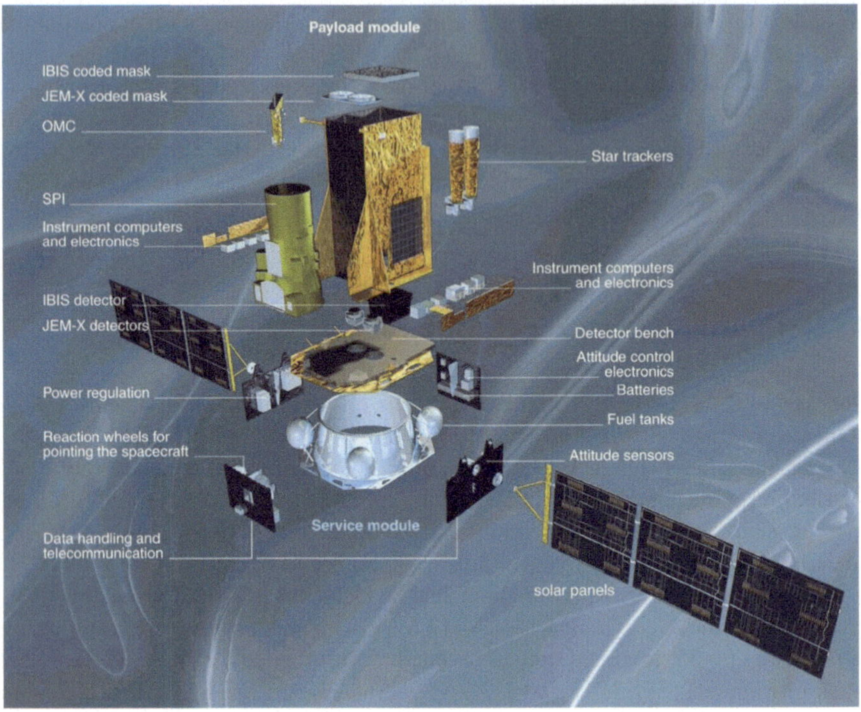

Fig. 4.33 Schematic diagram of INTEGRAL. [Credits: ESA]

An overview of the main scientific results of INTEGRAL are reported in [155, 245, 249, 257].

4.12 The Future of γ-Ray Astronomy

In recent years two important concepts have been promoted to establish a new mission in the medium- and high-energy γ-ray ranges. The main driver of new proposals is to develop a space observatory in which both the Compton and pair production regime are covered with the technology currently operating in space (the current proposal is based on a silicon tracker made of double-sided silicon microstrip detectors—DSSDs—instead of single side detectors). These instruments primarily make use of flight-proven technologies (e.g., Fermi/LAT, AGILE/GRID, PAMELA, and AMS) and mission configurations (Fermi and AGILE) and optimises the energy resolution for MeV line studies through the development of calorimeter configurations and accompanying dedicated electronics.

4.12.1 E-ASTROGAM

e-ASTROGAM [51] is a concept for a space mission observatory carrying a γ-ray telescope dedicated to the study of the non-thermal Universe in the energy range from 0.15 MeV to 3 GeV. The payload of the e-ASTROGAM satellite consists of a single γ-ray telescope operating over more than four orders of magnitude in energy by the joint detection of photons in both the Compton (0.15–30 MeV) and pair (>10 MeV) energy ranges. The telescope is made of three detection systems: a silicon tracker in which the cosmic γ-rays undergo a Compton scattering or a pair conversion; a calorimeter to absorb and measure the energy of the secondary particles and an anti-coincidence (AC) system to veto the background induced by charged particles. The telescope has a size of $120 \times 120 \times 78\,\text{cm}^3$. In one of the proposed configurations, the Si Tracker comprises 5,600 double-sided strip detectors (DSSDs) arranged in 56 layers. It is divided into four units of 5×5 DSSDs, the detectors being wire bonded strip to strip to form 2-D ladders. Each DSSD has a geometric area of $9.5 \times 9.5\,\text{cm}^2$, a thickness of 500 μm, and a strip pitch of 240 μm, for a total detection area of $9,025\,\text{cm}^2$. Such stacking of relatively thin detectors enables efficient tracking of the electrons and positrons produced by pair conversion, and of the recoil electrons produced by Compton scattering.

The calorimeter is a pixelated detector made of Thallium activated Cesium Iodide (CsI(Tl), a high-Z scintillation material) for efficient absorption of Compton scattered γ-rays and electron-positron pairs. It consists of an array of 33,856 parallelepiped bars of CsI(Tl) of 8 cm length and $5 \times 5\,\text{mm}^2$ cross-section, read out by silicon drift detectors (SDDs). The Calorimeter thickness is 8 cm of CsI(Tl), for 4.3 radiation-length detectors, that have an absorption probability of a 1-MeV photon on-axis of 88%.

The e-ASTROGAM detector is coupled with an anti-coincidence (AC) system made of segmented panels of plastic scintillators covering the top and four lateral sides of the instrument, and a Time of Flight (ToF) system, aimed at rejecting the particle background produced by the platform.

An updated version of the e-ASTROGAM mission concept, called All-Sky-ASTROGAM, a 100 keV to a few hundred MeV mission proposed as the "Fast" (F) mission of the European Space Agency an L2 orbit, can be found in [225].

4.12.2 AMEGO

AMEGO [148] is a concept for a space mission observatory for the study of the non-thermal Universe in the energy range from 200 keV to over 10 GeV. AMEGO has a configuration similar to e-ASTROGAM.

AMEGO consists of four subsystems: a DSSD tracker, a 3D position-sensitive virtual Frisch-grid Imaging Cadmium Zinc Telluride (CZT) calorimeter, a segmented thallium-activated Cesium Iodide (CsI) calorimeter and a plastic scintillator anti-coincidence detector (ACD). The silicon tracker consists of sixty layers of DSSDs. Each tower layer contains a 4×4 array of DSSDs, each 9.5×9.5 cm^2, with a thickness of 500 μm.

The calorimeter consists of an array of $8 \times 8 \times 40$ mm bars of CZT. One layer is under the silicon tracker, and another extends partially up the outer sides of the tracker. It provides a precise measurement of the location and energy of the scattered γ-ray. The CZT bars are operated in a drift mode that enables the 3-dimensional reconstruction of the location of the interaction in the detector, thus providing excellent positional resolution.

The CsI(Tl) calorimeter lies below the CZT and provides the depth to contain enough of a pair-conversion generated electromagnetic shower to extend the energy range to GeV energies. The design is similar to the Fermi/LAT CsI calorimeter, but it improves the low-energy performance by collecting the scintillation light with silicon photo-multipliers (SiPMs).

Acknowledgements Thanks to Nicolò Parmiggiani (nicolo.parmiggiani@inaf.it) for providing Figs. 4.30 and 4.32.

References

1. B.P. Abbott et al., ApJ **848**, L12 (2017)
2. F. Acero, M. Ackermann, M. Ajello, A. Albert et al., ApJS **218**, 2 (2015)
3. M. Ackermann et al., ApJ **793**, 1 (2014)
4. M. Ackermann et al., ApJ **799**, 86 (2015)
5. S.W. Allen, A.E. Evrard, A.B. Mantz, ARA&A **49**, 409 (2011)
6. R. Antonucci, ARA&A **31**, 473 (1993)
7. J.R. Arnold et al., JGR **67**, 4878 (1962)
8. W.B. Atwood, A.A. Abdo, M. Ackermann, W. Althouse, B. Anderson, M. Axelsson, L. Baldini, J. Ballet, D.L. Band, G. Barbiellini et al., ApJ **697**, 1071–1102 (2009)
9. H. Awaki, K. Koyama, H. Kunieda, Y. Tawara, Nature **346**, 544 (1990)
10. S.I. Babichenko et al., CosRe **14**, 878 (1977)
11. Badhwar et al., Astrophys. Space Sci. **27**, 147–155 (1974)
12. G. Barbiellini, G. Bordignon, G. Fedel, et al. in *American Institute of Physics Conference Series*, vol. 587, Gamma 2001: Gamma-Ray Astrophysics, ed. by S. Ritz, N. Gehrels, C.R. Shrader, 754 (2001)
13. P. Barr, Nature **328**, 785 (1987)
14. D. Barret, et al., SPIE, 106991G, SPIE10699 (2018)
15. S.D. Barthelmy et al., SSRv **120**, 143 (2005)
16. M. Bavdaz et al., SPIE, 106990X, SPIE10699 (2018)
17. E. Behar, ApJ **703**, 1346 (2009)
18. E. Behar, M. Sako, S.M. Kahn, ApJ **563**, 497 (2001)
19. T. Belloni, M. Klein-Wolt, M. Méndez, M. van der Klis, J. van Paradijs, A&A **355**, 271 (2000)
20. K. Bennett et al., A&A **56**, 469 (1977)
21. E. Berger, Ann. Rev. Astron. Astrophys. **52**, 43–105 (2014)
22. E. Berti, M. Volonteri, ApJ **684**, 822 (2008)
23. D.L. Bertsch et al., Nature **357**, 306 (1992)
24. G.F. Bignami et al., SSI **1**, 245 (1975)
25. G.F. Bignami et al., A&A **93**, 71–75 (1981)

26. G.F. Bignami, P. Caraveo, Nature **357**, 287 (1992)
27. G.F. Bignami, P. Caraveo, R.C. Lamb, ApJ **272**, L9 (1983)
28. G.F. Bignami, P. Caraveo, J.A. Paul, Nature **310**, 464 (1984)
29. J.A.M. Bleeker, A.J.M. Deerenberg, ApJ **159**, 215 (1970)
30. A.J. Blustin, M.J. Page, S.V. Fuerst, G. Branduardi-Raymont, C.E. Ashton, A&A **431**, 111 (2005)
31. H. Bohringer, R.A. Schwarz, U.G. Briel, ASPC, 368, ASPC...51 (1993)
32. T. Boller, M.J. Freyberg, J. Trümper, F. Haberl, W. Voges, K. Nandra, A&A **588**, A103 (2016)
33. J.A. Bowles, T.J. Patrick, P.H. Sheather, A.M. Eiband, JPhE **7**, 183 (1974)
34. H. Bradt, W. Mayer, S. Naranan, S. Rappaport, G. Spada, ApJ **150**, L199 (1967)
35. H.V.D. Bradt, T. Ohashi, K.A. Pounds, ARA&A **30**, 391 (1992)
36. W.N. Brandt, G. Hasinger, ARA&A **43**, 827 (2005)
37. L.I. Bratolyubova-Tsulukidze, N.L. Grigorov, L.F. Kalinkin et al., *Proceedings of the 11th International Cosmic Ray Conference*, vol. 1, issue 123 (1969)
38. J. Buff et al., ApJ **212**, 768 (1977)
39. A. Bulgarelli et al., A&A **627**, A13–36 (2019)
40. A. Bulgarelli et al., NIM A, **614**, 213–226 (2010)
41. P. Caraveo, Ann. Rev. Astron. Astrophys. **52**, 211–250 (2014)
42. P.W. Cattaneo et al., NIMA **630**, 251–257 (2011)
43. A.W. Chen, A. Argan, A. Bulgarelli et al., A&A **558**, A37 (2013)
44. G.W. Clark, G.P. Garmire, W.L. Kraushaar, ApJL **153**, L203 (1968)
45. G.W. Clark, W.H.G. Lewin, H.W. Schnopper, G. Sprott, J.E. McClintock, C. Canizares, ICRC **1**, 77 (1973)
46. Hitomi Collaboration et al., Nature **535**, 117 (2016)
47. Hitomi Collaboration et al., PASJ **70**, 12 (2018)
48. J.P. Conner, W.D. Evans, R.D. Belian, ApJ **157**, L157 (1969)
49. E. Costa et al., Nature **387**, 783 (1997)
50. J. Crummy, A.C. Fabian, L. Gallo, R.R. Ross, MNRAS **365**, 1067 (2006)
51. A. De Angelis et al., J. High Energy Astrophys. **19**, 1–106 (2018)
52. A. De Angelis, M. Mallamaci, Eur. Phys. J. Plus **133**, 663 (2018)
53. I. de la Calle Perez et al., Astron. Astrophys. **524**, A50 (2010), arXiv:1007.4762 [astro-ph.HE]
54. D. Dewey, S.A. Zhekov, R. McCray, C.R. Canizares, ApJ **676**, L131 (2008)
55. T. Di Matteo, V. Springel, L. Hernquist, Nature **433**, 604 (2005)
56. G. Dobler et al., ApJ **717**, 825 (2010)
57. C. Done, S.W. Davis, C. Jin, O. Blaes, M. Ward, MNRAS **420**, 1848 (2012)
58. M. Elvis, C.G. Page, K.A. Pounds, M.J. Ricketts, M.J.L. Turner, Nature **257**, 656 (1975)
59. M. Elvis, T. Maccacaro, A.S. Wilson, M.J. Ward, M.V. Penston, R.A.E. Fosbury, G.C. Perola, MNRAS **183**, 129 (1978)
60. I.N. Evans et al., ApJS **189**, 37 (2010)
61. A.C. Fabian, M.J. Rees, L. Stella, N.E. White, MNRAS **238**, 729 (1989)
62. A.C. Fabian, K. Iwasawa, C.S. Reynolds, A.J. Young, PASP **112**, 1145 (2000)
63. A.C. Fabian, J.S. Sanders, G.B. Taylor, S.W. Allen, C.S. Crawford, R.M. Johnstone, K. Iwasawa, MNRAS **366**, 417 (2006)
64. A.C. Fabian, M.L. Parker, D.R. Wilkins, J.M. Miller, E. Kara, C.S. Reynolds, T. Dauser, MNRAS **439**, 2307 (2014)
65. A.C. Fabian, A. Lohfink, E. Kara, M.L. Parker, R. Vasudevan, C.S. Reynolds, MNRAS **451**, 4375 (2015)
66. G. Fazio, Nature **225**, 905 (1970)
67. G.G. Fazio, E.M. Hafner, J. Geophys. Res. **72**, 2452 (1967)
68. E.D. Feigelson, E.J. Schreier, J.P. Delvaille, R. Giacconi, J.E. Grindlay, A.P. Lightman, ApJ **251**, 31 (1981)
69. H.C. Ferguson, M. Dickinson, R. Williams, ARA&A **38**, 667 (2000)
70. Fermi LAT Collaboration, arxiv:1902.10045
71. M. Feroci, E. Costa, P. Soffitta et al., NIMA **581**, 728 (2007)

72. C.E. Fichtel et al., ApJ, **198**, 163–182 (1975)
73. C.E. Fichtel, D.A. Kniffen, R.C. Hartman, ApJ, **186**, L99 (1973)
74. W. Forman et al., ApJ **665**, 1057 (2007)
75. D.J. Forrest et al., Solar Phys. **65**, 15–23 (1980)
76. H. Friedman, RSPSA **366**, 423 (1979)
77. H. Friedman, S.W. Lichtman, E.T. Byram, PhRv **83**, 1025 (1951)
78. T.J. Galama, P.M. Vreeswijk, J. van Paradijs et al., Nature **395**, 670 (1998)
79. N. Gehrels, Swift, AAS 205, 116.01 (2004)
80. R. Giacconi et al., ApJ **230**, 540 (1979)
81. R. Giacconi, H. Gursky, F.R. Paolini, B.B. Rossi, PhRvL **9**, 439 (1962)
82. R. Giacconi, E. Kellogg, P. Gorenstein, H. Gursky, H. Tananbaum, ApJ **165**, L27 (1971)
83. M. Giavalisco et al., ApJ **600**, L93 (2004)
84. I.M. Gioia, J.P. Henry, T. Maccacaro, S.L. Morris, J.T. Stocke, A. Wolter, ApJ **356**, L35 (1990)
85. J. Gofford, J.N. Reeves, F. Tombesi, V. Braito, T.J. Turner, L. Miller, M. Cappi, MNRAS **430**, 60 (2013)
86. P. Gorenstein, E.M. Kellogg, H. Gursky, ApJ **156**, 315 (1969)
87. M. Guainazzi, M.S. Tashiro (2018), arXiv:1807.06903
88. M. Guainazzi et al., A&A **341**, L27 (1999)
89. M. Guainazzi, S. Bianchi, MNRAS **374**, 1290 (2007)
90. H. Gursky et al., ApJ **223**, 973 (1978)
91. H. Gursky, E. Kellogg, S. Murray, C. Leong, H. Tananbaum, R. Giacconi, ApJ **167**, L81 (1971)
92. J.P. Halpern, ApJ **281**, 90 (1984)
93. J.P. Halpern, S.S. Holt, Nature **357**, 222 (1992)
94. F.A. Harrison et al., SPIE, 77320S, SPIE.7732 (2010)
95. F.A. Harrison et al., ApJ **770**, 103 (2013)
96. F.A. Harrison et al., ApJ **831**, 185 (2016)
97. R.C. Hartman et al., ApJ **385**, L1–L4 (1992)
98. R.C. Hartman, D.L. Bertsch, S.D. Bloom, A.W. Chen et al., ApJS **123**, 79H (1999)
99. G. Hasinger, M. van der Klis, A&A **225**, 79 (1989)
100. G. Hasinger, T. Miyaji, M. Schmidt, A&A **441**, 417 (2005)
101. W. Hermsen et al., A&A **175**, 141–150 (1987)
102. S.S. Holt, Ap&SS **42**, 123 (1976)
103. P.F. Hopkins, M. Elvis, MNRAS **401**, 7 (2010)
104. K. Hurley et al., Nature **372**, 652 (1994)
105. K. Iwamoto, P.A. Mazzali, K. Nomoto et al., Nature **395**, 672 (1998)
106. F. Jansen et al., A&A **365**, L1 (2001)
107. S. Jester, D.E. Harris, H.L. Marshall, K. Meisenheimer, ApJ **648**, 900 (2006)
108. T. Kallman, D.A. Evans, H. Marshall, C. Canizares, A. Longinotti, M. Nowak, N. Schulz, ApJ **780**, 121 (2014)
109. Kanbach, Rendiconti Lincei. Scienze Fisiche e Naturali, A DECADE OF AGILE (2019)
110. S. Kaspi et al., ApJ **574**, 643 (2002)
111. R.L. Kelley et al., SPIE, 99050V, SPIE.9905 (2016)
112. E. Kellogg, H. Gursky, C. Leong, E. Schreier, H. Tananbaum, R. Giacconi, ApJ **165**, L49 (1971)
113. T. Kii et al., ApJ **367**, 455 (1991)
114. A. King, ApJL **596**, L27 (2003)
115. A. King, ApJL **635**, L121 (2005)
116. A. King, K. Pounds, ARA&A **53**, 115 (2015)
117. A. Kinkhabwala et al., ApJ **575**, 732 (2002)
118. S. Kitamoto, H. Tsunemi, S. Miyamoto, K. Yamashita, S. Mizobuchi, Nature **342**, 518 (1989)
119. S. Kitamoto, H. Tsunemi, S. Miyamoto, K. Hayashida, ApJ **394**, 609 (1992)
120. R.W. Klebesadel, I.B. Strong, R.A. Olson, ApJ **182**, L85 (1973)

121. D.A. Kniffen et al., *Proceedings of the 14th International Cosmic Ray Conference (Munich)*, vol. 1, 100 (1975)
122. J. Knodlseder (2016), arXiv:1602.02728v1
123. W.L. Kraushaar et al., Astrophys. J. **177**, 341 (1972)
124. W.L. Kraushaar, G.W. Clark, Phys. Rev. Lett. **8**, 106 (1962)
125. Y. Krongold, F. Nicastro, M. Elvis, N. Brickhouse, L. Binette, S. Mathur, E. Jiménez-Bailón, ApJ **659**, 1022 (2007)
126. C. Labanti et al., NIM A **598**, 470–479 (2009)
127. S. Laha, M. Guainazzi, G.C. Dewangan, S. Chakravorty, A.K. Kembhavi, MNRAS **441**, 2613 (2014)
128. S. Laha, M. Guainazzi, S. Chakravorty, G.C. Dewangan, A.K. Kembhavi, MNRAS **457**, 3896 (2016)
129. R.C. Lamb et al., ApJ **212**, L63–L66 (1977)
130. G.B. Lansbury et al., ApJL **856**, L1 (2018)
131. A. Laor, ApJ **376**, 90 (1991)
132. A. Lawrence, M.G. Watson, K.A. Pounds, M. Elvis, Nature **325**, 694 (1987)
133. A.M. Levine et al., ApJS **54**, 581 (1984)
134. W.H.G. Lewin et al., ApJ **207**, L95 (1976)
135. R. Lin et al., Solar Phys. **219**, 3–32 (2002)
136. A.L. Longinotti et al., ApJL **813**, L39 (2015)
137. B. Luo et al., ApJS **228**, 2 (2017)
138. T. Maccacaro, I.M. Gioia, J.T. Stocke, ApJ **283**, 486 (1984)
139. G. Madejski, M. Sikora, Ann. Rev. Astron. Astrophys. **54** (2016)
140. F. Makino et al., ApJ **347**, L9 (1989)
141. H.L. Marshall et al., ApJS **156**, 13 (2005)
142. A. Martocchia, G. Matt, V. Karas, T. Belloni, M. Feroci, A&A **387**, 215 (2002)
143. J.L. Masnou et al., *Proceedings of the 17th International Cosmic Ray Conference (Munich)*, vol. 1 (1981), p. 100
144. M. Matsuoka, T. Ikegami, H. Inoue, K. Koyama, PASJ **38**, 285 (1986)
145. M. Matsuoka, L. Piro, M. Yamauchi, T. Murakami, ApJ **361**, 440 (1990)
146. J.R. Mattox et al., ApJ **401**, L23–L26 (1992)
147. N.J. McConnell, C.-P. Ma, ApJ **764**, 184 (2013)
148. J. McEnery et al., arxiv:1907.07558
149. I. McHardy, B. Czerny, Nature **325**, 696 (1987)
150. I.M. McHardy, A. Lawrence, J.P. Pye, K.A. Pounds, MNRAS **197**, 893 (1981)
151. B.R. McNamara, F. Kazemzadeh, D.A. Rafferty, L. Bîrzan, P.E.J. Nulsen, C.C. Kirkpatrick, M.W. Wise, ApJ **698**, 594 (2009)
152. C. Meegan et al., ApJ **702**, 791 (2009)
153. J.F. Meekins, K.S. Wood, R.L. Hedler, E.T. Byram, D.J. Yentis, T.A. Chubb, H. Friedman, ApJ **278**, 288 (1984)
154. N. Meidinger et al., SPIE, 103970V, SPIE10397 (2017)
155. S. Mereghetti et al., Proceedings of the THESEUS workshop 2017. J. Italian Astron. Soc. (Mem.SAIt) (2018)
156. A. Merloni et al., (2012), arXiv:1209.3114
157. A. Merloni et al., MNRAS **437**, 3550 (2014)
158. J.M. Miller, ARA&A **45**, 441 (2007)
159. J.M. Miller, J. Raymond, C.S. Reynolds, A.C. Fabian, T.R. Kallman, J. Homan, ApJ **680**, 1359 (2008)
160. I.F. Mirabel, V. Dhawan, S. Chaty, L.F. Rodriguez, J. Marti, C.R. Robinson, J. Swank, T. Geballe, A&A **330**, L9 (1998)
161. S. Miyamoto, S. Kitamoto, K. Mitsuda, T. Dotani, Nature **336**, 450 (1988)
162. P. Morrison, Il Nuovo Cimento, **VII**(6) (1958)
163. R.F. Mushotzky, AdSpR **3**, 157 (1984)
164. K. Nandra et al., (2013), arXiv:1306.2307

165. K. Nandra, K.A. Pounds, MNRAS **268**, 405 (1994)
166. E. Nardini et al., Science **347**, 860 (2015)
167. M. Oda, P. Gorenstein, H. Gursky, E. Kellogg, E. Schreier, H. Tananbaum, R. Giacconi, ApJ **166**, L1 (1971)
168. K. Oh et al., ApJS **235**, 4 (2018)
169. T. Ohashi, K. Makishima, H. Inoue, K. Koyama, F. Makino, M.J.L. Turner, R.S. Warwick, PASJ **41**, 709 (1989)
170. A. Owens et al., IEEE Trans. Nuclear Sci. **39**(2), 559–567 (1991)
171. P. Padovan et al., A&A Rev. **25**, 2 (2017)
172. F.B.S. Paerels, S.M. Kahn, ARA&A **41**, 291 (2003)
173. F. Paerels, J. Cottam, M. Sako, D.A. Liedahl, A.C. Brinkman, R.L.J. van der Meer, J.S. Kaastra, P. Predehl, ApJ **533**, L13 (2000)
174. G.C. Perola et al., ApJ **306**, 508 (1986)
175. F. Perotti, M. Fiorini, S. Incorvaia, E. Mattaini, E. Sant Ambrogio, NIMA **556**, 228 (2006)
176. Peterson et al., ApJ **145**, 962 (1966)
177. Peterson et al., Phys. Rev. Let. **16**, 142 (1966)
178. L.E. Peterson, ARA&A **13**, 423 (1975)
179. L.E. Peterson, J.R. Winckler, Phys. Rev. Lett. **1**, 205 (1958)
180. G. Piccinotti, R.F. Mushotzky, E.A. Boldt, S.S. Holt, F.E. Marshall, P.J. Serlemitsos, R.A. Shafer, ApJ **253**, 485 (1982)
181. A.M.T. Pollock et al., A&A **94**, 116–120 (1981)
182. G. Ponti et al., MNRAS **453**, 172 (2015)
183. G. Ponti, R.P. Fender, M.C. Begelman, R.J.H. Dunn, J. Neilsen, M. Coriat, MNRAS **422**, L11 (2012)
184. K.A. Pounds, K.L. Page, MNRAS **372**, 1275 (2006)
185. K.A. Pounds, J.N. Reeves, MNRAS **397**, 249 (2009)
186. K.A. Pounds, S. Vaughan, MNRAS **423**, 165 (2012)
187. K.A. Pounds, K. Nandra, G.C. Stewart, I.M. George, A.C. Fabian, Nature **344**, 132 (1990)
188. M. Prest, G. Barbiellini, G. Bordignon et al., NIMA **501**, 280 (2003)
189. W.C. Priedhorsky, J. Terrell, S.S. Holt, ApJ **270**, 233 (1983)
190. W. Priedhorsky, G. Hasinger, W.H.G. Lewin, J. Middleditch, A. Parmar, L. Stella, N. White, ApJ **306**, L91 (1986)
191. D. Psaltis, T. Belloni, M. van der Klis, ApJ **520**, 262 (1999)
192. H.-O. Radecke, G. Kanbach, *Data Analysis in Astronomy IV* (Plenum Press, New York, 1992)
193. J.N. Reeves et al., ApJ **701**, 493 (2009)
194. P. Reig, T. Belloni, M. van der Klis, A&A **412**, 229 (2003)
195. R.C. Reis, M.T. Reynolds, J.M. Miller, D.J. Walton, Nature **507**, 207 (2014)
196. R.A. Remillard, J.E. McClintock, ARA&A **44**, 49 (2006)
197. R.A. Remillard, B. Grossan, H.V. Bradt, T. Ohashi, K. Hayashida, Nature **350**, 589 (1991)
198. R.A. Remillard, J.E. McClintock, C.D. Bailyn, ApJ **399**, L145 (1992)
199. C.S. Reynolds, SSRv **183**, 277 (2014)
200. G.R. Ricker, G.W. Clarke, R.E. Doxsey, R.G. Dower, J.G. Jernigan, J.P. Delvaille, G.M. MacAlpine, R.M. Hjellming, Nature **271**, 35 (1978)
201. S.R. Rosen et al., A&A **590**, A1 (2016)
202. R. Rothschild et al., SSI **4**, 269 (1979)
203. R.E. Rothschild, E.A. Boldt, S.S. Holt, P.J. Serlemitsos, ApJ **189**, L13 (1974)
204. J.S. Sanders, A.C. Fabian, S.W. Allen, R.G. Morris, J. Graham, R.M. Johnstone, MNRAS **385**, 1186 (2008)
205. P.W. Sanford, K.O. Mason, F.J. Hawkins, P. Murdin, A. Savage, ApJ **190**, L55 (1974)
206. R.D. Saxton, A.M. Read, P. Esquej, M.J. Freyberg, B. Altieri, D. Bermejo, A&A **480**, 611 (2008)
207. V. Schonfelder et al., IEEE Trans. Nuclear Sci. **NS-31**(1) (1984)
208. V. Schonfelder et al., ApJS **86**, 657–692 (1993)
209. V. Schönfelder et al., ApJS **86**, 657 (1993)

210. V. Schonfelder et al., A&A Suppl. Ser. **143**, 145–179 (2000)
211. E.J. Schreier, P. Gorenstein, E.D. Feigelson, ApJ **261**, 42 (1982)
212. D.A. Schwartz, H.S. Hudson, L.E. Peterson, ApJ **162**, 431 (1970)
213. T. Shanks, I. Georgantopoulos, G.C. Stewart, K.A. Pounds, B.J. Boyle, R.E. Griffiths, Nature **353**, 315 (1991)
214. A. Siemiginowska, Ł. Stawarz, C.C. Cheung, D.E. Harris, M. Sikora, T.L. Aldcroft, J. Bechtold, ApJ **657**, 145 (2007)
215. J.F. Smith, G.M. Courtier, RSPSA **350**, 421 (1976)
216. R.K. Smith, N.S. Brickhouse, D.A. Liedahl, J.C. Raymond, ApJ **556**, L91 (2001)
217. P. Sreekumar et al., ApJ **400**, L67 (1992)
218. M. Su, D.P. Finkbeiner, ApJ **753**, 61 (2012)
219. M. Su, T.R. Slatyer, D.P. Finkbeiner, ApJ **724**, 1044 (2010)
220. B.N. Swanenburg et al., Nature **275**, 298 (1978)
221. B.N. Swanenburg et al., ApJ **234**, L69–L73 (1981)
222. Y. Tanaka et al., PASJ **36**, 641 (1984)
223. Y. Tanaka et al., Nature **375**, 659 (1995)
224. M. Tashiro et al., SPIE, 1069922, SPIE10699 (2018)
225. V. Tatischeff et al., *Proceedings of the 12th INTEGRAL conference "INTEGRAL looks AHEAD to Multimessenger astronomy")* (Geneva (Switzerland), 2019)
226. M. Tavani, *Multiple Messengers and Challenges in Astroparticle Physics* (2018), p. 145. ISBN 978-3-319-65423-2
227. M. Tavani, G. Barbiellini, A. Argan et al., A&A **502**, 995–1013 (2009)
228. B.G. Taylor, R.D. Andresen, A. Peacock, R. Zobl, SSRv **30**, 479 (1981)
229. D.J. Thompson et al., ApJ **200**, 79–82 (1975)
230. D.J. Thompson et al., ApJ **213**, 252–262 (1977)
231. D.J. Thompson et al., ApJ Supp. **86**, 629–656 (1993)
232. F. Tombesi et al., MNRAS **443**, 2154 (2014)
233. F. Tombesi, M. Cappi, J.N. Reeves, G.G.C. Palumbo, T. Yaqoob, V. Braito, M. Dadina, A&A **521**, A57 (2010)
234. F. Tombesi, M. Cappi, J.N. Reeves, V. Braito, MNRAS **422**, L1 (2012)
235. F. Tombesi, M. Cappi, J.N. Reeves, R.S. Nemmen, V. Braito, M. Gaspari, C.S. Reynolds, MNRAS **430**, 1102 (2013)
236. A. Tortosa, S. Bianchi, A. Marinucci, G. Matt, P.O. Petrucci, A&A **614**, A37 (2018)
237. G. Trinchieri, G. Fabbiano, C.R. Canizares, ApJ **310**, 637 (1986)
238. J. Trümper et al., Nature **349**, 579 (1991)
239. M. Tsujimoto et al., PASJ **70**, 20 (2018)
240. H. Tsunemi, S. Kitamoto, M. Manabe, S. Miyamoto, K. Yamashita, M. Nakagawa, PASJ **41**, 391 (1989)
241. H. Tsunemi, S. Kitamoto, S. Okamura, D. Roussel-Dupre, ApJ **337**, L81 (1989)
242. M.J.L. Turner et al., MNRAS **244**, 310 (1990)
243. T.J. Turner, K.A. Pounds, MNRAS **240**, 833 (1989)
244. M.J.L. Turner, A. Smith, H.U. Zimmermann, SSRv **30**, 513 (1981)
245. Ubertini et al., Nuclear Phys. B Proc. Suppl. **243**, 52 (2013)
246. M. van der Klis, F. Jansen, J. van Paradijs, W.H.G. Lewin, E.P.J. van den Heuvel, J.E. Trumper, M. Szatjno, Nature **316**, 225 (1985)
247. G. Villa, C.G. Page, M.J.L. Turner, B.A. Cooke, M.J. Ricketts, K.A. Pounds, D.J. Adams, MNRAS **176**, 609 (1976)
248. W. Voges et al., A&A **349**, 389 (1999)
249. Walter et al., A&ARv **23**, 2 (2015)
250. R.S. Warwick et al., MNRAS **197**, 865 (1981)
251. R.S. Warwick et al., ApJ **470**, 349 (1996)
252. M.G. Watson, G.C. Stewart, W. Brinkmann, A.R. King, MNRAS **222**, 261 (1986)
253. B.J. Wilkes, M. Elvis, ApJ **323**, 243 (1987)
254. O.R. Williams et al., ApJ **389**, 157 (1992)

255. P.F. Winkler Jr., A.E. White, ApJ **199**, L139 (1975)
256. C. Winkler et al., A&A **411**, L1–L6 (2003)
257. C. Winkler et al., Space Sci. Rev. **161**, 149 (2011)
258. D.M. Worrall, A&ARv **17**, 1 (2009)
259. D.M. Worrall, E.A. Boldt, S.S. Holt, R.F. Mushotzky, P.J. Serlemitsos, ApJ **243**, 53 (1981)
260. X.-P. Wu, Y.-J. Xue, L.-Z. Fang, ApJ **524**, 22 (1999)
261. F. Yuan, W. Cui, R. Narayan, ApJ **620**, 905 (2005)
262. A.C. Zoglauer, First light for the next generation of Compton and pair telescopes, Ph.D. Thesis, Garching: Max-Planck-Institut für Extraterrestrische Physik, 2006, MPE Report, No. 289 (2006)

Chapter 5
From Raw Data to Scientific Products: Images, Light Curves and Spectra

Jiachen Jiang and Dheeraj R. Pasham

5.1 Overview

This chapter will describe the main steps and some of the common intricacies involved in the process of getting from the raw data to the most basic scientific products, i.e., images, light curves and energy spectra. The outline of this process is shown in Fig. 5.1 and each step will be discussed in separate sections throughout the chapter.

5.1.1 Astrophysical Photons to Raw Data

The first step in any experimental science is to take data. Unfortunately, the Earth's atmosphere is not transparent to X-ray and γ-rays. Hence to detect high-energy radiation from astrophysical phenomena telescopes need to be operated from above the Earth's atmosphere. We refer interested readers to Chap. 3 for more details on X-ray and γ-ray detectors. Traditionally, the high-energy sky has been accessed either via rocket experiments, balloon missions, or from spacecrafts that orbit our planet. With rocket and balloon experiments the data is typically recovered when the detectors return to ground at the end of their mission. On the other hand, the data acquired by orbiting satellites is transmitted back to Earth on a semi-regular basis when they pass over a ground station. For example, the main ground station for the

J. Jiang
Department of Astronomy, Tsinghua Univerisity, 30 Shuangqing Road, Beijing 100084, China
e-mail: jcjiang@tsinghua.edu.cn

D. R. Pasham (✉)
MIT Kavli Institute for Astrophysics and Space Research, MIT, 77 Massachusetts Avenue, Cambridge, MA 02139, USA
e-mail: drreddy@mit.edu

C. Bambi (ed.), *Tutorial Guide to X-ray and Gamma-ray Astronomy*,
https://doi.org/10.1007/978-981-15-6337-9_5

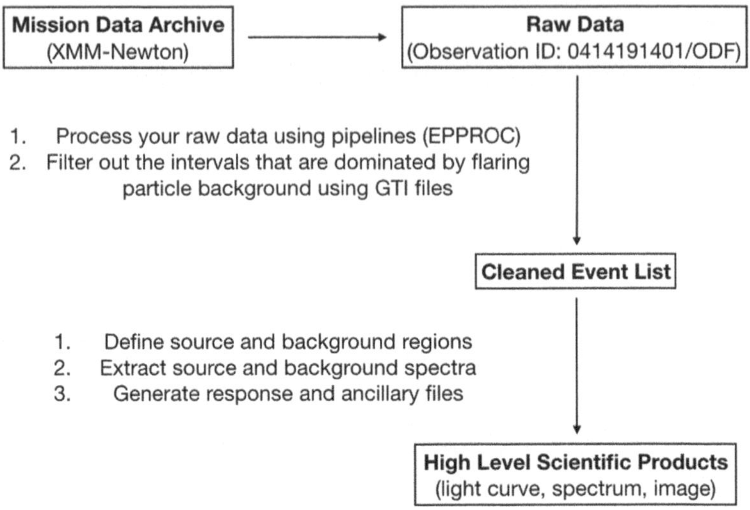

Fig. 5.1 Outline of the process of getting from raw data to scientific products

Neil Gehrels Swift observatory is the Malindi Ground Station located in Kenya. The data received at ground stations is then passed on to the mission operations center (MOC) which often either validates the data or performs some minimal data sorting before sending it to the mission's data center.

The organized data from the data centers is eventually stored in the HEASARC (High Energy Astrophysics Archive) or in a location that is accessible via http or ftp from the HEASARC. In some cases, the data access is exclusively given to either the proposing team or to the detector development team for a certain fixed period of time. In these cases, the idea is to allow the proposing team to carry out their research program or, in the case of detector team, to validate the preliminary data. In all cases, the datasets eventually turn up on the archive. This is because astrophysical X-ray missions are funded by national budgets and thus they require the data acquired by them to be made public.

5.1.2 Raw to Cleaned Data

The working philosophy of a typical X-ray or a γ-ray instrument is to register properties of individual photon events that hit the detector material and produce a threshold-crossing amount of current/charge. Thus the raw (or level-1) data is essentially a table (or a set of tables) where each column describes a different attribute of an event. Several of these raw event attributes can be common among many different detectors. These include basic properties like the time of the event, the amount of charge generated by the event (known as the Pulse Height Amplitude or PHA), and in the case of detectors with imaging capability, the raw coordinates of the event in units of detector pixels. A major portion of data processing/cleaning is to convert these event attributes into astrophysically relevant quantities that can be used to extract scientific products which inform us about the underlying physical processes.

For example, the quantity PHA does not directly tell us anything meaningful about the astrophysical object that was observed. However, the PHA value is related to the energy of the incoming photon by a relation that itself depends on the energy and the position of the event on the detector. This information is encoded in a calibration file of the detector known as the response matrix. The data cleaning programs/pipelines use this information to convert the PHA column into something called a Pulse Invariant (PI) which is linearly related to the energy of the incoming X-ray photon. Similarly, the raw positions of the events in the detector's reference frame have to be transformed into the sky coordinates to be useful. This is usually done with the aid of a small optical star tracker telescope on board. By knowing the pointing direction of the optical telescope and by cross-matching its images with previously known star catalogs, and combining that knowledge with the information about the relative orientation between the optical and the X-ray detector plane, the data reduction programs are able to derive the sky coordinates (right ascension and the declination values) of the individual pixels in the image.

Furthermore, the raw data includes events resulting from high-energy (X-ray or γ-ray) photons as well as high-energy particles interacting with the detector. These particle events can be from various sources including cosmic rays, trapped charged particles in Earth's magnetosphere, etc. They can interact with the detector, generate charge and mimic the behavior of high-energy photons. In addition to the above mentioned event properties, there are often several instrument-specific event attributes like the flag, grade or the pixel pattern produced by the event. These can be used to separate the source events from the background particle events as part of the data cleaning process.

5.2 Getting the Cleaned Eventlist

We demonstrate the end-to-end process from data acquisition to scientific products using a sample dataset of the quasar 3C 273 from *XMM-Newton* (XMM hereafter). The XMM observation used in this chapter is part of a cross-calibration campaign between various X-ray telescopes from 2018 (XMM observation ID 0414191401). XMM is equipped with three X-ray CCD cameras (pn, MOS1, and MOS2) which are all part of the European Photon Imaging Camera (EPIC). Although we will only consider the EPIC-pn observation in this chapter, the data reduction for EPIC-MOS1 and EPIC-MOS2 data are very similar.

5.2.1 Accessing the Raw Data

HEASARC[1] is currently the largest repository for high-energy astrophysical data. One can access public observations taken recently by currently active missions like *Chandra*, *Swift*, *XMM-Newton*, *NICER*, and *Fermi* to name a few, and also data acquired by some of the first high-energy missions such as *HEAO-1* and *HEAO-2/Einstein* observatory that were operational in the late 1970s to early 1980s. The three main steps to access data are as follows (see Fig. 5.2 for the three webpage layouts).

1. HEASARC's home page shows a data query form which can be used to search for any publicly-available observations. You can start by entering `3C 273` for `Object Name or Coordinates`. Alternatively, you may enter the coordinates of the target in one of the formats shown below the query box. For `What missions and catalogs do you want to search?` option, choose `XMM-Newton`. Leave the rest to default values and click `Start Search`. Readers are strongly recommended to use the `Detailed Mission` option to try different combinations of search keywords.
2. A list of catalogs will be returned once the search is done. Different catalogs show relevant observations in different archives of the mission. In this example, we only consider the public available pointing observations of 3C 273. Therefore, choose `XMM-Newton Master Log and Public Archive`.
3. Choose the observation you want to analyse, and retrieve the data. Here we focus on the observation ID 0414191401. Click `XMM ODF Basic Data`, which contains all the raw data obtained during this observation, and then download the datasets. The raw event files will be included in the downloaded folder together with the corresponding spacecraft, housekeeping and radiation monitor files for each instrument.

5.2.2 Software Prerequisites: HEAsoft, XMM-SAS, XMM Calibration Files and ds9

Several programs are necessary to reduce and calibrate the raw data and to extract and visualize the scientific data products. A large collection of such tools is provided by NASA's HEASARC software or HEAsoft. Instructions for HEAsoft installation can be found here: https://heasarc.gsfc.nasa.gov/docs/software/heasoft/.

The Science Analysis System (SAS) is a set of programs, scripts and libraries, specifically designed to reduce and analyze XMM data. It can be installed by following the instructions outlined here: https://www.cosmos.esa.int/web/xmm-newton/download-and-install-sas.

[1] https://heasarc.gsfc.nasa.gov/cgi-bin/W3Browse/w3browse.pl.

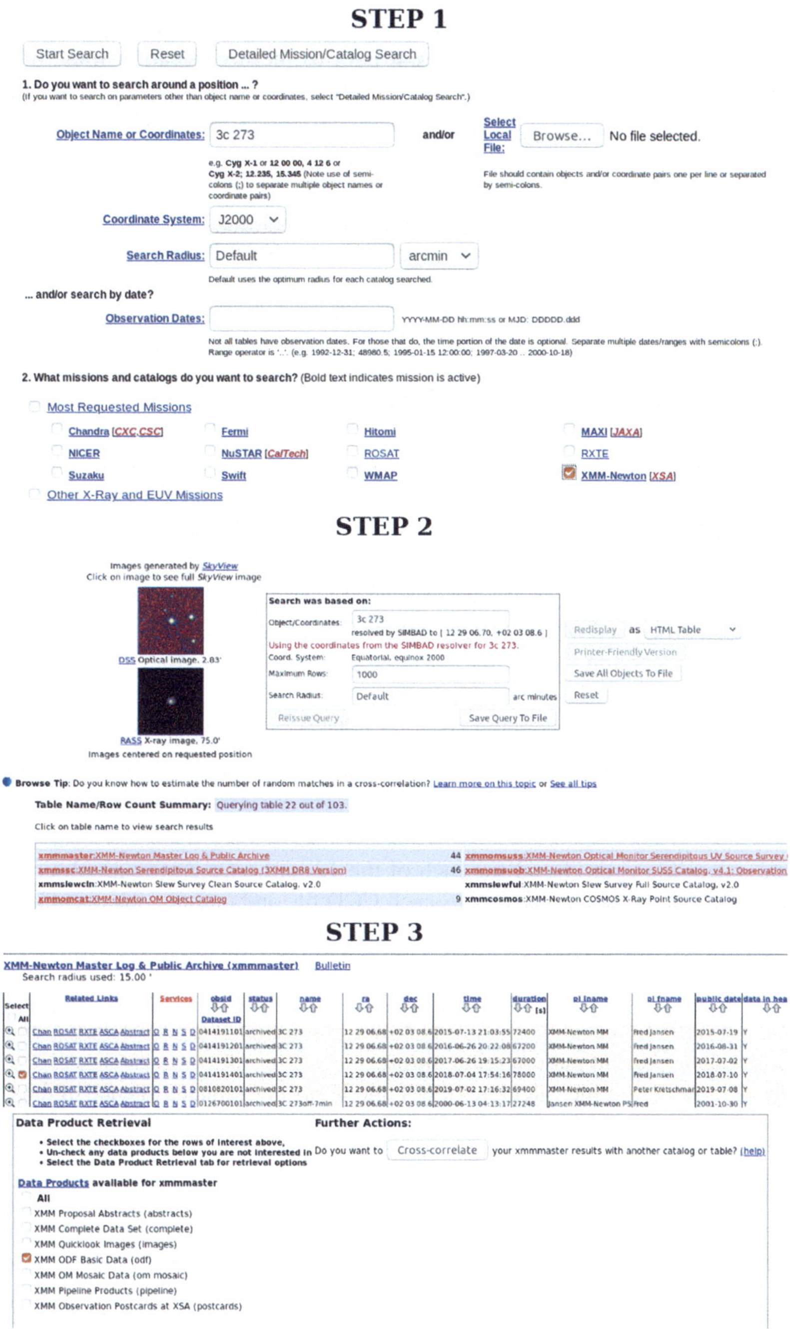

Fig. 5.2 Three steps to download raw data from HEASARC (See text for details)

XMM SAS requires calibration files to reduce the raw data. Thus, some of SAS tasks require you to specify a local path to these calibration files. Instructions on how to setup a local up-to-date calibration database can be found here: https://www.cosmos.esa.int/web/xmm-newton/current-calibration-files. It is very important to keep the calibration files updated in order to generate accurate scientific products in the end.

Finally, ds9 is an imaging and visualization tool used frequently in high-energy astrophysics data analysis. Installation instructions can be found here: http://ds9.si.edu/site/Home.html.

5.2.3 Reprocessing the Raw Data

Most of the high-energy data stored on HEASARC contain both the raw data as well as the cleaned data that has been processed with a recent version of calibration files. If a particular mission has been be decommissioned then it will likely have been processed with the most up-to-date calibration data. However, if a particular observatory is still active it is recommended to check and make sure that the cleaned data has been processed with the latest version of the calibration data. If that is not the case, then re-processing the raw data may be necessary using, for example, mission specific pipeline tools as described below.

In the case of XMM reducing the raw datasets is pretty straightforward. For EPIC-pn it is accomplished using the XMM SAS task `epproc`. The `epproc` pipeline includes the following subroutines `epframes`, `badpixfind`, `badpix`, `epevents` and `attcalc`. They are for the following purposes:

1. One image file is produced for each of the 12 pn CCDs during an observation. The `epframes` routine processes each CCD file and outputs raw eventlists.
2. The `badpixfind` routine finds bad pixels in the CCDs, which are later labelled in an updated eventlist by `badpix`.
3. `epevents` processes the updated eventlist created in the second step and generates a calibrated eventlist with PHA and PI columns by performing gain shift correction and event pattern recognition.[2]
4. The `attcalc` routine calculates the X and Y sky coordinates of each of the CCD pixels.

The resulting eventlist is saved to
`'3401_0414191401_EPN_S003_ImagingEvts.ds'`.
The number '3401' is the revolution number of this XMM observation. Each orbit of XMM has a particular revolution number, and may contain more than one observation; '0414191401' is the observation number; 'EPN' stands for the instrument

[2]We refer interested readers to the following website for the introduction of different event patterns: https://xmm-tools.cosmos.esa.int/external/xmm_user_support/documentation/uhb/epic_evgrades.html.

name; S003 is the exposure number. For example, S003 is the EPIC-pn exposure that is taken in the small window mode[3] and S027 is the other EPIC-pn exposure that is taken during this observation but in the large window mode. Here we consider exposure S003.

5.2.3.1 Filtering Flaring Particle Background

In addition to the standard pipeline, it is also necessary to check if your EPIC observations are dominated by flaring particle background. To do so, you need to define good time intervals (GTIs) by identifying intervals of low-background during an observation and save that information to a GTI file. The `evselect` tool is used to generate a light curve to visualise the background level and the `tabgtigen` tool is used to create a corresponding GTI file that will be used in the next step.

The commands are as follows:

```
evselect table=3401_0414191401_EPN_S003_ImagingEvts.ds \

withrateset=Y rateset=rateEPIC_PN_0414191401_S003_tol.fits \

maketimecolumn=Y timebinsize=100 makeratecolumn=Y

expression='#XMMEA_EP && (PI>10000 && PI<12000) && (PATTERN==0)'
```

The parameters `maketimecolumn` and `makeratecolum` are set to yes in order to make a light curve. By running this command, we create an EPIC-pn light curve with events in the energy range of 10 and 12 keV that are dominated by particles. Events above 12 keV are excluded to avoid hot pixels being wrongly classified as high energy events.

The output light curve file '`rateEPIC_PN_0414191401_S003_tol.fits`' can be viewed by running the `lcurve` tool as follows:

```
> lcurve
lcurve 1.0 (xronos5.22)
Number of time series for this task[1]
Ser. 1 filename +options (or file of filenames +options)[]
Series 1 file 1:rateEPIC_PN_0414191401_S003_tol.fits

  Selected FITS extensions: 1 - RATE TABLE;

  Source ........... Start Time (d) .... 18303 18:21:
52.468
```

[3]https://xmm-tools.cosmos.esa.int/external/xmm_user_support/documentation/uhb/epicmode.html.

```
FITS Extension .... 1 - 'RATE ' Stop Time (d) ..... 18304
11:55:06.210
No. of Rows ....... 632 Bin Time (s) ...... 100.0
Right Ascension ... Internal time sys.. Converted
to TJD
Declination ....... Experiment ........ XMM EPN
Filter ............ Thick
Corrections applied: Vignetting - No ; Deadtime - No ;
Bkgd - No ; Clock - Yes

  Selected Columns: 3- Time; 1- Y-axis; 2- Y-error;

  File contains binned data.

  Name of the window file ('-' for default window)[-]

  Expected Start ... 18303.76519059666 (days) 18:21:52:
468 (h:m:s:ms)
Expected Stop .... 18304.49659965848 (days) 11:55: 6:
210 (h:m:s:ms)

  Minimum Newbin Time 100.00000 (s)
for Maximum Newbin No.. 632

  Default Newbin Time is: 123.57009 (s) (to have 1
Intv. of 512 Newbins)
Type INDEF to accept the default value

  Newbin Time or negative rebinning[] 100.

  Newbin Time ...... 100.00000 (s)
Maximum Newbin No. 632

  Default Newbins per Interval are: 512
(giving 2 Intervals of 512 Newbins each)
Type INDEF to accept the default value

  Number of Newbins/Interval[512] 512
Maximum of 2 Intvs. with 512 Newbins of 100.000 (s)
Name of output file[default]
Do you want to plot your results?[yes]
Enter PGPLOT device[/XW]

  512 analysis results per interval
```

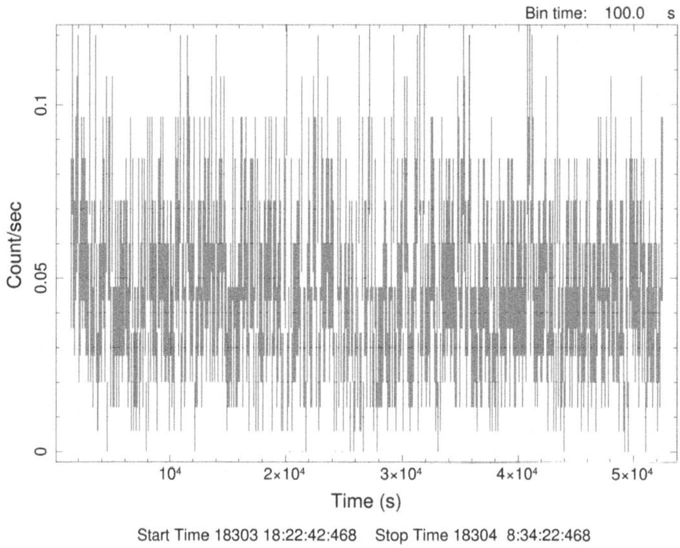

Fig. 5.3 The 10–12 keV band light curve extracted from the EPIC-pn observation of 3C 273

```
  81% completed
Intv 1 Start 18303 18:22:42
Ser.1 Avg 0.4672E-01 Chisq 441.1 Var 0.4025E-03 Newbs.
512
Min 0.000 Max 0.1200 expVar 0.4672E-03 Bins 512
```

A light curve as in Fig. 5.3 will be shown in a new plotting window. The 10–12 keV band count rate of this EPIC-pn observation is consistently below 0.12 cts s^{-1} during the entire observation. A reference EPIC-pn count rate for the threshold of a high background count rate is 0.4 cts s^{-1}. The threshold number can be different for different observations but 0.4 cts s^{-1} is considered to be a good reference number of most standard observations. If there are any intervals during your observation that show more than 0.4 cts s^{-1} GTI files should be generated as follows.

```
  tabgtigen table=rateEPIC_PN_0414191401_S003_tol.fits \

  expression='RATE<=0.4' gtiset=pn_gti.fits
```

This command will generate a GTI file `'pn_gti.fits'`. A screenshot of this file is in Fig. 5.4. A GTI file only has two columns, the start time and the end time of a good time interval, and each row represents a good interval.

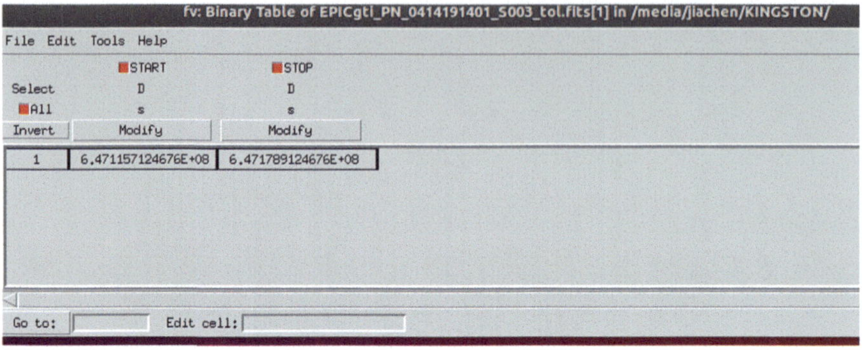

Fig. 5.4 An example of a user-defined good time interval (GTI) file

The last step is to apply the user-defined GTI file `'pn_gti.fits'` to the calibrated event file `'3401_0414191401_EPN_S003_ImagingEvts.ds'`. To do so, we can use the `evselect` tool again.

```
evselect table=3401_0414191401_EPN_S003_ImagingEvts.ds \

withfilteredset=Y filteredset=EPICclean_pn.fits \

destruct=Y keepfilteroutput=T \

expression='#XMMEA_EP&&gti(pn_gti.fits,TIME)&&(PI>150)'
```

The output file 'EPICclean_pn.fits' is the final cleaned event file, which can used to extract high level scientific products.

5.2.3.2 Cleaned Event File

A screenshot of the various columns in the cleaned event file 'EPICclean_pn.fits' is shown in Fig. 5.5. There can be several columns for each event but some of the main parameters are:

- Time of the event in units of seconds,
- Raw pixel position (RAWX/RAWY), camera coordinates (DETX/DETY) and physical sky coordinate (X/Y),
- The energy of the event in PHA (channel number) and PI (eV),
- Event pattern and CCD information (PATTERN, CCDNR).

File Edit Tools Help

	TIME	RAWX	RAWY	DETX	DETY	X	Y	PHA	PI	FLAG	PATTERN	PAT_ID	PAT_SEQ	CCDNR
Select	D	I	I	I	I	J	J	I	I	J	B	I	B	B
All	s	pixel	pixel	0.05 arcsec	0.05 arcsec	0.05 arcsec	0.05 arcsec	channel	eV					
Invert	Modify	Modify	Modify	Modify	Modify	Modify	Modify	Modify	Modify	Modify	Modify	Modify	Modify	Modify
1	6.471157124739E+08	36	191	248	-309	26461	28087	177	975	0	0	0	0	4
2	6.471157124973E+08	38	192	61	-372	26264	28072	34	305	0	3	5121	1	4
3	6.471157125068E+08	37	191	141	-306	26364	28042	582	3033	0	0	0	0	4
4	6.471157125886E+08	37	191	112	-292	26343	28018	100	942	0	7	5121	1	4
5	6.471157126643E+08	37	192	137	-395	26326	28123	65	351	0	0	0	0	4
6	6.471157126782E+08	37	192	156	-329	26369	28069	53	282	0	0	0	0	4
7	6.471157127220E+08	38	187	42	19	26399	27704	203	1113	0	0	0	0	4
8	6.471157127858E+08	36	196	200	-654	26283	28386	454	2528	0	0	0	0	4
9	6.471157129288E+08	38	191	43	-259	26292	27961	1176	9746	0	3	5121	1	4
10	6.471157129513E+08	31	191	656	-263	26855	28203	88	463	0	0	0	0	4
11	6.471157129619E+08	64	165	-2072	1837	25160	25207	26	259	5	3	1	1	4
12	6.471157129623E+08	37	191	97	-261	26341	27984	33	175	0	0	0	0	4
13	6.471157130182E+08	38	190	4	-229	26268	27918	228	1255	0	0	0	0	4
14	6.471157131101E+08	39	188	-52	-61	26282	27742	415	2146	0	0	0	0	4
15	6.471157131406E+08	41	189	-182	-131	26135	27755	195	1019	0	0	0	0	4
16	6.471157131478E+08	36	189	220	-130	26505	27911	95	512	0	0	0	0	4
17	6.471157131763E+08	38	191	20	-252	26273	27945	541	2965	0	0	0	0	4
18	6.471157131888E+08	36	191	212	-298	26432	28063	92	497	0	0	0	0	4
19	6.471157132210E+08	35	195	268	-592	26369	28355	174	942	0	0	0	0	4
20	6.471157132494E+08	39	193	-28	-450	26152	28109	108	590	0	0	0	0	4

Fig. 5.5 An example of a calibrated and cleaned event file

5.3 Generate Scientific Products

5.3.1 Images: Selecting Source and Background Regions

In order to generate scientific products, source and background regions need to be defined. Note that the 'source' region includes events from both the source 3C 273 and the background. Therefore, an additional 'background' region nearby has to be defined in order to estimate the background properties. `ds9` software can be used to view the event file as an image as follows.

```
ds9 EPICclean_pn.fits
```

In order to obtain an image as shown in Fig. 5.6, choose `Scale>log` and `Color>heat` on the top menu bar of the window. Choose `Edit>region` to define a region. A circular region with an optimal radius of 35 arcsec is used to extract source products. This aperture encloses roughly 90% of light from a point source. Making the radius larger usually does not help as you also increase the background contribution. The background region is defined as the polygon region shown in Fig. 5.6. To enable better statistics much larger region is required to extract the background products. Care should be taken to ensure that the background region does not include any contaminating sources.

If there are several point sources in the field of view and you are not sure which one is your target source, you may click `Analysis>Name resolution`. Type in your source name in the query form, and click `Retrieve`. This tool requires internet connection in order to obtain the coordinates of your target source from online services, e.g. `Simbad`. Click `File>Crosshair to`, and the corresponding camera position of your target source will be marked in the image. An example of

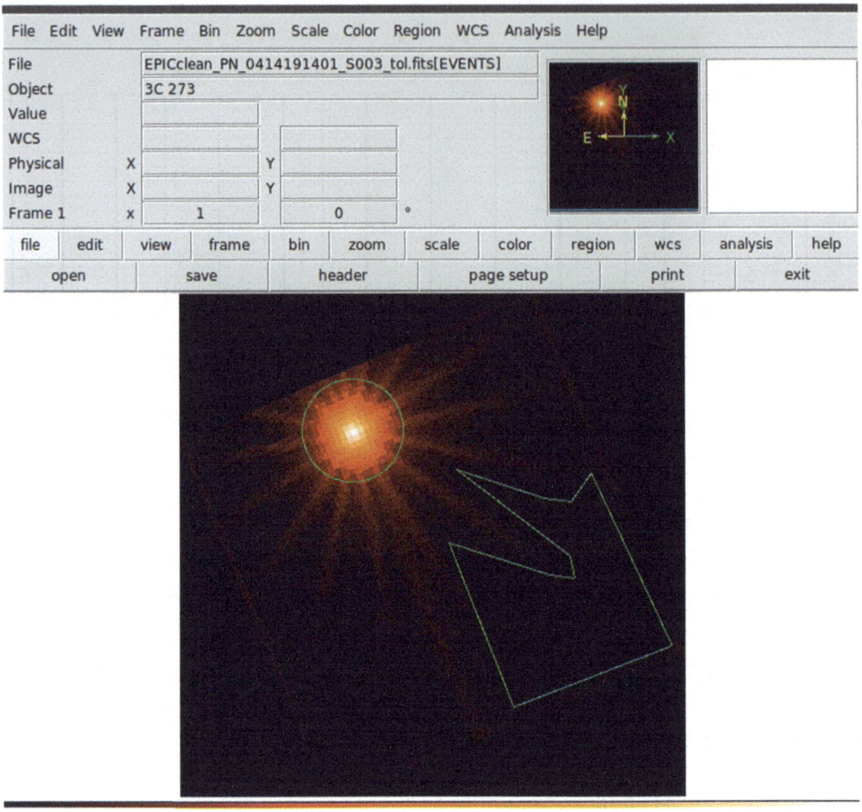

Fig. 5.6 The source and background regions used to extract the source and background spectra of 3C 273 from our example XMM observation

using the `Name resolution` tool is shown in Fig. 5.7, where the green crossbar shows the position of 3C 273 on the detector.

5.3.2 *Light Curves*

Again, `evselect` can be used to extract the source light curve. First, we extract the source+background light curve by considering the source region defined above:

```
evselect table=EPICclean_pn.fits energycolumn=PI \

expression='#XMMEA_EP&&(PATTERN<=4 \
```

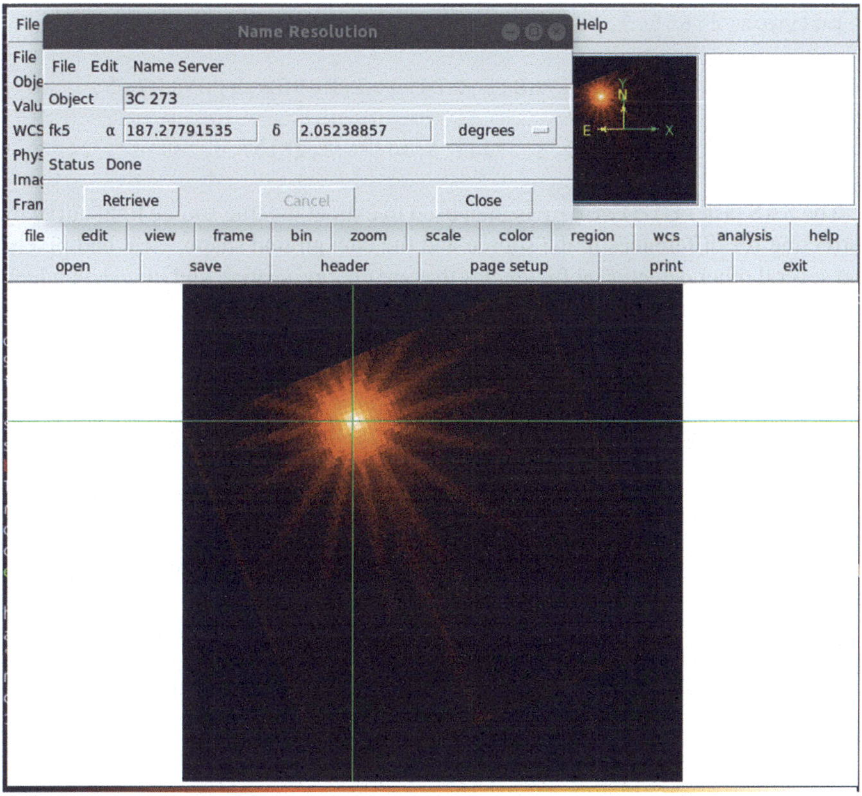

Fig. 5.7 An example of locating our target source (3C 273) on the detector plane

```
((X,Y) IN circle(26344.403,28019.495,700)) \
&&(PI in [300:10000])' \
withrateset=yes rateset="PN_src_lightcurve_raw.lc" \
timebinsize=100 maketimecolumn=yes makeratecolumn=yes
```

This command will generate a 'source'+'background' light curve 'PN_src_lightcurve_raw.lc' using events between 0.3–10 keV (PI: 300–10000) in 100 s bins. Users can define their preferred energy range and bin size by changing the range of `PI` and the value of `timebinsize` respectively.

Then we use the same command above but change the circular region to a polygon region as following to generate the background light curve 'PN_bkg_lightcurve_raw.lc':

```
polygon 27656.5,26499.5,29032.5,26067.5,29368.5,26019.5

29304.5,26323.5,27768.5,27491.5,28952.5,27107.5,

29320.5,27059.5,29592.5,27443.5,30696.5,25107.5,28536.5,24275.5
```

The SAS task `epiclccorr` is then used to extract the true source light curve by correcting for differences in the source and the background shape, area, bad pixels, and several other detector inefficiencies that are energy, spatial, and time dependent. The command is as follows:

```
epiclccorr srctslist=PN_src_lightcurve_raw.lc \

eventlist=EPICclean_pn.fits outset=PN_lccorr.lc \

bkgtslist=PN_bkg_lightcurve_raw.lc \

withbkgset=yes applyabsolutecorrections=yes
```

The output file 'PN_ccorr.lc' is the final corrected source light curve. By using the `lcurve` tool as above, you will be able to visualise the final background-corrected light curve as in Fig. 5.8. Further studies of light curves can be found in Sect. 7.2.

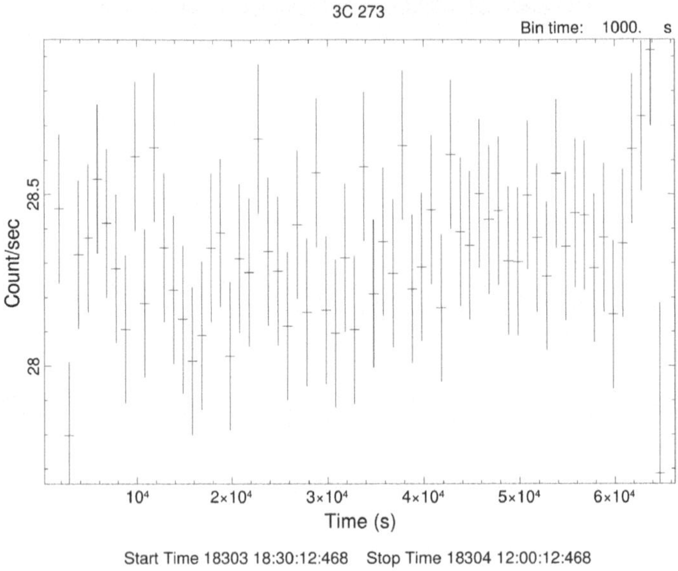

Fig. 5.8 The 0.3–10 keV band light curve of 3C 273 during our example observation

5.3.3 *Spectra*

The `evselect` tool can also be used to generate spectra. Use the following command to extract a source+background spectrum from the same source region defined above:

```
evselect table=EPICclean_pn.fits withspectrumset=yes \

spectrumset=PNsrc_spectrum.fits \

energycolumn=PI spectralbinsize=5 withspecranges=yes \

specchannelmin=0 specchannelmax=20479 \

expression='(FLAG==0) && (PATTERN<=4) \

&& ((X,Y) IN circle(26344.403,28019.495,700))'
```

The output file 'PNsrc_spectrum.fits' is the source+background spectrum. Similarly, the background spectrum 'PNbkg_spectrum.fits' can be extracted by using a similar command and the polygon region defined above.

The second step is to calculate the area of the source and background regions by using the `backscale` tool:

```
backscale spectrumset=PNsrc_spectrum.fits badpixlocation=EPICclean_
pn.fits

backscale spectrumset=PNbkg_spectrum.fits badpixlocation=EPICclean_
pn.fits
```

Note that the bad pixels must be given when calculating the area of the regions. The results will be written in the header of the `fits` files of the spectra as `BACKSCAL`.

The third step is to generate corresponding redistribution matrix and ancillary files using the `rmfgen` and `arfgen` tools.

To generate the redistribution matrix file:

```
rmfgen spectrumset=PNsrc_spectrum.fits rmfset=PN.rmf
```

To generate the ancillary file:

```
arfgen spectrumset=PNsrc_spectrum.fits arfset=PN.arf \

withrmfset=yes rmfset=PN.rmf \

badpixlocation=EPICclean_pn.fits detmaptype=psf
```

The last step is to rebin the spectrum to make sure each energy spectral bin of the spectrum matches the distribution for your preferred goodness of fit statistics, e.g. χ^2. Obviously, rebinning is only necessary if χ^2 statistics are to be used. Here is an example where the spectrum is grouped to have minimum 20 counts per bin and not to oversample the intrinsic energy resolution by a factor of more than 3:

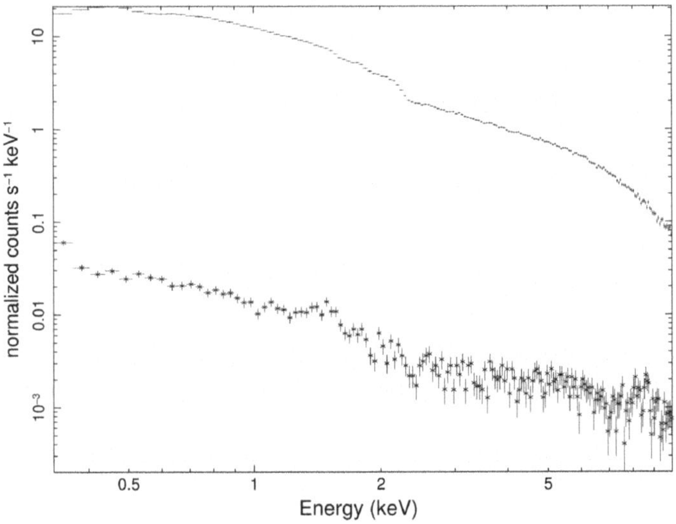

Fig. 5.9 The data (upper) and background (lower) spectra shown in XSPEC

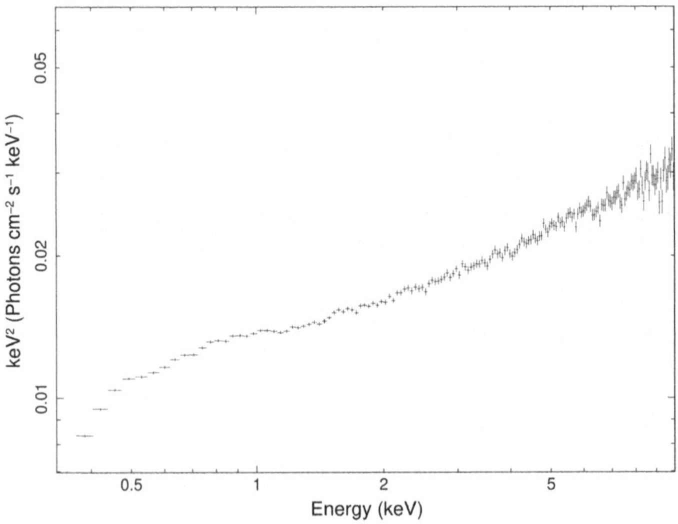

Fig. 5.10 The unfolded spectrum of 3C 273 using a power-law model with $\Gamma = 0$

```
specgroup spectrumset=PNsrc_spectrum.fits \

mincounts=20 oversample=3 rmfset=PN.rmf arfset=PN.arf \

backgndset=PNbkg_spectrum.fits groupedset=PN_spec_grp.fits
```

We refer interested readers to Sect. 6.4 for more introduction of goodness of fit statistics.

The output file 'PN_spec_grp.fits' is ready to be analysed in spectral analysis software, such as XSPEC. Figure 5.9 shows the source and background spectra. Note that the spectra in Fig. 5.9 are in units of normalized counts. In order to have a *quick* look at the intrinsic spectral shape, one can "unfold" the spectrum by using a power-law model with photon index $\Gamma = 0$, which is a constant model, to remove the impact of the instrumental effective area. As an example, the unfolded νF_{ν} spectrum of 3C 273 is shown in Fig. 5.10, and shows a power law-shaped continuum. Interested readers can find more instruction for spectral analysis using XSPEC in Sect. 7.3.

Chapter 6
Basics of Astrostatistics

Vinay L. Kashyap

6.1 Introduction

The term Statistics is used to describe both compilations and mathematical descriptions of data. The former may include summaries of the data, like the median weight of newborn babies, variances in stock prices, amortization tables for mortgages, etc. The latter describes connections across the data: either correlations (is cigarette smoking injurious to health?) or dependencies (how injurious is it?); provide a framework for making decisions (should you stop smoking?); and is necessary to understand the processes that generate the data and the certainty with which we can draw actionable conclusions. The former is related to the latter in the same way that astrometry is relevant to astrophysics: it is necessary, but not sufficient. In this chapter, we will focus on the latter aspect, and assume that the reader is familiar with the usual ways of summarizing their data.

In astronomy, where most events are one-off, it is critical to understand how different a measurement could have been, to understand the spread, and prevent us from over-interpreting an observation and fooling ourselves. Astrostatistics, in particular, is the field dedicated to studying the mathematical underpinnings of astronomical data, to obtain *estimates* and *uncertainties* on quantities useful for astrophysical inference, while taking into account instrument sensitivities, random fluctuations, the circumstances of the observations, and avoid the pitfalls of making incorrect inferences. Importantly, it is used as a guide in asking the right question of the data and to obtain the best possible answer.

For instance, consider two observations of two sources, one which yields 100 counts in 10 ks, and one which yields 10 counts in 1 ks. The estimated count rates are 0.10 ± 0.00316 and 0.10 ± 0.01. If we were to ask which is the brighter source (ignoring complications due to background), the answer will depend on how well we understand the data: if we assume the errors are well described by a Gaussian and are symmetric, we would claim that no difference will be discerned with repeated observations; if instead we account for the skew in the Poisson likelihood, then the

V. L. Kashyap (✉)
Center for Astrophysics, Harvard & Smithsonian, 60 Garden Street,
Cambridge, MA 02138, USA
e-mail: vkashyap@cfa.harvard.edu

C. Bambi (ed.), *Tutorial Guide to X-ray and Gamma-ray Astronomy*,
https://doi.org/10.1007/978-981-15-6337-9_6

source with the shorter observation will be more likely to be brighter in more of the repeated observations.

Knowing how the uncertainties are distributed gives us a powerful lever to obtain better estimates of measurables that more precisely reflect the physics that generates them. The important part to note here is that astrostatistics is not just about computing means and variances: the mathematics of uncertainty characterization allows us to detect sources, develop and fit models, compare competing models, group and classify objects, etc.

The purpose of this chapter is to provide a framework for astronomers to understand statistical issues that are relevant to astronomical analysis and place them in context. In particular, we will describe the basic statistical tool-set needed for contemporary analysis of high-energy astronomical data. Thus, we will first discuss the Poisson distribution, in the context of several others that are relevant, in Sect. 6.2. Next, in Sect. 6.3, we will provide a guideline to how error bars and uncertainties are evaluated, and how uncertainty intervals are set. We will also briefly discuss Bayesian analysis in Sect. 6.3.2 in the context of uncertainty intervals. Then, in Sect. 6.4, we will discuss the underpinnings of the fitting process, introducing the concept of likelihoods and parametric curve fitting. In Sect. 6.5 we will then discuss the basics of decision making, via hypothesis tests, p-value thresholds, goodness-of-fit tests, and model comparisons, and point out some important limitations in the process. Finally, in Sect. 6.6, we will point the reader to resources for more in depth study.

6.2 Distributions

When an observable quantity is measured, it can be considered to be *sampled* from amongst several possible values that it could take due to natural fluctuations. The underlying set from which this value is sampled is called a *distribution*. Distributions put precise probabilities on obtaining a particular value in an experiment. For example, when a fair coin is flipped, it can land on either the head or the tail with equal probability. When that coin is flipped repeatedly (say 20 times), what are the chances that it will land heads 10 times? 15 times? 20 times? The probability of these occurrences are described by the *Binomial* distribution (see below).

Note that most useful distributions, whether defined over a continuous or discrete variable, are invariably *proper*. That is, a distribution $f(.)$ over a continuous variable x or a discrete variable k is normalizable such that

$$\int_x dx\, f(x) = 1 \quad \text{or} \quad \sum_k f(k) = 1 \,.$$

In contrast, higher order moments like the mean

$$E[x] = \int_x dx\, x\, f(x) \quad \text{or} \quad E[k] = \sum_k k\, f(k)\,,$$

variance

$$V[x] = \int_x dx\, x^2\, f(x) - E[x]^2 \quad \text{or} \quad V[k] = \sum_k k^2\, f(k) - E[k]^2\,,$$

etc., are not necessarily defined.

In principle, there are an infinite number of possible distributions, limited in their usefulness only by the application for which they are most suited for. There are, however, a small number of distribution families that are often used in, or are directly applicable to, astronomical analyses, and we describe their relevance briefly below.

Uniform

The simplest of all distributions, it has a uniform probability of generating a number between two specified values. It is supported over the entire real number line $\mathbb{R}$, but without loss of generality can be defined to be unity in the range $x \in [0, 1]$. Any arbitrary continuous range can be obtained by a translation and scaling linear transformation,

$$\begin{aligned} U(x; a, b) &= \frac{1}{b-a} \quad a \le x \le b \\ &= 0 \text{ otherwise}\,, \\ &\quad \forall x, a, b \in \mathbb{R}\,, \\ E[x] &= \frac{a+b}{2}\,, \\ V[x] &= \frac{1}{12}(b-a)^2\,. \end{aligned} \tag{6.1}$$

Sampling from it forms the first step in all numerical Monte Carlo methods, and algorithms to obtain high-fidelity draws from it are widely used, in fields ranging from cryptography to ray tracing. Most pseudorandom number generators in codes used in astronomy analyses (everything from Matlab, IDL, Python, R, etc) uses the Mersenne Twister method.[1] It is not cryptographically secure, but it has a period of $2^{1937} - 1$ for 32-bit integers and is thus adequate for numerical simulation purposes.

Gaussian

Also called the **Normal** distribution, it is one of the most common distributions encountered in descriptions of data. It is defined over the full real line, with mean and variance the sole determining parameters, and all higher moments identically 0,

[1] Matsumoto 1997; http://www.math.sci.hiroshima-u.ac.jp/~m-mat/MT/emt.html.

$$N(x;\mu,\sigma^2) = \frac{1}{\sigma\sqrt{2\pi}} e^{-\frac{(x-\mu)^2}{2\sigma^2}} ,$$
$$\forall x, \mu, \sigma \in \mathbb{R} ,$$
$$E[x] = \mu ,$$
$$V[x] = \sigma^2 . \quad (6.2)$$

The primary reason for its ubiquity is that it is the natural distribution that results for summaries of data. This is a consequence of the Central Limit Theorem, which holds for large sample sizes for well-behaved samples, i.e., for samples which are drawn from distributions with well-defined means and variances. The Gaussian distribution has several mathematical properties that make it useful in astronomical analyses: (1) it is symmetric and defined over the full real line; (2) its log-form is parabolic, which means first-order Taylor expansions are interpretable as being distributed as Gaussians; (3) its Fourier transform is also Gaussian in form in frequency space; (4) it is easily generalized to multiple dimensions; (5) it is the Mother wavelet for the Mexican Hat wavelet; and (6) the product or convolution of two Gaussians is also a Gaussian, all of which makes it a convenient way to characterize error bars. Figure 6.3 (left) shows some examples of the Gaussian distribution, centered at $\mu = 0$, but for different values of σ. The area enclosed between $\pm\{1, 2, 3\}\sigma$ is $\{0.683, 0.954, 0.997\}$ respectively for a one-dimensional Gaussian. The corresponding values for the 2-D case are $\{0.393, 0.865, 0.989\}$.

Log-Normal
Astronomical data often cover a large dynamic range. For instance, stellar bolometric luminosities range from $\approx 10^{-3}$ $L_\odot$ for cool dM9 dwarfs to $>10^5$ $L_\odot$ for blue supergiants. Most distributions are not optimized to represent such broad ranges of data, and suitable distributions are best defined over the log scale. The log-Normal serves this purpose, as it is essentially the Gaussian distribution, defined over the transformed variable $\ln x$,

$$f(x;\mu,\sigma^2) = \frac{1}{\sigma\sqrt{2\pi}} \frac{1}{x} e^{-\frac{(\ln x-\mu)^2}{2\sigma^2}} ,$$
$$\forall x \in \mathbb{R}_{>0}, \forall \mu, \sigma \in \mathbb{R} ,$$
$$E[x] = e^{\mu+\frac{\sigma^2}{2}} ,$$
$$V[x] = (e^{\sigma^2} - 1)\, e^{2\mu+\sigma^2} . \quad (6.3)$$

Note that despite the similarity to the Gaussian (the $\frac{1}{x}$ factor is absorbed in the differential measure to form $d \ln x$), the parameters are not as simply defined. The mean has an additional correction term of $\sigma^2/2$, and the variance includes corrections based on the center μ. Despite these complications, the log-Normal is often used to model luminosity functions.

Binomial
This is often considered the baseline distribution, since it can be built up from first principles as a combinatorics problem of selecting k objects out of a sample of N

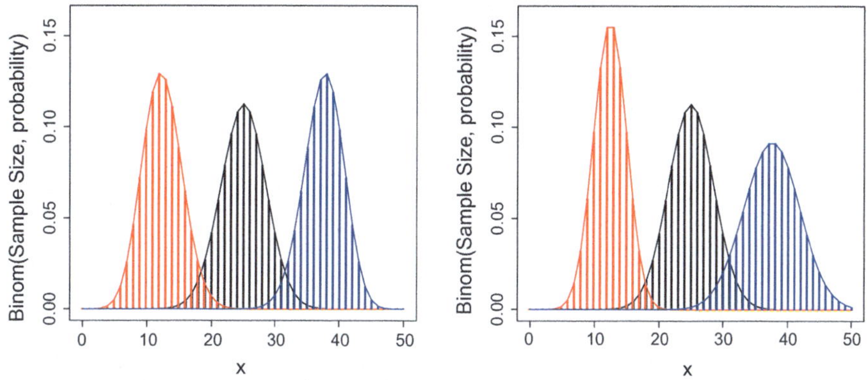

Fig. 6.1 *Left:* Binomial distribution for $p = 0.25, 0.5, 0.75$ and sample size $N = 50$ (red, black, blue respectively). *Right:* Binomial distribution for $p = 0.5$ and different sample sizes of $N = 25, 50, 75$ (red, black, blue respectively)

when the probability of picking it out of one is p,

$$\begin{aligned} Binom(k; N, p) &= {}^{N}C_{k}\; p^{k}(1-p)^{N-k}\,, \\ &\quad \forall k \in \mathbb{N}_0, \forall p \in [0,1]\,, \\ E[k] &= Np\,, \\ V[k] &= Np(1-p)\,. \end{aligned} \tag{6.4}$$

Several other distributions can be constructed as asymptotic extensions (e.g., the form of the Poisson is derived in the limit $N \rightarrow \infty$ keeping the count rate fixed). Unlike the Gaussian, it is defined only for whole numbers $k, N \in \mathbb{N}_0$, while $p \in [0, 1]$. It is useful to describe problems where the distribution of the selection of one of a binary outcome (heads or tails, 0 or 1) is being described. For instance, it can be used to set the error bars on enclosed energy ($p = EE$) radii for point sources: for a source with N events, the fractional error on the enclosed energy is $\sqrt{\frac{EE\,(1-EE)}{N}}$, which then can be projected against the cumulative distribution function $F(< r_{EE})$ to obtain the corresponding error on the radius r_{EE} (note that this error is approximate, as it relies on symmetry, thus making it invalid for $EE \approx 0, 1$ and interpolation in $F(< r_{EE})$, thus requiring large N). There are also versions where more than one type of object can be selected, called the Multinomial distribution. Figure 6.1 shows some examples of the Binomial distribution, demonstrating how the choices of p and N affect its shape.

Poisson

This describes the probability of observing k *counts* when a value λ is expected,[2]

$$\begin{aligned} Pois(k;\lambda) &= \frac{\lambda^k}{\Gamma(k+1)} \cdot e^{-\lambda}\,, \\ &\quad \forall k \in \mathbb{N}_0, \forall \lambda \in \mathbb{R}_{\geq 0}\,, \\ E[k] &= \lambda\,, \\ V[k] &= \lambda\,. \end{aligned} \tag{6.5}$$

As such, it is the fundamental underlying distribution that is used to describe all of high-energy photon counts data. As in the case of the Binomial distribution, the Poisson is also defined over the whole number line, $k \in \mathbb{N}_0$, and the governing parameter $\lambda \geq 0$. Unlike the Gaussian, the range of the Poisson is bounded, and it is thus strongly skewed as $\lambda \to 0$. This skew has important ramifications for astronomical analyses: best-fit model parameter estimates will be biased, and computed error bars will be incorrect if the wrong distribution is assumed. Note that a feature of the Poisson distribution is that the variance is equal to the estimate; this is the origin of the $\sqrt{N}$ error typically used in counting statistics. An illustration of what the Poisson distribution looks like for different values of λ is shown in Fig. 6.2, which bins the data from an X-ray source known to be unvarying at different time bins; smaller bins leads to a highly skewed distribution of counts, while large bins lead to a close approximation to a Gaussian.

Gamma

The continuous variable form of the Poisson is the Gamma distribution. It has the same mathematical form as the Poisson (the product of a power and an exponential), but is defined over $x \in \mathbb{R}_{\geq 0}$, and is governed by two parameters (α, β) that control its location and shape.

$$\begin{aligned} \gamma(x;\alpha,\beta) &= \frac{\beta^\alpha}{\Gamma(\alpha)} \cdot x^{\alpha-1}\, e^{-x\beta}\,, \\ &\quad \forall x, \alpha, \beta \in \mathbb{R}_{\geq 0}\,, \\ E[x] &= \frac{\alpha}{\beta}\,, \\ V[x] &= \frac{\alpha}{\beta^2}\,. \end{aligned} \tag{6.6}$$

It is often used as a so-called conjugate prior in Bayesian analyses that involve the Poisson distribution. It is a highly flexible functional form, able to mimic a large

---

[2] Statisticians use Greek letters for variables that describe model parameters and Roman letters for quantities that describe the data. In particular, they use λ as the symbol to represent brightness or strength of a source. This is sometimes also called intensity, but always has units [count].

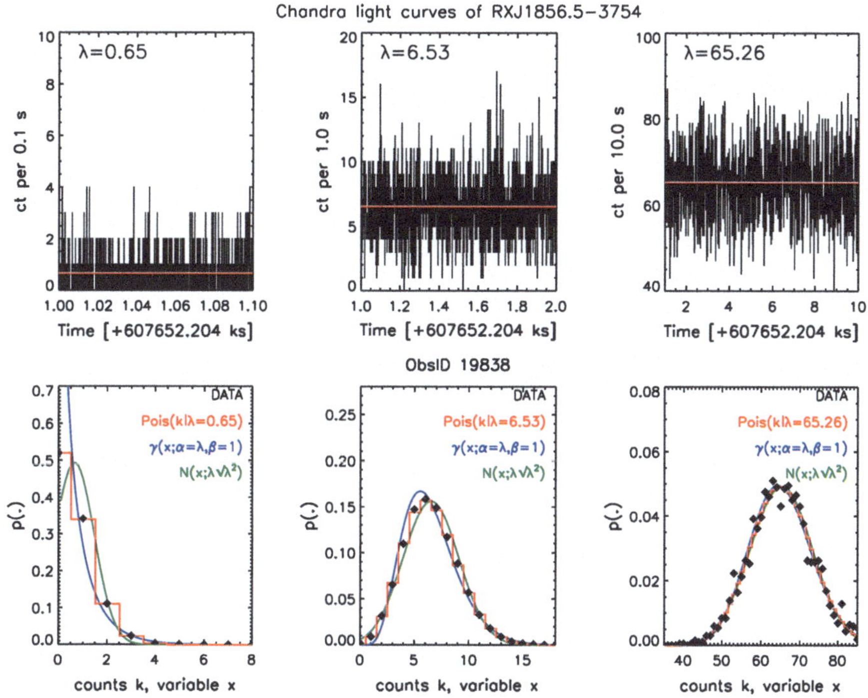

Fig. 6.2 Illustrating the Poisson and Gamma distributions by comparison to a steady X-ray source. The top panels show the light curves of counts in each bin for the isolated neutron star RXJ 1856.5-3754 from a 20 ks *Chandra* observation, made with different time bins: 0.1 s (*left*), 1 s (*middle*), and 10 s (*right*). The time range is set such that each panel has roughly similar number of bins shown. The source is known to be steady over timescales of several years, and is thus a good example of a constant source. The average value of the counts per time bin, computed over the whole observation, is marked with a horizontal red line. In the bottom panel, the distribution of the counts in the corresponding light curves is shown as the green diamonds. For comparison, a Poisson distribution (Eq. 6.5) constructed for the same model mean is shown as a red histogram, an equivalent Gamma distribution (Eq. 6.6) is shown as the blue curve, and a corresponding Gaussian (Eq. 6.2 is shown as the green curve. Note that the abscissa in the lower panels are used in two ways, both for integer counts k (for the stepped histograms), as well as for a continuous variable x (for the continuous curves)

variety of unimodal distributions encountered in astronomical data.[3] It also surfaces in several instances as special cases (e.g., see Chi-squared below). Notice that as with the Gaussian, the parameters are determinable from the mean and variance of an observed sample that obeys the γ-distribution, as $\alpha = \frac{E[x]^2}{V[x]}$ and $\beta = \frac{E[x]}{V[x]}$. The Gamma distribution is also illustrated in Fig. 6.2 as the blue curve overlaid on the red histogram representing the equivalent Poisson distribution.

[3]In this, it is similar to the Weibull distribution, which has an $\frac{1}{\alpha}\, x^{\beta}$ form in the exponential instead of $x\beta$.

Chi-Squared

This is a special case of the γ distribution, with $\alpha = \nu/2$ and $\beta = 1/2$, where ν are the degrees of freedom,

$$\begin{aligned}
\gamma\left(\chi^2; \frac{\nu}{2}, \frac{1}{2}\right) &= \frac{1}{2^{\nu/2}\Gamma(k/2)}(\chi^2)^{\nu/2-1}e^{-\chi^2/2}\,, \\
&\quad \forall\chi^2 \in \mathbb{R}_{\geq 0}, \forall\nu \in \mathbb{Z}^+\,, \\
E[\chi^2] &= \nu\,, \\
V[\chi^2] &= 2\nu\,,
\end{aligned} \tag{6.7}$$

where $\chi^2 \geq 0$ and $\nu \in \mathbb{Z}$. It is the solution to the question that asks, if ν independent Gaussians are combined together, what is the probability that the sum of their squared deviations, weighted by the reciprocal of their variances, add up to the given χ^2. This phrasing anticipates model fitting described below in Sect. 6.4, with χ^2 and $(-2 \times \text{exponent})$ of the Gaussian playing the role of the weighted squared deviations.

Student's t

This is a versatile distribution that is encountered in several vastly different situations. It is also called the Cauchy, or the Lorentzian, or the Beta-profile distribution, and is characterized by tails that cover larger areas than the Gaussian (Fig. 6.3).

$$\begin{aligned}
t_\nu(x) &= \frac{\Gamma\left(\frac{\nu+1}{2}\right)}{\sqrt{\nu\pi}\,\Gamma(\nu/2)}\left(1+\frac{x^2}{\nu}\right)^{-\frac{\nu+1}{2}}\,, \\
&\quad \forall x \in \mathbb{R}, \forall\nu \in \mathbb{Z}^+\,, \\
E[x] &= 0 \quad \nu > 1\,, \\
&= \text{undefined} \quad \nu \leq 1\,, \\
V[x] &= \frac{\nu}{\nu-2} \quad \text{for } \nu > 2\,, \\
&= \text{undefined} \quad \nu \leq 2\,.
\end{aligned} \tag{6.8}$$

Formally, it is derived as the ratio of a Normal and a $\sqrt{\chi^2}$ distribution. In a statistical context, it describes the uncertainty with which the mean of a sample can be determined when the variance is also determined from the same sample; in that situation, ν represents the size of the sample.

Pareto

One of the most common distributions encountered in astrophysics is the power-law, which arises whenever a physical process operates over a large range of scales. In such situations, when the physical process is essentially scale-free, or there is no preferred scale, the energy in the system cascades such that the distribution looks self-similar everywhere, and can be described as a function with a power-law index α. The statistical analogue of the power-law is the Pareto distribution, which is defined such that it has a well-defined lower bound to prevent it from becoming improper, i.e., so that the integral stays finite.

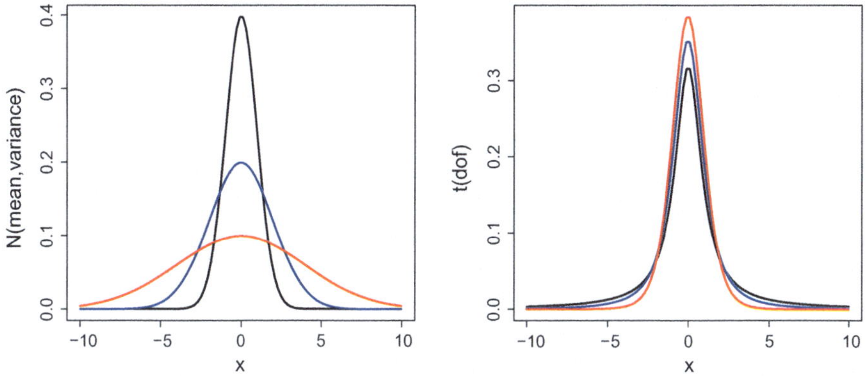

Fig. 6.3 Comparing the Gaussian and t_ν distributions. *Left:* Normal distribution for $\mu = 0, \sigma = 1, 2, 4$ (black, blue, red). *Right:* t distribution for $\nu = 1, 2, 7$ (black, blue, red)

$$\begin{aligned} P(x; \alpha, x_0) &= \alpha x_0^{\alpha}\, x^{-(\alpha+1)} \,, \\ &\quad \forall x_0, \alpha \in \mathbb{R}_{>0}, \forall x > x_0 \,, \\ E[x] &= \frac{\alpha x_0}{\alpha - 1} \quad \alpha > 1 \,, \\ &= \text{undefined} \quad \alpha \leq 1 \,, \\ V[x] &= \frac{x_0^2 \alpha}{(\alpha - 1)^2 (\alpha - 2)} \quad \alpha > 2 \,, \\ &= \text{undefined} \quad \alpha \leq 2 \,. \end{aligned} \tag{6.9}$$

6.2.1 Distributions Versus Functions

It is important to understand the difference between functions and distributions. Even though distributions are defined using functional forms, they each represent fundamentally different quantities. A function is a deterministic locus of points that satisfy the mathematical form. In contrast, a distribution represents a sampling of a variable conditioned on the defined parameters. Samples from distributions are indicated with a special symbol "$\sim$", as e.g.,

$$X \sim f(x; \theta) \,.$$

6.3 Error Bars

6.3.1 Propagation of Errors

One of the first things a researcher has to do with a measurement is to scale, shift, and transform the signal that comes out of the detector to a form that is physically relevant. A simple example is to count the photons registered in a certain time and compute the count rate as the number of photons registered per second (see, e.g., the light curves shown in the top row of Fig. 6.2). We expect the counts to follow a Poisson distribution (see above). How, then, can we use that information to place an error bar on the count rate?

If the quantity of interest is distributed as a Gaussian, uncertainty intervals can be propagated through any number of transformations $g = f(x_1, x_2, .., x_K)$ using the chain rule,

$$\sigma_g^2 = \sum_{i=1}^{K} \left(\frac{\partial f}{\partial x_i}\right)^2 \sigma_i^2 . \tag{6.10}$$

This expression is a consequence of doing a Taylor expansion of the Gaussian around the mean and computing the variance of the difference, and ignoring terms $o(x^2)$ and higher. Some common transformations are shown below:

$$\begin{aligned}
g &= \text{constant} \cdot x &&\Rightarrow \sigma_g = \text{constant} \cdot \sigma_x && \text{(errors scale)}\\
g &= \ln x &&\Rightarrow \sigma_g = \tfrac{\sigma_x}{x} && \text{(fractional error)}\\
g &= \frac{1}{x} &&\Rightarrow \tfrac{\sigma_g}{g} = \tfrac{\sigma_x}{x} && \text{(fractional error preserved)}\\
g &= x + y &&\Rightarrow \sigma_g = \sqrt{\sigma_x^2 + \sigma_y^2} && \text{(add variances)}\\
g &= \frac{x}{y} &&\Rightarrow \sqrt{\left(\tfrac{\sigma_x}{x}\right)^2 + \left(\tfrac{\sigma_y}{y}\right)^2} && \text{(add fractional variances)}
\end{aligned} \tag{6.11}$$

That is, a multiplication by a constant scales the error on the resultant value by the same factor; a log-transformation assigns the fractional error to the new variable; a reciprocal relation preserves the fractional error across the transformation; for an additive combination of two independent variables the variances are added together; and for the ratio of independent variables the fractional variances are added together. In the count rate example we were considering above, a Poisson count N also has variance N. If N is sufficiently large (say > 30), the Poisson distribution becomes similar to the Gaussian (as in the third column of Fig. 6.2), and we can then set $\sigma = \sqrt{N}$. The exposure time τ is usually measured with high precision, and if we ignore the measurement uncertainty on it, it can be effectively treated as a constant. Then, because rate $= \frac{\mathrm{N}}{\tau}$, the error on the rate, $\sigma_{\text{rate}} = \frac{\sqrt{N}}{\tau}$. If this rate is then converted to a flux at the telescope by dividing with a (known) effective area factor A_{eff} and then to luminosity by multiplying by a distance modulus $4\pi D^2$, where D is the distance to the source, and the log of the luminosity is computed as $\log L = \log_{10}\left(\frac{\text{rate } 4\ \pi\ D^2}{\tau\ A_{\text{eff}}}\right)$, then the error on $\log L$ is, by the chain rule, $\sigma_{\log L} = \frac{1}{\ln(10)}\frac{\sigma_L}{L} = \frac{0.4343}{\sqrt{N}}$.

This method of propagating errors assumes that the Gaussian distribution is appropriate at every stage, and that the transformations in question are differentiable, and that the variance is well-defined in all cases. These are highly restrictive assumptions. Note that they formally break down for the ratios in the examples shown in Eqs. 6.11 when the denominators approach zero. The estimates and uncertainties for such ratios are unstable, and subject to large fluctuations; it is for this reason that the fractional hardness ratio $\left(\frac{Hard\ counts - Soft\ counts}{Hard\ counts + Soft\ counts}\right)$ and the color ($\log(Soft\ counts) - \log(Hard\ counts)$) are used extensively to track spectral changes in preference to the simple ratio $\frac{Soft\ counts}{Hard\ counts}$.

6.3.2 *Digression: Frequentist Versus Bayesian Analysis*

There are two major approaches in Statistics theory. The Frequentist viewpoint treats the observed data as just one realization amongst an ensemble that is obtainable, with the ultimate physical quantity that the data are describing to be immutable, i.e., that there is one truth. The Bayesian viewpoint is that the data at hand are what are available, and cannot be changed, and they predict a variety of plausible values for the parameter, which are described via a distribution. Both approaches give the same answers for the same setups, but expose different assumptions and work through different pathways to get to the result. For astronomers, who tend to obtain one dataset at a time, Bayesian analysis may seem a more natural approach. Nevertheless, it is best to use whichever technique is best suited to the particular question being asked of the data.

Bayesian analysis relies on probability calculus, and on conditional probabilities in particular. The main axioms of probability theory are that

$$\begin{aligned}
\text{probability}(A\ or\ B) &= \text{probability}(A)\ and\ \text{probability}(B)\ less\ \text{probability}(A\ and\ B) \\
p(A+B) &= p(A) + p(B) - p(AB) \qquad (6.12)\\
\text{probability}(A\ and\ B) &= \text{probability}(A\ given\ B)\ times\ \text{probability}(B) \\
p(AB) &= p(A|B) \cdot p(B) \\
&\equiv p(B|A) \cdot p(A) \qquad (6.13)
\end{aligned}$$

where A, B, etc. are statements that can take truth values with probability $p()$. There is an equivalent axiom that can be derived from the above, which states that the sum of the probability of A and its negation $\overline{A}$ is 1, i.e.,

$$p(A) + p(\overline{A}) = 1\,.$$

The second axiom of Eq. 6.13 describes conditional probability using the notation $A|B$. It leads directly to Bayes' Theorem,

$$p(A|B) = \frac{p(B|A) \cdot p(A)}{p(B)} . \tag{6.14}$$

Despite the almost trivial simplicity of the Theorem, it is a powerful tool that underlies probabilistic analysis. Its power comes from how the statements A, B, etc. are interpreted. Consider the case where A represents a model with parameter θ, and B represents the data that the model seeks to describe. Then, Bayes' Theorem (Eq. 6.14) becomes

$$p(\theta|D) = \frac{p(D|\theta) \cdot p(\theta)}{p(D)} . \tag{6.15}$$

The term $p(\theta)$ is called the *prior* probability of the model parameter, i.e., the range of values it can be expected to have before the data are acquired. The term $p(D|\theta)$ is called the *Likelihood* (it is sometimes written as $L(\theta|D)$), and denotes the probability that the data can be described with the particular model parameter. The term $p(D)$ in the denominator is called the *evidence* of the data, and acts as the normalization factor for the probability, ensuring that the sum of the probabilities adds up to 1. The term $p(\theta|D)$ is called the *posterior* probability of the model parameter θ, which is an update to the prior probability after an observation is made.

As an example, let us consider a case where 10 counts are observed in a pixel in an image. We may seek to evaluate the probabilities of the brightness values for the source that produced this many counts. A priori, we do not know what θ can be; but it is reasonable, for astronomical sources, that the brightness can cover a large dynamic range, going several 100s of counts to small fractions of a count, so we adopt $p(\theta) = \frac{1}{\theta}$ as the prior. The likelihood is described by a Poisson distribution, $p(D = 10|\theta) = \frac{\theta^{10}\, e^{-\theta}}{\Gamma(11)}$. Specifically now, let us consider the case $\theta = 5$. For this value, $p(\theta = 5) = 0.2$ and $p(D = 10|\theta = 5) = 0.018$. We can evaluate the numerator of Eq. 6.15 similarly for several values of θ, and evaluate the integral $\int_0^{\infty} d\theta p(D = 10|\theta) \cdot p(\theta) = 0.1$, which defines the value of the denominator $p(D)$ since the probability that θ has some value between 0 and ∞ is 1. Thus, the posterior probability density at $\theta = 5$ can be evaluated as $p(\theta = 5|D = 10) = 0.0018$.

In a similar manner as above, complex problems can be handled by repeated applications of Bayes' Theorem, reducing a problem with several interconnected variables into separated factors that can each be evaluated.

6.3.3 Uncertainty Intervals

An uncertainty interval can be thought of as a measure of the width of the distribution. Frequentist and Bayesian analyses approach this differently. In the Frequentist paradigm, a *confidence interval* is defined by a process: a calculated interval can be expected to contain the true value a certain fraction of the times that an observation is performed. In the Bayesian paradigm, a *credible region* is the range in the parameter values which encloses a certain fraction of the area under the probability distribution. We will refer to either of these in general as *uncertainty intervals*, and use the paradigm specific terms when referring to a particular case.

A Credible Interval is defined as the range $[a, b]$ that corresponds to a specified area $\int_{x=a}^{x=b} dx f(x) = c$ under the curve. It is possible to find multiple values of a and b that lead to the same value of the integral. For instance, while 68% of the area under a Gaussian $N(0, \sigma^2)$ is enclosed within the interval $[-1\sigma, +1\sigma]$, the same fraction is also enclosed between $[-\infty, +0.17\sigma]$, between $[-0.17\sigma, +\infty]$, etc. In fact, there are an infinite number of intervals which enclose 68% of the area of a Gaussian. Thus, there is no unique instance of "a" credible interval, and some other factor of importance must be stated. The degeneracy is typically broken by stating either equal-tail (EQT) intervals, which sets $[a, b]$ such that equal areas are left out at either end of the support of the distribution, i.e., $\int_{x<a}^{x-a} dx f(x) = \int_{x=b}^{x>b} dx f x = \frac{1-c}{2}$, or using the highest posterior density (HPD) intervals, which enclose the all the highest possible values of the probability density. An EQT is invariant even under non-linear coordinate transformations, and an HPD, which always includes the mode of the distribution, guarantees the shortest uncertainty interval by design (for unimodal distributions; HPDs for multimodal distributions can be split into several segments).

Different techniques are used to estimate the uncertainties in the case of non-linear weighted least-squares fitting. These methods rely on the magnitude of variation of a particular statistic (often related to the likelihood), and are described below (see Sect. 6.4.1.3).

6.4 Fitting

Typically, one has an astrophysical model that is a function of several variables (aka parameters). The model is used to predict the incident flux, and is then further modified by instrumental effects such as the effective area and spectral response to be put into the same form as the data. The task is to determine which values of the parameters is the best description of the data. This is usually done by finding the extremum of a suitable metric. There are several choices we can make for the metric. The absolute deviation of the model from the data is a popular choice, and is also called the L_1-norm. The sum of the squared of the deviations is also called the L_2-norm. The common example of the simple linear regression fit uses the L_2 norm: if there are pairs of data (x_i, y_i), the sum squared deviation can be written as

$$L_2 = \sum_i (y_i - m \cdot x_i - c)^2 ,$$

where m is the slope of the fitted straight line and c is its y-intercept. Setting $\partial L_2/\partial m = \partial L_2/\partial c = 0$ results in the well-known solutions for the slope and intercept, $m = \frac{\mathrm{cov}(x,y)}{V[x]}$, and $c = E[y] - m \cdot E[x]$.

6.4.1 χ^2

While the above is an adequate solution for when the errors on x_i are all identical (this is called homoskedasticity), in typical astronomical data the errors vary considerably amongst the x_i (this is called heteroskedasticity). In such a case, it is advantageous to use the sum of squared deviations inversely weighted by the variances, which leads us to the familiar χ^2 statistic, which can be written in a generalized form as

$$\chi^2 = \sum_i \frac{(\mathrm{Data}_i - \mathrm{Model}_i)^2}{\mathrm{Error}_i^2} . \tag{6.16}$$

Notice here that the χ^2 value is simply the exponent of the Gaussian density; it can in fact be written without loss of generality as

$$\chi^2 = -2 \sum_i \ln N(x_i; \mu_i, \sigma_i^2) .$$

Furthermore, the Gaussian density is a measure of the likelihood of the data for the specified model, and thus, minimizing the χ^2 is equivalent to maximizing the likelihood. This ensures that the optimal solution is found in the Gaussian regime, since by definition the best-fit has the highest likelihood of explaining the data. Thus, the χ^2 is the appropriate statistic to use to obtain fits to data whose errors are Normally distributed. This method of fitting has several useful properties: (1) the fitting is done as L_2 minimization, for which several well-established numerical algorithms exist; (2) the quality of the fit can be estimated (see Sect. 6.4.1.2 below); and (3) uncertainty ranges on the best-fit parameter estimates can be computed by measuring the changes in χ^2 over the parameter space (see Sect. 6.4.1.3).

6.4.1.1 Digression: Degrees of Freedom

Degrees of freedom is a loosely defined concept that gives a sense of the number of independent quantities or variables needed to describe the system under consideration. Its precise value is context dependent, and is influenced by the question being addressed. As an example, when a model with m parameters is fit to a dataset with N bins (which could be the number of spectral bins, or time bins in a light curve, or pixels in an image), the relevant degrees of freedom $\nu = N - m$. A fit is only possible if $N > m$, and the problem is overdetermined and the solution will be overfit if the reverse is true. In contrast, when the uncertainty interval of a single parameter is being estimated by computing the change in χ^2 from the best-fit at fixed parameter values, the degrees of freedom is the number of parameters being held fixed. When the standard deviation of a sample of size N is estimated, because it also requires that the mean be computed from the sample beforehand, the degrees of freedom is reduced by 1 to $\nu = N - 1$, and the sum of the squared deviations from the mean is divided by this factor rather than by N.

6.4.1.2 Goodness of Fit

The χ^2 statistic has an additional useful property that the statistic obtained from an ensemble of good fits is distributed as the χ^2 distribution (Eq. 6.7) with $\nu = N - m$ degrees of freedom, where data size is N, and the model has m parameters that are allowed to be free. Thus, when the observed χ^2 statistic lies in the range $\nu \pm \sqrt{2\nu}$, that is an indication that the derived best-fit belongs to the set of good fits to the data. Sometimes ν is divided into χ^2 to form the so-called *reduced* χ^2, and extreme values of this, $\frac{\chi^2}{\nu} \equiv \chi^2_\nu \gg 1 + \sqrt{\frac{2}{\nu}}$, are seen as an indication that the fit is not good. However, there are several reasons that χ^2_ν can be large, and being a bad fit is only one of them; other possibilities include: defining χ^2 with a different denominator; underestimation of errors; not accounting for systematic errors; and applying it to data not distributed as a Gaussian.

6.4.1.3 Error Bars on Parameters

The similarity of χ^2 minimization to the Gaussian likelihood provides a means by which uncertainties intervals on the best-fit parameter values can be determined. The idea is that any change away from the best-fit values will lead to an increase in the χ^2 and a corresponding decrease in the likelihood. This change can be mapped to the χ^2_ν distribution, and thresholds can be identified where successively larger areas under the distribution are included. Thus, to compute the 1σ confidence interval on (say) the jth parameter, we will have to locate the value of the parameter θ_j where the integral of the distribution from $\chi^2_{\rm min}|\theta_j$ to $+\infty$ equals 0.16 (that is, the central portion of the distribution includes 68% of the area, leaving 32% outside, which are split into equal 16% segments on either side of the distribution). Since there is one variable that is being varied (θ_j), this corresponds to computing the appropriate quantiles of $\chi^2_{\nu=1}$. The 1σ bound is reached when $\Delta\chi^2 = \chi^2|\theta_j - \chi^2_{\rm best-fit} = +1$. Similarly, the 90% bound is reached when $\Delta\chi^2 = +2.7$ (a common choice for users of `XSPEC`) and a 3σ equivalent bound when $\Delta\chi^2 = +9$. When the error bounds on several parameters n are being considered simultaneously, the threshold values should be obtained using the integrated percentile values of the χ^2_n distribution with n degrees of freedom. Specifically, for the so-called banana plots in 2-D, $\Delta\chi^2 = +4.6$ for a 90% bound.

6.4.1.4 χ^2 Variants

There are several variants of the χ^2 statistic, which mainly differ in how the variance in the denominator is defined. These include using the Gehrels estimate [9] of the Poisson 84th percentile as the Gaussian 1σ equivalent, or the Primini method of using the model estimate from the previous iteration [14]. It is important to note that while such variants are often adequate as a tool to obtain the best fit solution, they cannot be used to compute error bars as described above if the computed statistic is not a χ^2 distribution (Eq. 6.16).

6.4.2 cstat

While the χ^2 is an appropriate statistic to minimize for Gaussian distributed data, using it in other situations will lead to biased estimation of parameters. This is the case in the vast majority of high-energy datasets, which are based on photon counts, and are governed by the Poisson distribution (see Eq. 6.5). Note that even though a Gaussian is an adequate approximation to a Poisson for large N, parameter estimates will remain biased. It is therefore necessary to use the Poisson likelihood in places where counts data are involved.

Because there is a significant amount of software and theoretical methods available that is built upon χ^2 based fitting, it is advantageous to cast the Poisson likelihood in the same form. This is the origin of the `cstat` statistic,

$$
\begin{aligned}
\texttt{cstat} &= -2\sum_i \ln Pois(D_i, M_i) \\
&= -2\sum_i [D_i \ln M_i - M_i - \ln \Gamma(D_i + 1)] \\
&\rightarrow -2\sum_i [D_i \ln M_i - M_i - (D_i \ln D_i - D_i)] \\
&\Longrightarrow 2\sum_i [(M_i - D_i) + D_i \cdot (\ln D_i - \ln M_i)] \ , \qquad (6.17)
\end{aligned}
$$

where D_i are the counts in bin i and M_i are the predicted model intensities in units of [counts], and the $\Gamma(.)$ factor is expanded using Stirling's approximation.

In the asymptotic limit of large datasets, the `cstat` is distributed as the χ^2 distribution, so all the techniques used to determine error bars and goodness of fit can be applied. Even though such cases are rare, this is still a useful result, because using `cstat` eliminates the bias in the parameters that results when the wrong distribution is used. In the low counts case, which is more common, a full description of the expected distribution of `cstat` is not yet available. However, recent work by J. Kaastra [12] has brought forth a useful approximation where the expected value C_μ and the variance C_σ^2 of the `cstat` for the given model intensities $\{M_i\}$ can be estimated. Thus, if we know what value of `cstat` is expected for the best-fit model intensities $\{\hat{M}_i\}$, we can compare it to the observed value $\texttt{cstat}_{\text{best-fit}}$, and evaluate how many standard deviations away it is from the expected value. If, say, $\texttt{cstat}_{\text{best-fit}} > C_\mu + 3\,C_\sigma$, or $\texttt{cstat}_{\text{best-fit}} < C_\mu - 3\,C_\sigma$, then it can be inferred that the model fit is improbable at the <0.3% level.

6.5 Hypothesis Tests and Model Comparison

Obtaining best-fit parameters and error bars on them is often insufficient. In many instances, astronomers must decide amongst several competing theories. A decision must be made to *choose* one hypothesis or model over another, identify a model that works well, or eliminate a model that does not. This is the realm of hypothesis testing and model comparison.

In comparison to estimation problems, testing is fraught with misinterpretation. It is necessary to understand both what a comparison means and what it does not. The underlying cause of much of the confusion is the so-called p-value. We will discuss the p-value and its use in Null-Hypothesis significance tests in Sect. 6.5.1, and then discuss some errors that emerge as a consequence of their use. We will suggest some schemes to work around these issues.

6.5.1 *p-Values and Hypothesis Testing*

Two crucial concepts underlie the mechanism of hypothesis testing:

p-Value
In any distribution, the area under it over a range starting from a particular value, and extending to the end of the domain over which the distribution is defined, is the p-value. As an example, the area under a Gaussian $N(0, 1^2)$ ranging from the $+1\sigma$ point to $+\infty$ represents $p = 0.16$. Similarly, the area from $+3\sigma$ to $+\infty$ represents $p = 0.003$. Often, the problem is reversed such that the point that corresponds to a specified p-value is of interest, and is used as a threshold for detection. For a given distribution $f(\mathbb{S})$,

$$p(\mathbb{S}_c) = \int_{\mathbb{S}>\mathbb{S}} d\mathbb{S}\; f(\mathbb{S})\,, \tag{6.18}$$

with a summation replacing the integral for discrete distributions. The p-value represents the probability that a chance fluctuation results in observed values of $\mathbb{S} > \mathbb{S}_c$.

Null Distribution
In order to be able to say that a given distribution is different, or preferred, we must first specify a distribution that it should be different from. This is a default distribution, which we would expect to see if there were no signal in the data. It is

also sometimes called the "Null Model", or the "Null Hypothesis", and is denoted with the symbol H_0, in contrast to the alternate, interesting hypothesis or model that is tested, H_1. For example, if we were interested in testing whether a coin was biased, the null distribution would be the Binomial with the probability of heads and tails being equal, $B(k; N, 0.5)$. In the `cstat` goodness-of-fit check described above (Sect. 6.4.2), the null distribution is that distribution of `cstat` statistic values one would obtain in repeated experiments if the best-fit model were indeed a good representation, i.e., $N(\texttt{cstat}; C_\mu, C_\sigma^2)$.

A typical hypothesis test is carried out by first defining a statistic, $\mathbb{S}$, that summarizes the model and the data. This could be the number of heads in repeated coin tosses, or the χ^2, or the `cstat`, etc. A null distribution $f_0(\mathbb{S})$ is then constructed, and a $p_{\text{threshold}}$ value is set, corresponding to a critical value $\mathbb{S}_c$ that will be used in decision making. Note that it is important to set the threshold *before* the analysis takes place, in order to guard against wishful thinking playing a role in the subsequent analysis. Typically, statisticians use $p = 0.05$ as a standard choice of threshold. Note that this corresponds to a chance fluctuation of 1 in 20 that the Null distribution can generate values beyond the stated threshold. Astronomers have historically tended to use stricter thresholds, typically set at 3σ, corresponding to $p = 0.003$.

Next, the same statistic of interest $\mathbb{S}'$ is computed for the alternate model, and is compared against $\mathbb{S}_c$. If $\mathbb{S}' > \mathbb{S}_c$, this is taken as evidence that H_1 is preferred over H_0 at significance $p_{\text{threshold}}$. This is often described as "rejecting the Null". If $\mathbb{S}' \leq \mathbb{S}_c$, then it is considered that there is no evidence to prefer H_1 over H_0 at significance $p_{\text{threshold}}$.

Note that the former condition does not guarantee that H_1 is true, nor does the latter condition constitute proof that H_0 is true. Null hypothesis tests can only reject the null, as in, the measured statistic $\mathbb{S}'$ is deep in the tail of the null distribution, and hence is unlikely to have originated from it. But it is not proof that the Null is "false". Nor is it the case that if the Null cannot be rejected then it is "true". The results of such tests must therefore be interpreted with care. There are difficulties that arise both when the data quality is poor as well as when it is good. We will discuss the problems that arise at weak signals in Sect. 6.5.2 below. Counter-intuitively, when the signal is strong, the Null distribution is often exposed as being inapplicable, either due to uncorrected systematic errors which become non-ignorable relative to statistical fluctuations, or due to model approximations which fail to account for real-world complexities. This will lead to almost all tests rejecting the Null. This is the reason why high-counts low-resolution spectra which are fit with models with $\chi^2/\nu \gg 1$ are often published in the literature.

6.5.2 Threshold Based Errors

As described above, statistical decisions are made by appealing to how much of the area of a distribution falls beyond a previously set threshold. While this mechanism leads to precision in how a result is described, it is important to note that the result so obtained may be inaccurate in several ways. The art of specifying thresholds is often one of trading off the different errors that arise due to the inevitable fluctuations that arise with any measurement and inference.

Type I Errors
The area of a distribution $\alpha = \int d\mathbb{S}\ p(\mathbb{S} > \mathbb{S}^*)$, representing the probability that samplings of the statistic $\mathbb{S}$ can have values larger than $\mathbb{S}^*$, is the probability of obtaining false positives from the Null distribution, and is called the Type I error. It can be thought of as an occurrence rate, signifying the fraction of times that a false detection is obtained when samples $\{\mathbb{S}\}$ drawn from $p(\mathbb{S})$ exceed $\mathbb{S}^*$. It is fundamentally equivalent to the p-value at the threshold. When the p-value is sufficiently small, it is interpreted as being so far in the tail of the Null distribution that it is unlikely to be a draw from it, and thus cause for the rejection of the Null hypothesis. As emphasized above, it behooves us to be careful about what this means exactly: it does not mean that the Null is false, only that the probability of observing such a signal is $<\alpha$, and thus cause to consider alternative explanations. It is illustrated in the upper panel of Fig. 6.4 as the shaded region to the right of the vertical line representing the threshold.

Type II Errors
In contrast to a false positive, it is possible that there truly exists a signal whose distribution has area $1 - \beta$ *below* the threshold (i.e., it has $p = \beta$). Then, with probability $1 - \beta$, the observed signal will fail to reject the Null, and the signal will be deemed to be not detected. This is called a false negative, or the Type II error. It is essentially the mirror of the Type I error, in that it represents the probability that a

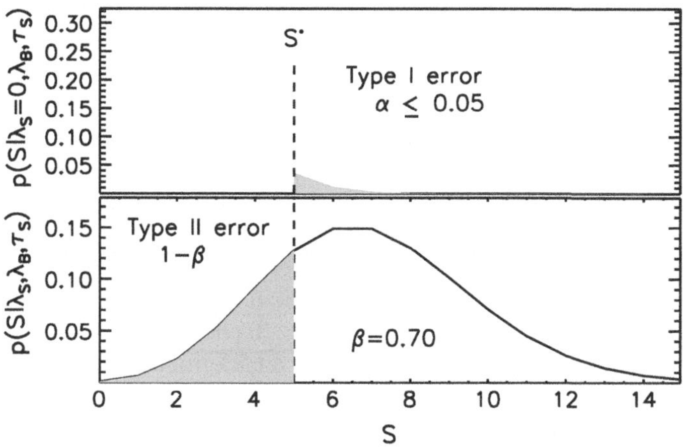

Fig. 6.4 Illustrating Type I and Type II errors [13]. The upper panel shows the example of a Null distribution arising, e.g., from a background, along with the Type I error α representing the area of the curve beyond the threshold $\mathbb{S}^*$. The lower panel shows the example of an alternate distribution, e.g., resulting from a source, along with the Type II error $1 - \beta$ representing the fraction of the distribution that would remain below the threshold

draw from an alternate distribution is classified as indistinguishable from that from the Null distribution. It is illustrated in the bottom panel of Fig. 6.4 as the shaded region to the left of the threshold. The p value in this case is also called the *statistical power* of a test that uses as its threshold $\mathbb{S}^*$. The higher the power, the better the test is at finding the signal; but note that there is a trade-off, in that the lower false negative rate would be naturally accompanied by a higher false positive rate.

Upper Limits

The combination of Type I and Type II errors also leads to a statistically rigorous definition of an upper limit. A typical problem encountered in astronomy is the case where a source is not detected, and we seek to establish the limiting brightness it could have beyond which it would have been detected. Datasets with such points are usually called 'censored', and can be dealt with in a non-parametric fashion using Survival Analysis (see works by E.Feigelson and coauthors [7, 11]). But the statistical description of an *upper limit* is more subtle. It is worth pointing out that defining an uncertainty interval on the source intensity does not provide a solution to this problem: as discussed above, an uncertainty interval precisely defines the bounds on a parameter for a specified area under the distribution, but cannot be defined in a unique manner. Consider an uncertainty interval defined with the lower bound set at the same point as the corresponding p-value. The upper bound is then (if the distribution is defined over $\mathbb{R}$) at $+\infty$, and while that would be a true statement, it is not a useful one. However, the power of the detection test provides a way to define a useful upper limit [13]. Consider the pair of distributions in Fig. 6.4, where the statistic of interest $\mathbb{S} = N$, the number of counts: the distribution in the upper panel represents the Null, or the distribution of the background, and the bottom panel represents the source. In a given observation, the background distribution is generally well known, and can be considered to be fixed. In contrast, the brightness of an undetected source is not known at all. When the true brightness is small, the probability that the source will not be detected, $1 - \beta$, will be large, and vice versa. If, then, in addition to the detection threshold α, we require that an existing source should also be detected with probability β^*, the upper limit to the source brightness is that value which achieves a power of β^*. A source with higher brightness would be detected more often than β^*, and vice versa. Note that when $\beta^* = 0.5$ the source brightness coincides with the nominal brightness value of the threshold (if the distributions are not skewed), allowing for an easy interpretation. Thus, specifying an upper limit requires two significance levels, and both the Type I and Type II errors are needed.

As an example, consider a case where $N_C = 20$ photons are counted in a region that is believed to contain a source along with a background. Suppose further that $N_B = 100$ counts are collected in a source-free area $\rho = 10$ times the area of the source region. The expected background under the source region is $\hat{\lambda_B} = N_B/\rho = 10$, and the estimated source strength $\hat{\lambda_S} = N_C - N_B/\rho = 10$, and the nominal propagated uncertainty (see Sect. 6.3) in the estimate is $\hat{\sigma_S} = \sqrt{N_C + N_B/\rho^2} \approx 4.6$. The source would be considered undetected by either the signal-to-noise criterion ($\frac{\hat{\lambda_S}}{\hat{\sigma_S}} < 3$) or considering the p-value of the Poisson likelihood for the background intensity, $p = \sum_{k=20}^{\infty} Pois(k; \lambda_B) = 0.00345$ (whereas the detection threshold set at the usual 3σ-equivalent would be $p \leq \alpha = 0.003$). Computing an error bar on the source brightness λ_S is not helpful in this case, partly because there is no guarantee that the source *exists*, and partly because there is no way to uniquely set an uncertainty interval. For instance, computing $\hat{\lambda_S} \pm N\hat{\sigma_S}$ gives [5.4, 14.6] and $[-3.7, 23.7]$ for $N = 1$ and $N = 3$ respectively. The negative lower bound is an indication that the Gaussian approximation breaks down. Computing the Bayesian posterior distribution $p(\lambda_S|N_C, N_B, \rho)$ (see, e.g., [21]), we can compute an equal-tail 68% interval of [6.7, 15.9] ([0.5, 27.7] at 99.7%), or *one-sided* 68% intervals of [0, 13.2] *or* [8.9, $+\infty$) ([0, 26.3] or [0.9, $+\infty$) at 99.7%). This is clearly untenable, and we must instead compute the upper limit of the detectable source brightness, i.e., determine that λ_S at which we can be reasonably certain that the threshold criterion is set. This latter criterion is the power of the test, β (see above). Just as one has to decide the level at which error bars are reported (1σ, 2σ, etc.) a choice must be made as to what value of β to report. For the sake of simplicity, we choose $\beta = 0.5$, as signifying the case when the source has a 50% chance of being detected at the given threshold α. This is equivalent to computing when the counts in the source region exceed the criterion for detection, i.e., when the hypothesis that the counts in the source region are entirely drawn from the background can be rejected. The threshold for this is achieved if $\geq$21 counts are observed, and the upper limit is set by computing the smallest value of λ_S where the probability of obtaining 21 or more counts exceeds $\beta = 0.5$, which can be calculated numerically as $\lambda_S < \mathbb{UL}(N_B = 100, \rho = 10, \alpha = 0.003, \beta = 0.5) = 11.6$. If the source strength were greater than that, the source would be detected more than half the time it is observed. Notice that the number of counts observed in the source region, N_C, is irrelevant to this calculation because the detection threshold is set based only on the background distribution.

False Discovery Rate

A relatively recent innovation in statistical methods is the False Discovery Rate (FDR), which combines aspects of both Type I and Type II errors. It represents the fraction of those tests where the Null is rejected where it is falsely rejected. This is useful to devise tests where the sample from the alternate is small compared to the sample from the Null. Tests that control for FDR (i.e., ensure small values of FDR) account for large disparities in sample sizes. An example of how it can be used is illustrated by the `wavdetect` algorithm in `CIAO` that is used to detect sources in X-ray images. The threshold for detection is set by requiring a wavelet correlation strength that would result in one false detection on average over the entire image, i.e., an $\mathbb{S}^*$ corresponding to a $p \equiv \alpha = \frac{1}{N_{\rm pixel}}$, where $N_{\rm pixel}$ are the number of pixels in the image, and also the number of independent hypothesis tests that are carried out within the image.

Type M Errors

One of the consequences of a threshold-based selection of alternatives is that when the signal, also called the *effect size*, is small, the times when the Null is rejected also require large fluctuations in the signal. These fluctuations can be so strong that the estimated signal strength is strongly biased, and leads to clearly incorrect inferences. This is illustrated in the left two panels of Fig. 6.5. Consider a signal described by $H_1 : N(x; \mu, 1^2)$, compared to a Null distribution $H_0 : N(x; 0, 1^2)$. Let us consider tests where the Null is rejected at thresholds corresponding to $p = \{0.1, 0.05, 0.01\}$ (equivalent to $\sigma > \{+1.6, +2, +2.6\}$; and represented with red, blue, and green colors respectively). That is, if a sample is drawn from H_1, and it exceeds the threshold set based on H_0, the hypothesis that the draw is from H_0 can be rejected. The middle panel shows what happens when the thresholds are applied to a true signal of various strengths, $\mu \in [0, 4]$ (the case of $\mu = 0.1$ is illustrated in the left panel). The instances when the Null is rejected all require samples at large p-values in the alternate distribution, and the resulting sample estimates are invariably larger than the true signal. This effect is well known in astronomy, and is encountered in all cases where automated source detection is used to detect weak sources.[4]

Type S Errors
Just as in the case of one-sided thresholding that can lead to a signal being detected with the wrong magnitude, two-sided thresholds can lead to an even more spectacular failure of the test, with the signal being detected with the wrong sign. This is illustrated in the left and right panels of Fig. 6.5. As above, consider a signal described by $H_1 : N(\mu, 1^2)$, compared to a Null distribution $H_0 : N(0, 1^2)$. Let us consider tests where the observed signal exceeds a threshold on either side of zero, with $|\sigma| > \{1.6, 2, 2, 6\}$. The area under H_1 that exceeds these deviations are shown in the left panel, shaded in red, blue, and green respectively. This is a situation one might encounter if searching for emission and absorption lines in a spectrum. The right panel shows the fraction of observations where the Null would be rejected with the signal strength estimate being negative. As the figure demonstrates, there is a non-zero chance that a weak emission line source can produce a "detection" of an absorption line.

The basic takeaway from this discussion is that statistical tests that decide between alternatives should not be treated as black boxes. The results of the tests should be considered in the context of the different ways that they could go wrong, and thresholds should be set to minimize these errors. Most importantly, decisions of choice should be made as late as possible in the process in order to avoid introducing unaccounted and uncalibrated biases into subsequent analyses.

6.5.3 Likelihood Ratio Tests

It is often the case that two distinct models must be compared and one chosen as being a better descriptor of the data. We can use the machinery of hypothesis tests to do this (see Sect. 6.5.1 above). The optimal statistic would be one that quantifies

[4]This bias is sometimes called the Eddington Bias, though strictly speaking the Eddington Bias also includes the effects of population characteristics. That is, the measured source strengths are affected by both the Type M bias as well as there being more weaker sources that have upward fluctuations in the measurements than stronger sources that have downward fluctuations.

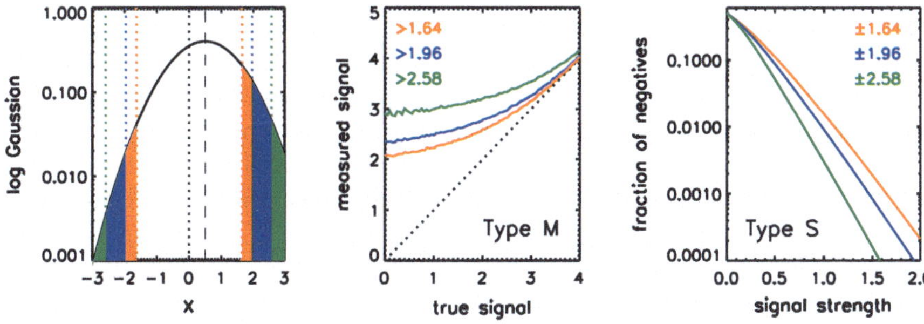

Fig. 6.5 Illustrating Type M and Type S errors. *Left:* Example distribution $N(x; 0.1, 1^2)$ of a small effect size compared to the Null distribution $N(x; 0, 1^2)$, showing how it exceeds different thresholds set on the Null at the two-sided ±90% (red), ±95% (blue), and ±99% (green) two-sided levels. *Middle:* The expected value of samples drawn from the alternate distribution with different effect sizes that exceed the thresholds at $p = 0.9$ (red), $p = 0.95$ (blue), and $p = 0.99$ (green). *Right:* The fraction of samples drawn from the alternate distribution with different effect sizes that are detected with the opposite sign at the same thresholds as in the *left* figure. Notice that even at a signal strength of 1.5, there is a probability of 10^{-4} that an observation will return a fluctuation of the wrong sign at $p < 0.005$

how well a model describes the data, and thus should be dependent on the likelihood. The natural quantity to consider is the ratio of the likelihoods R of the two models,

$$R = \frac{\mathbb{L}(\hat{\Theta}_1)}{\mathbb{L}(\hat{\Theta}_2)}, \tag{6.19}$$

where $\mathbb{L}(\hat{\Theta}_k)$ represents the likelihood of model k at the best-fit parameter values $\hat{\Theta}_k$. Note that Θ_k may be comprised of different parameters, and indeed different numbers of parameters, for different k. Often, the negative of twice the natural log of this quantity,

$$LR = -2 \ln R \tag{6.20}$$

is used instead, as a guard against significant differences being hidden near the lower bound as $R \to 0$. In the Gaussian regime, LR is also easily computed as the χ^2 (see Sect. 6.4.1). Conventionally, the numerator comes from the simpler model, and the denominator from the more complex model, so small values of R or large values of LR are interpreted as favoring the complex model.

Likelihood Ratio Tests (LRTs) work by considering how the distribution of the likelihoods are affected by the quality of the model fit. If there is nothing to choose between the two models, then the R and LR distributions should be consistent with that expected from statistical fluctuations. If one or the other model is superior, then R will be expected to differ from 1, and LR from 0. But since there is no explicit Null distribution to compare with, these distributions are not known in all cases and must often be calibrated using Monte Carlo methods. However, in the limit of large data sizes, LR is distributed as a χ^2 distribution with degrees of freedom equal to the difference in dimensionality between the parameter spaces of the two models. This is known as *Wilks' Theorem*, and it is invoked whenever a decision must be made to add a component to a model or not. For instance, if an optically thin thermal emission model is being used to fit to a coronal spectrum, and a choice must be made whether to thaw the metallicity or not, the value of LR is calculated for models with the metallicity frozen and thawed, and this value is compared against the χ_1^2 distribution to decide whether the p-value is small enough that the more complex model, with the metallicity thawed, is required. Note that, as discussed above in Sect. 6.5.2, if the p-value is not small, this does not forbid the metallicity being thawed, but merely states that the simpler model is good enough.

It is important to understand the regime of applicability of Wilks' Theorem. As alluded to above, it is asymptotically valid for large data sets, as the size of the sample $\rightarrow \infty$. There are two additional conditions that are crucial: first, the simpler model must be *nested* within the complex model, and second, the simpler model *should not fall on the edge of the parameter space* spanned by the complex model. The first condition precludes direct comparisons between, e.g., power-law and blackbody spectral models. The second indicates that the existence of emission (or absorption) lines cannot be searched for in this manner, because the simpler model (one with no emission line) is identical to the boundary of the complex model where line intensity is zero. In such cases, LR is *not* well described by the χ^2 distribution, and the computed p-value could be either an underestimate or an overestimate. This situation was explored in depth by the CHASC AstroStatistics group [22]. They prescribed a general method based on Monte Carlo simulations to calibrate the LRT when Wilks' Theorem is inapplicable:

1. First compute best-fit parameter values and error distributions $p(\Theta_{1,2}|\text{data})$ for the two models;
2. From the best-fit parameter values, compute LR_{observed};
3. Draw N sets of samples of Θ_1, the simpler model's parameters, from this distribution;
4. Create N simulated data sets from the sample parameter values;
5. Fit both the models to the simulated data sets, and compute the LR for each simulated sample;
6. Construct the distribution $f_{\text{sim}}(LR)$ as the sampling distribution for when the simpler model is the correct descriptor of the data;
7. Compare LR_{observed} against $f_{\text{sim}}(LR)$, and compute the approximate p-value.

6.6 Further Reading

In this chapter, we have described the foundational statistics necessary to understand and analyze high-energy astronomy data. Astrostatistics is an old field, arguably dating back to Pierre Laplace and certainly to Arthur Eddington, but is also an active field of research where new methods and techniques are being developed to handle the numerous problems that are encountered. Here we have focused specifically on concepts dealing with errors and uncertainties. The literature is vast and constantly growing. The papers, books, and monographs that were used, or implicitly or explicitly referred to here, are listed below, along with several others that can point the reader to more details and a greater depth of understanding. This list is not designed to be complete, but is rather expected to be representative.

References

Papers and Monographs

1. Y. Avni, Energy spectra of X-ray clusters of galaxies. ApJ **210**, 642 (1976), https://ui.adsabs.harvard.edu/abs/1976ApJ...210..642A/abstract
2. L. Bretthorst, Bayesian Fourier analysis (1988), https://bayes.wustl.edu/glb/book.pdf
3. W. Cash, Parameter estimation in astronomy through application of the likelihood ratio. ApJ **228**, 939 (1979), https://ui.adsabs.harvard.edu/abs/1979ApJ...228..939C/abstract
4. D.W. Hogg, Data analysis recipes: probability calculus for inference (2012), arXiv:1205.4446, https://arxiv.org/pdf/1205.4446.pdf
5. D.W. Hogg, J. Bovy, D. Lang, Data analysis recipes: fitting a model to data (2010), arXiv:1008.4686, https://arxiv.org/pdf/1008.4686.pdf
6. D.W. Hogg, D. Foreman-Mackey, Data analysis recipes: using Markov Chain Monte Carlo. ApJS **236**, 11 (2018), https://ui.adsabs.harvard.edu/abs/2018ApJS..236...11H/abstract
7. E.D. Feigelson, P.I. Nelson, Statistical methods for astronomical data with upper limits. I. Univariate distributions. ApJ **293**, 192 (1985), https://ui.adsabs.harvard.edu/abs/1985ApJ...293..192F/abstract
8. P.E. Freeman, V. Kashyap, R. Rosner, D.Q. Lamb, A wavelet-based algorithm for the spatial analysis of Poisson data. ApJS **138**, 185 (2002), https://ui.adsabs.harvard.edu/abs/2002ApJS..138..185F/abstract
9. N. Gehrels, Confidence limits for small numbers of events in astrophysical data. ApJ **303**, 336 (1986), https://ui.adsabs.harvard.edu/abs/1986ApJ...303..336G/abstract
10. A. Gelman, J. Carlin, Beyond power calculations: assessing type S (sign) and type M (magnitude) errors. Perspect. Psychol. Sci. **9**, 641 (2014), https://doi.org/10.1177/1745691614551642
11. T. Isobe, E.D. Feigelson, P.I. Nelson, Statistical methods for astronomical data with upper limits. II. Correlation and regression. ApJ **306**, 490 (1986), https://ui.adsabs.harvard.edu/abs/1986ApJ...306..490I/abstract
12. J.S. Kaastra, On the use of C-stat in testing models for X-ray spectra. A&A **605**, 51 (2017), https://ui.adsabs.harvard.edu/abs/2017A&A...605A..51K/abstract
13. V.L. Kashyap, D.A. van Dyk, A. Connors, P.E. Freeman, A. Siemiginowska, X. Jin, A. Zezas, On computing upper limits to source intensities. ApJ **719**, 900 (2010), https://ui.adsabs.harvard.edu/abs/2010ApJ...719..900K/abstract
14. K. Kearns, F. Primini, D. Alexander, Bias-free parameter estimation with few counts, by iterative chi-squared minimization. ASPC **77**, 331 (1995), https://ui.adsabs.harvard.edu/abs/1995ASPC...77..331K/abstract
15. B.C. Kelly, Some aspects of measurement error in linear regression of astronomical data. ApJ **665**, 1489 (2007), https://ui.adsabs.harvard.edu/abs/2007ApJ...665.1489K/abstract
16. T. Loredo, From Laplace to supernova SN 1987 A: Bayesian inference in astrophysics (1990), http://hosting.astro.cornell.edu/staff/loredo/bayes/L90-LaplaceToSN1987A-scan.pdf
17. M. Matsumoto, T. Nishimura, Mersenne twister: a 623-dimensionally equidistributed uniform pseudo-random number generator. ACM Trans. Model. Comput. Simul. **8**, 1 (1998), https://doi.org/10.1145/272991.272995

18. R. Neal, Probabilistic inference using Markov Chain Monte Carlo methods. Techincal Report CRG-TR-93-1 (Department of Computer Science, University of Toronto, 1993), https://www.cs.toronto.edu/~radford/review.abstract.html
19. S. Nieuwenhaus, B.U. Forstmann, E.-J. Wagenmakers, Erroneous analyses of interactions in neuroscience: a problem of significance. Nat. Neurosci. **14**, 1105 (2011), https://doi.org/10.1038/nn.2886
20. T. Park, V.L. Kashyap, A. Siemiginowska, D.A. van Dyk, A. Zezas, C. Heinke, B. Wargelin, Bayesian estimation of hardness ratios: modeling and computations. ApJ **652**, 610 (2006), https://ui.adsabs.harvard.edu/abs/2006ApJ...652..610P/abstract
21. F.A. Primini, V.L. Kashyap, Determining X-ray source intensity and confidence bounds in crowded fields. ApJ **796**, 24 (2014), https://ui.adsabs.harvard.edu/abs/2014ApJ...796...24P/abstract
22. R. Protassov, D.A. van Dyk, A. Connors, V.L. Kashyap, A. Siemiginowska, Statistics, handle with care: detecting multiple model components with the likelihood ratio test. ApJ **571**, 545 (2002), https://ui.adsabs.harvard.edu/abs/2002ApJ...571..545P/abstract
23. J.D. Scargle, J.P. Norris, B. Jackson, J. Chiang, Studies in astronomical time series analysis. VI. Bayesian block representations. ApJ **764**, 167 (2013), https://ui.adsabs.harvard.edu/abs/2013ApJ...764..167S/abstract
24. J.H.M.M. Schmitt, Statistical analysis of astronomical data containing upper bounds: general methods and examples drawn from X-ray astronomy. ApJ **293**, 178 (1985), https://ui.adsabs.harvard.edu/abs/1985ApJ...293..178S/abstract
25. J.H.M.M. Schmitt, T. Maccacaro, Number-counts slope estimation in the presence of Poisson noise. ApJ **310**, 334 (1986), https://ui.adsabs.harvard.edu/abs/1986ApJ...310..334S/abstract
26. D.A. van Dyk, A. Connors, V.L. Kashyap, A. Siemiginowska, Analysis of energy spectra with low photon counts via Bayesian posterior simulation. ApJ **548**, 224 (2001), https://ui.adsabs.harvard.edu/abs/2001ApJ...548..224V/abstract

Books

27. J. Babu, E. Feigelson, *Astrostatistics* (1996), https://www.crcpress.com/Astrostatistics/Babu-Feigelson/p/book/9780412983917
28. P.R. Bevington, D.K. Robinson, *Data Reduction and Error Analysis for the Physical Sciences*, 3rd edn. (2003), http://hosting.astro.cornell.edu/academics/courses/astro3310/Books/Bevington_opt.pdf
29. L. Wasserman, *All of Non-Parametric Statistics* (2006), http://www.stat.cmu.edu/~larry/all-of-nonpar/
30. C.K. Rasmussen, C.E. Williams, *Gaussian Processes for Machine Learning* (2006), http://www.gaussianprocess.org/gpml/
31. W.H. Press, S.A. Teukolsky, W.T. Vetterling, B.P. Flannery, *Numerical Recipes: The Art of Scientific Computing*, 3rd edn. (2007), http://numerical.recipes
32. E. Feigelson, J. Babu, *Modern Statistical Methods for Astronomy with R Applications* (2012), https://astrostatistics.psu.edu/MSMA/
33. K. Arnaud, R. Smith, A. Siemiginowska, *Handbook of X-Ray Astronomy* (2011), http://hea-www.cfa.harvard.edu/~rsmith/xrayastronomyhandbook/
34. P. Gregory, *Bayesian Logical Data Analysis for Physical Sciences* (2012), https://doi.org/10.1017/CBO9780511791277
35. A. Gelman, J.B. Carlin, H.S. Stern, D.B. Dunson, A. Vehtari, D.B. Rubin, *Bayesian Data Analysis*, 3rd edn. (2013), http://www.stat.columbia.edu/~gelman/book/
36. E. Robinson, *Data Analysis for Scientists and Engineers* (2016), https://press.princeton.edu/titles/10911.html

Chapter 7
Data Analysis

William Alston, Peter Boorman, Andrea Bulgarelli, and Michael Parker

7.1 Spatial Study: Exploring a Sky Image

In this section, different techniques and methods are introduced for exploring an X-ray image. First, techniques used to identify the presence of an X-ray emitting source are presented in Sect. 7.1.1, focusing on: manually altering the image to find a source ('by eye'; Sect. 7.1.1.1); a rigorous method for statistically quantifying the confidence associated to the detection of a source (the 'no-source' probability; Sect. 7.1.1.2) and algorithms for source detection (Sect. 7.1.1.3). Techniques for quantifying and studying extended emission are then detailed in Sect. 7.1.2, before finally detailing sources of multi-wavelength data in Sect. 7.1.3 to compliment X-ray imaging analysis.

7.1.1 Source Presence

The first step in exploring an X-ray image encompasses an understanding of the target of interest. In some cases, the target is known (e.g., for *pointed* observations)

W. Alston
Institute of Astronomy, Madingley Road, Cambridge CB3 0HA, UK
e-mail: wna@ast.cam.ac.uk

P. Boorman (✉)
Astronomical Institute, Academy of Sciences, Bǒcní II 1401, 14131 Prague, Czech Republic
e-mail: peter.boorman@asu.cas.cz

Faculty of Physical Sciences and Engineering, Department of Physics & Astronomy, University of Southampton, Southampton SO17 1BJ, UK

A. Bulgarelli
INAF/OAS Bologna, Via Piero Gobetti 93/3, 40129 Bologna, Italy
e-mail: andrea.bulgarelli@inaf.it

M. Parker
European Space Agency (ESA), European Space Astronomy Center (ESAC), Villanueva de la Cañada, 28691 Madrid, Spain
e-mail: mparker@sciops.esa.int

C. Bambi (ed.), *Tutorial Guide to X-ray and Gamma-ray Astronomy*,
https://doi.org/10.1007/978-981-15-6337-9_7

and in others unknown. For unknown targets, users typically rely on source detection algorithms designed to disentangle individual X-ray sources down to a certain flux threshold. In addition, a priori unknown X-ray targets can be identified in the surrounding field-of-view of pointed observations—known as serendipitous surveys. To-date there have been several published serendipitous surveys for X-ray instruments (e.g., *Chandra*; [8], *NuSTAR*; [16], *XMM-Newton*; [26]). In both targeted and untargeted observations, the source can be so faint that the *significance* of the source presence in the image is required.

For these reasons, this section is organised as follows. First, in Sect. 7.1.1.1, methods are discussed to detect faint sources in X-ray images. Section 7.1.1.2 then details a statistical framework to quantify the confidence associated with detecting a source. Finally, some popular source detection algorithms are described in Sect. 7.1.1.3, together with their advantages and disadvantages.

7.1.1.1 Detection by Eye

In the event that a target is comparable in brightness to the background, it can be difficult to assess the detection of the source. A preliminary technique to explore is identifying the presence of an X-ray source by eye. Three specific methods can be very useful to achieve this: spatial smoothing, spatial binning and energy binning, details of which are enumerated below. To illustrate these three processes, SAOImage DS9[1] is used with a *NuSTAR* observation of the AGN UGC 9944. The calibrated & cleaned events file (after running `nupipeline`) is shown as an image in the upper left panel of Fig. 7.1.

1. **Energy binning:** Each X-ray event in an events list has a corresponding energy associated to it, and it can be very useful to filter an image to only display events in a specific range of energies. The range chosen ultimately depends on the X-ray spectrum of the target, and predictions of the target's emission as a function of energy as well as the instrument's sensitivities could reveal the energy band that the source emission would be strongest as compared to the background emission, and thus most likely to show up in an X-ray image. By including less-sensitive energy regions or weak emission energy bands associated to the source can give the fainter portions of the source's spectrum more weight in the overall image, which is then harder to distinguish from the background by eye.
   Energy can be listed in a number of different ways in an events list, and thus it is very important to filter on the correct column in deriving an energy-binned events file. The raw energy recorded by the instrument of a detected event is called the Pulse Height Amplitude (PHA), given as an integer 'channel'. The corresponding energy of each PHA value is proportional to the channel number, with the constant of proportionality being the gain. However, the gain is a function of detected event location and time of detection, such that the gain can be different for every event

[1] http://ds9.si.edu/site/Home.html.

in the PHA column. For this reason, many instruments have their own pipelines that produce calibrated event lists with a Pulse Invariant (PI) column. This PI channel number is a gain-corrected PHA value, and thus all PI channel values are (in general) directly proportional to energy with some uniform dependence. Often the process of producing calibrated events lists also includes creating a column in units of energy also (e.g., the `energy` column for *Chandra* after reprocessing). To check the different columns that are available, use the `ftool fv` to view the header of the events file. The units of each column are also displayed, showing whether the columns are PHA or gain-corrected PI, as well as the units—e.g., `chan` or energy (e.g., eV, keV). Due to the potential confusion associated with binning an events file by energy, the following two methods are examples of unambiguous techniques for accomplishing this.

- *With SAOImage DS9:* To bin an events list on an arbitrary column in SAOImage DS9, use the 'Binning Parameters' option under the 'Bin' menu. As an example, in Fig. 7.1 the events file for UGC 9944 has been filtered by using the syntax `pi={pilow}:{pihigh}` in the 'Bin Filter' text box. Here `{pilow}` and `{pihigh}` are the lower and upper bounds of the PI channel column in the corresponding *NuSTAR* events file, with the channels 35:160 corresponding to 3–8 keV for *NuSTAR*.
- *With Chandra Interactive Analysis of Observations* (`ciao`[2]): The `dmcopy`[3] command inside CIAO is used to bin events files on arbitrary columns. The syntax for binning on energy is:
  `dmcopy "{evt_file}[{value}={low}:{high}]" {out_evt_file}`,
  where `{evt_file}` is the input events file, `{value}` is the column of the events file being binned, `{low}` & `{high}` are the lower & upper bounds of the filter and `{out_evt_file}` is the output events filename to be created.

2. **Spatial binning:** Spatial binning involves creating new larger pixels by combining individual adjacent pixels in the original image. This results in a gain of image sensitivity, potentially detecting faint features and sources. However this is at the cost of spatial resolution, potentially losing information on smaller size scales in the image. To do this in SAOImage DS9, select 'Bin' and 'Binning Parameters', again. Then in the 'Bin' column, type the spatial binning factor you would like for the X and Y directions and click 'Apply'. In the lower right panel of Fig. 7.1, the events file is spatially binned by a factor of 4 in the X and Y directions. This is seen to increase the sensitivity of individual pixels as the count rate increases per pixel (note the colour bar at the bottom of the panel), but spatial information is lost associated to size scales less than 4 pixels as a result.
3. **Spatial smoothing:** Spatial smoothing involves convolving a kernel of fixed shape with each pixel of the image, and can help the human eye to associate multiple X-ray events to a single source. This can help for faint targets with a deficit of

[2] https://cxc.harvard.edu/ciao/.

[3] https://cxc.harvard.edu/ciao/ahelp/dmcopy.html.

detected photons, by helping to associate a collection of photons with a single common origin. However, this process effectively distributes a single pixel event to multiple pixels, which decreases the overall observed brightness of the event in an image. Thus, over-smoothing can result in an over-prediction of the source-emitting region or even the complete disappearance of the source in the image. To smooth an image in SAOImage DS9, select 'Smooth' and 'Smooth Parameters'. In the menu box that appears, select the function you would like to use for smoothing. Here, the default Gaussian smoothing is used with a radius of 3 pixels. The resulting image is shown in the lower left panel of Fig. 7.1, and the source is circled in green for clarity. The source is clearly identifiable after applying these methods, but one method on its own does not reveal the source significantly. Note a caveat to altering images by eye: the filtering method can increase the visible sensitivity of sources other than the target of interest. For example, in the lower left panel of Fig. 7.1, the background emission in the lower region of the image has also been enhanced by the procedures, such that a robust knowledge of the source location is essential.

Different combinations of these methods may be optimum for a particular source, and it is recommended to try a selection to determine an efficient strategy to detect an individual source by eye.

7.1.1.2 A Statistical Test of Detection Significance

Detecting by eye is useful for re-centering source extraction regions and extracting useful science products for analysis (e.g., spectra and light curves). However, sometimes a significance of the detection of a source in an X-ray image is useful. X-ray photons are emitted with probability dictated by the Poisson distribution given in Eq. 7.1. Here, λ is the average number of events in a time interval (not necessarily an integer) and k is the integer number of events in the time interval.In the high counts regime, the Poisson distribution tends to a Gaussian distribution (see Fig. 7.2). However, in the low counts regime when sources are faint, the Poisson nature of the source emission is very different from Gaussian. This is why a statistical test of significance must use the Poisson distribution inherently.

$$\mathscr{P} = P(k \text{ events in interval}) = \frac{\lambda^k e^{-\lambda}}{k!} \tag{7.1}$$

Here, the method outlined in [29][4] for the 'simplest' case that the background is known with zero uncertainty is followed as an example. The strategy for determining source significance is to work under the assumption that there is no source, and that any events detected are sampled solely from the background. Then, a statistical test is devised to calculate the confidence with which this null hypothesis can be rejected—known as the 'no-source probability'. To do this, consider an X-ray image

[4] Also see the GitHub codebase available at: https://github.com/giacomov/gv_significance.

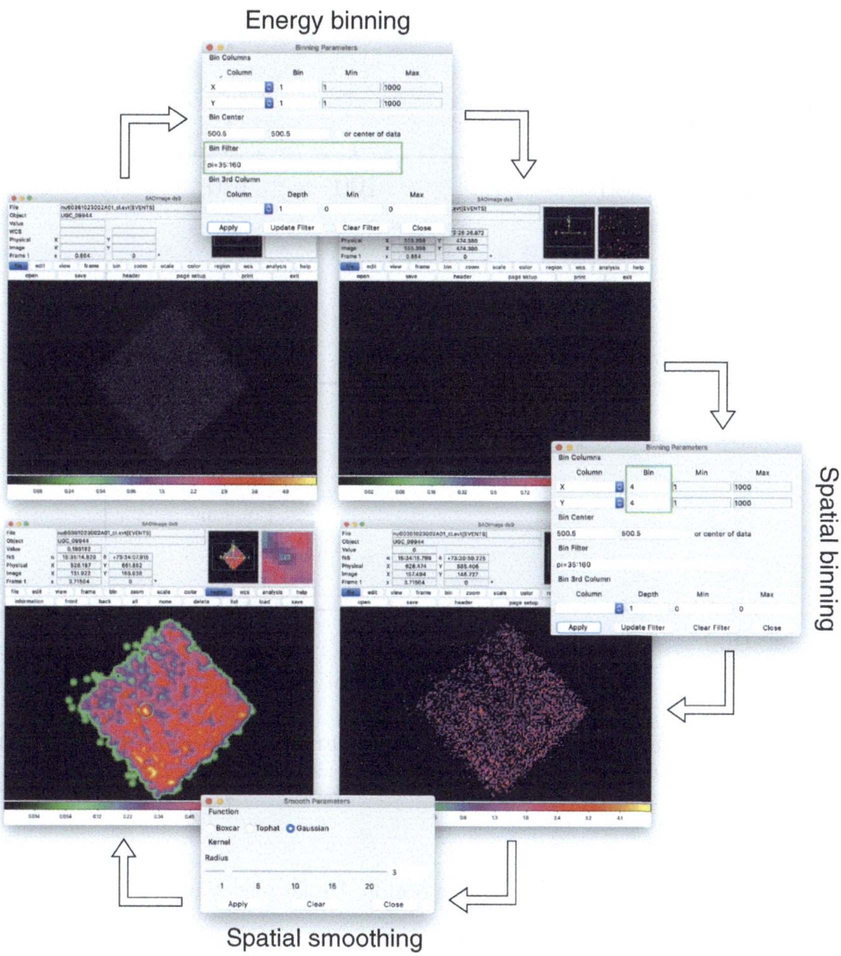

Fig. 7.1 Faint source detection by eye in SAOImage DS9. (*Upper left*) Original unaltered X-ray image of UGC 9944. (*Upper right*) Image filtered by energy to only show photons with $3 < E < 8$ keV. (*Lower right*) Image spatially binned by a factor of 4 in the X and Y directions. (*Lower left*) Image smoothed with a Gaussian kernel of radius 3 pixels. The source is circled in green, and can be seen clearly, whereas could not be seen in any intermediate step

with source counts extracted from a source region and background counts extracted from a background region, completely devoid of any source counts. Parameters are defined as follows.

- T: Total counts (source + background) extracted from the source extraction region.
- A_S: Area of the source extraction region.
- A_B: Area of the background extraction region.
- B: Background counts extracted from the background extraction region.

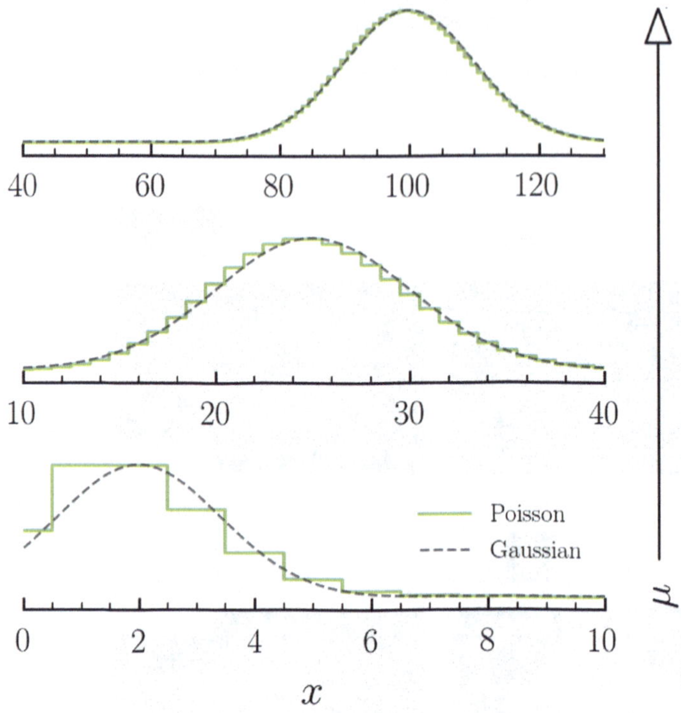

Fig. 7.2 Three instances of the Poisson and Gaussian distributions for increasing values of the mean, μ. As the mean increases, the Poisson distribution is seen to resemble the Gaussian distribution in increased accuracy

- $\boldsymbol{B_S}$: Background counts in the source region. Under our assumptions, this is just $B_S = B \times A_S/A_B$.
- $\boldsymbol{S}$: Total number of source counts in the source extraction region, $S = T - B_S$.

Following this setup, the probability of detecting S or more background counts in the source extraction region is given by Eq. A10 of [30], which written in our defined notation is:

$$\mathscr{P}(B \geqslant S) = \sum_{i=S}^{\infty} \frac{(B_S)^i}{i!} \times e^{-B_S} \tag{7.2}$$

This is the sum of the Poisson distribution, and can be calculated analytically as a function of the Γ and incomplete Γ functions (e.g., [29]) as:

$$\mathscr{P}(B \geqslant S) = 1 - \frac{\Gamma(S+1, B_S)}{\Gamma(S+1)} \tag{7.3}$$

This equation is calculated in Python with the `scipy.special.gammaincc` function. For example, suppose that 15 source + background counts have been extracted from a source extraction region 5 times smaller than a background extraction region which separately yielded 25 background counts. Therefore $B_S = B \times A_S/A_B = 25/5 = 5$ and $S = T - B_S = 15 - 25/5 = 10$. Equation 7.3 then yields a no-source probability of 6.9×10^{-5}. By assuming a number of background counts, one can then calculate via Monte Carlo simulations the percentage of time a source would be detected above some significance level for a given expectation value of counts in the Poisson distribution (see Fig. 7.1 [29]). For an alternative no-source probability incorporating uncertainties into the background counts, see [17, 29]. In some scenarios, the source is not detected and it is very useful in such cases to extract an upper limit to the count rate at the source location. For this, the work of [14] provide a detailed comparison of frequentist and Bayesian techniques for establishing confidence limits and upper limits to source count rates in the low counts regime.

7.1.1.3 Searching for Sources in an X-Ray Image

The Point Spread Function (PSF) is the image produced by a Delta function of light incident on an X-ray detector. The Encircled Energy Fraction is then the two-dimensional integral of the PSF, and can be very useful for parameterising and analysing dependencies of the PSF (see Fig. 7.12 of [18] for an example with *NuSTAR*). Due to the complex configuration of focusing X-ray optics used by modern high-resolution X-ray imaging facilities, the PSF can vary substantially with the following parameters:

- Position of the source on the detector field of view (described by the 'off-axis' angle, or the distance of the source from the focus of the X-ray optics on the detector).
- The energy of the detected photon (and hence the X-ray spectrum of the emitting source).
- The number of photons being emitted (and hence the count rate of the source).

These effects are highlighted in Fig. 7.3, taken from the *Chandra* PSF help page.[5]

The complex nature of X-ray imaging PSFs can dramatically effect the outcome of source detection algorithms, and thus some knowledge of an individual instrument's PSF is typically always required pre-analysis. A selection of algorithms for source detection implemented exclusively by the *Chandra* team into `ciao` are presented below.

- **Sliding-cell** (`celldetect` in `ciao`[6]): Two coaligned cells of different size are placed at each pixel in the image. By using the number of counts in the corresponding annular region, the relative number of counts in the source region is used

[5] https://cxc.harvard.edu/ciao/PSFs/psf_central.html.

[6] https://cxc.harvard.edu/ciao/ahelp/celldetect.html.

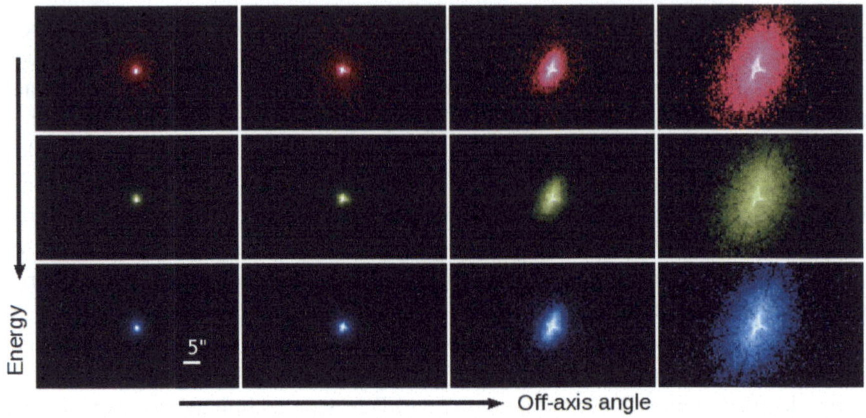

Fig. 7.3 The dependence of the *Chandra* PSF is shown as a function of off-axis angle and detected photon energy from the *Chandra* help page dedicated to the PSF (See footnote 5). From left to right, the off-axis angles are 0, 2.4, 4.7 and 9.6 arcmin, and from top to bottom, the photon energies are 0.92, 1.56 and 3.8 keV

to determine if there is an excess of counts above some predefined threshold. The cell size is also matched to the PSF, or can be fixed by the user.

  - *Advantages:* the algorithm is well understood, and works well for point sources in uncrowded fields.
  - *Disadvantages:* struggles to separate sources in crowded fields, and often divides extended sources into multiple point sources.

- **Wavelet basis functions** (`wavdetect` in `ciao`[7]): An improvement on sliding-cells was to replace the cells with wavelet basis functions, such as the 'Mexican Hat' [12]. Such functions are designed to better match the PSF shape than with sliding-cell methods, and can work over arbitrary scales defined by the user. The correlation of a well-chosen wavelet function with an events image will give strong correlations around concentrations of counts.

  - *Advantages:* more computationally efficient and capable of detecting extended emission.
  - *Disadvantages:* the scale of the wavelets is set by the user, and the algorithm can take a much longer time than `celldetect` to compute (the size of the image being searched is an important factor to consider).

- **Voronoi–Tesselation–Percolation** ([9], e.g., `vtpdetect` in `ciao`[8]): For an X-ray image of individual photon events, a Voronoi tessellation is a unique set of cells, each enclosing one of the points in the image. The algorithm determines

[7]https://cxc.harvard.edu/ciao/ahelp/wavdetect.html.

[8]https://cxc.harvard.edu/ciao/ahelp/vtpdetect.html.

individual densities for each pixel and analyses the distribution of densities to reveal source candidates.

- *Advantages:* finds extended sources, no scale is required from the user and can find low surface brightness features.
- *Disadvantages:* struggles with crowded fields and computationally expensive for large numbers of photons.

7.1.2 Source Extension

Broadly speaking, for modern X-ray focusing telescopes detected sources can appear point-like or extended on a detector. Here, extended refers to spatially resolved emission on scales larger than that of the instrument PSF. Studying the distribution of extended X-ray emission is a powerful technique for a wide range of astronomical studies. For example, spatially resolving the ^{44}Ti present in the supernova remnant Cassiopeia A enabled [13] to rule out symmetric explosion mechanisms for core-collapse supernovae. Another example is provided with the Fe Kα emission from obscured AGN found to be extended over hundreds of parsec surrounding the central AGN (e.g., [10]).

A common method used to determine if an X-ray source is extended is to construct the radial profile of the source, and compare it to the radial profile of the equivalent PSF predicted at the same off-axis angle and spectral energy distribution as the source (e.g., [10, 15]). Each instrument additionally provides a way to predict the PSF for a given observation. For example, the *Chandra* team recommends to use the *Chandra* Ray Tracer (ChaRT; [6]) and MARX [7] to simulate a bespoke PSF per source. Papers dedicated to the PSF of other instruments include *Swift* [20], *XMM-Newton* [25] and *NuSTAR* [3]. The process for generating a radial profile of a source using SAOImage DS9 and `ciao` is enumerated below for a *Chandra* observation of Mrk 573 (summarised in Fig. 7.4).

1. Load the image in SAOImage DS9 and create a new annular region, centered on the source.
2. Select the inner and outer most radii of your radial profile as well as the number of annuli (data points) you would like, and click 'Generate'. Save this region with a specific name, e.g. 'annuli_src.reg'.
3. Create a new annular region with inner radius $\geqslant$ the outermost radius of the source annuli, and arbitrary outer radius. Save this region separately with a different name, e.g. 'annuli_bgd.reg'.
4. Using the `ciao` command `dmextract`, extract the corresponding radial profile and save in fits file format before plotting the radial profile.

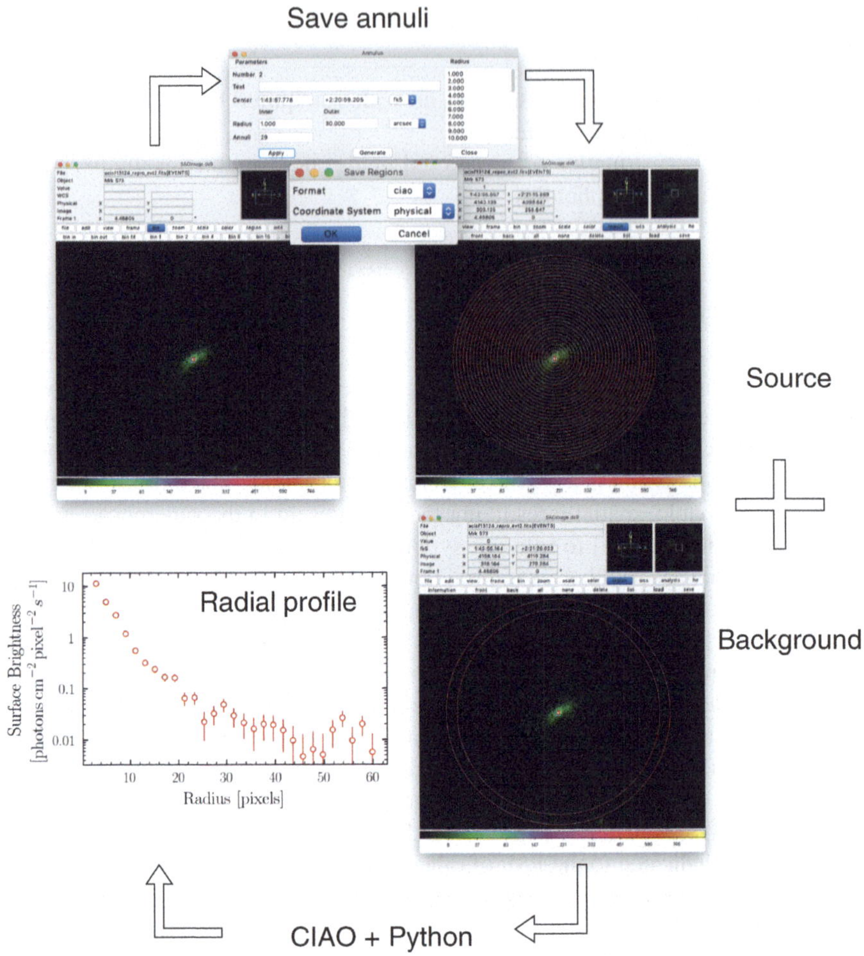

Fig. 7.4 Radial profile extraction in SAOImage DS9 and `ciao`. (*Upper left*) X-ray image of Mrk 573 in the 0.3–8 keV band. (*Upper right*) Image with radial profile annuli overplotted. (*Lower right*) Single background annulus chosen to be outside the source annuli. (*Lower left*) Resulting radial profile extracted with `ciao` and plotted in Python

7.1.3 Multi-wavelength Identifiers

There are multiple resources and instruments available to analyse an X-ray source identified in an image across the electromagnetic spectrum. This can be extremely useful for a wide array of applications such as: establishing different emission regions in a source; identifying robust source positions; multi-wavelength classifications or emission processes and assessing contaminants in the X-ray source extraction region.

A small selection of powerful resources are summarised below for accomplishing these tasks.

- **NED**[9]: The NASA/IPAC Extragalactic Database represents a comprehensive and ambitious effort to catalogue every known extragalactic object. Per source, there is a wide array of available multi-wavelength information, including a useful tool to view the broadband Spectral Energy Distribution of the source. In addition, users can browse the archived spectra for a source, as well as any multi-wavelength images that are available (see Fig. 7.5).
- **IRSA**[10]: The NASA/IPAC Infrared Science Archive provides a multitude of infrared-based survey data covering the entire sky. A key resource is from the *Wide-field Infrared Survey Explorer* (*WISE*; [32]), which provides 4-band infrared photometry for the majority of the sky, allowing exquisite source localisation for cross-referencing in X-ray images. The four photometric filters available on *WISE* enable additional infrared colour classification of objects (see Fig. 7.12 of [32]), and can be very useful for inferring the nature of a source in conjunction with X-ray analysis.
- **SDSS**[11]: The Sloan Digital Sky Survey provides unprecedented optical photometry and spectral constraints on a vast number of sources over the northern sky. The most recent Data Release 15 is detailed in [2], and multiple tutorials and examples can be found on the website for optimum data access techniques.
- ***Hubble* Legacy Archive**[12]: This website provides access to *Hubble Space Telescope* products, such as images & spectra, and even offers Jupyter notebooks to browse for data via Python. The high spatial resolution and separate wavebands to X-ray offered by *Hubble* gives a unique comparison to X-ray images for possible contaminant identifications, for instance.

7.2 Time Series Analysis

Time-series analysis or *signal processing* is a branch of statistics routinely applied to time-dependent processes throughout economics, engineering and the sciences. Many astrophysical sources vary in brightness over time, or are transient in nature, providing an extra dimension by which to study the underlying physics (e.g. [57, 111]). The terms *time series* and *light curve* are used interchangeably in astronomy to mean a finite sequence $x(t_i)$ of N data points of some measured quantity (e.g. photon counts, flux, etc.). In this section we will review the basics of time series analysis and some commonly applied methods in X-ray astronomy, with a focus on

[9]https://ned.ipac.caltech.edu.

[10]https://irsa.ipac.caltech.edu/frontpage/.

[11]https://www.sdss.org.

[12]https://hla.stsci.edu.

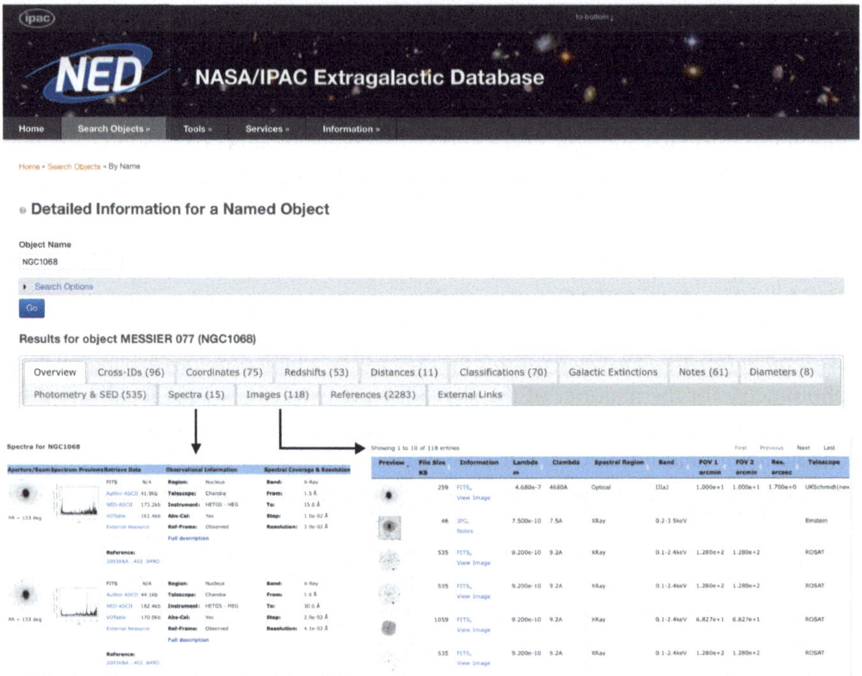

Fig. 7.5 Example search results shown for the source NGC 1068 on NED. The lower left and right panels illustrate the available spectra and images listed

accretion onto black holes. Example code and data analysis can be found at https://github.com/coherentnoise.

7.2.1 Astrophysical Time Series

The observed time series from an astrophysical source follows a stochastic (random) noise process, where each recorded observation is one realization of the stochastic process (e.g. [95, 97]). A stochastic process is considered stationary when its probabilistic structure is time invariant. An implicit assumption in many of the methods used for analysing astrophysical data detailed below, is that the process generating the time series under consideration is stationary. In practice, the stochastic process is typically only *weakly stationary*, where only the second order moments are time invariant, e.g. the variance or rms (root mean square amplitude), power spectral density, cross-correlation function and cross-spectrum. The effects of genuine *strong* non-stationarity is discussed in detail in Sect. 7.2.11.

X-ray detectors are typically photon counting devices, recording a discrete number of counts c_i in the time interval $[t_i, t_{i+1}]$, hence $x(t_i) = c_i/\delta t$, where $i \epsilon [0, N-1]$

and the length of the light curve is given by $T = N\delta t$. In the following sections the subscript i will be dropped from much of the notation. The production of source light curves from raw detector data is discussed in detail in Chap. 5.

7.2.1.1 Poisson Noise

The photon counting nature of X-ray detectors means the observed signal in a given energy band, $x(t)$, comprises the 'true' underlying physical signal $s(t)$ plus some Poisson counting noise $n(t)$, such that

$$x(t) = s(t) + n(t) \tag{7.4}$$

where n(t) has mean 0 and sigma (rms) of $\sqrt{s(t)}$. If the true flux is $s(t)$ (in time interval dt) then we will observe random values $X \sim Pois(s)$, i.e. each measurement is drawn from a Poisson distribution. But in the limit of moderately high counts/interval we approximate the Poisson distribution as a Normal distribution and treat the observed signal as the sum of the true signal s(t) plus a noise term, n(t). We can thus approximate variance$(n) \sim s(t)$. The noise is uncorrelated with the signal, however, since the amplitude of Poisson noise is $\sqrt{flux}$, it is not independent of it.

7.2.2 *Quantifying Variability*

For a time series, $x(t)$, variability can be quantified using a number of statistical tests. In an observed light curve, some of the total variance, S^2, will be intrinsic to the source σ_{XS^2}, and some will come from the measurement uncertainties, $\overline{\sigma^2_{\mathrm{err}}}$ [115]. The 'excess variance' can be used to estimate the intrinsic source variance for N_x data points using:

$$\sigma^2_{\mathrm{XS}} = S^2 - \overline{\sigma^2_{\mathrm{err}}} \tag{7.5}$$

where $S^2 = \sum_{i=1}^{N}(x_i - \bar{x})^2/(N-1)$ is the unbiased sample variance, and $\overline{\sigma^2_{\mathrm{err}}} = \frac{1}{N}\sum_{i=1}^{N}\sigma^2_{err,i}$ is the mean square error (mse). $\sqrt{\sigma_{\mathrm{XS}^2}}$ gives the root mean squared (rms) variability amplitude. The fractional rms variability amplitude (F_{var}; see e.g. [53]) is the normalised version of the excess variance defined as:

$$F_{\mathrm{var}} = \sqrt{\frac{S^2 - \overline{\sigma^2_{\mathrm{err}}}}{\bar{x}^2}}. \tag{7.6}$$

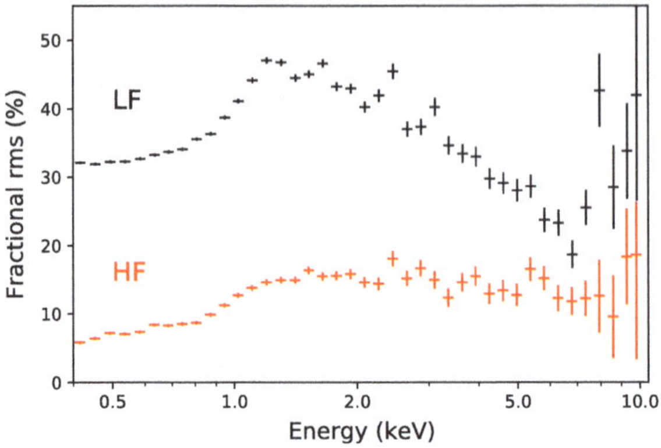

Fig. 7.6 The rms spectrum in fractional units for *XMM-Newton* observations of the AGN, IRAS 13224-3809. The long timescale (low-frequency) spectra is shown in black, and the short timescale (high frequency) spectra in red. The normalisations are different due to the amplitude of the PSD being different at different timescales (rms spectra are equivalent to integrating the PSD between two frequency bounds

Practically, $\sigma^2_{\rm XS}$ is estimated using $n = M \cdot N$ data points, where M and N are typically > 20, resulting in $n > 400$. This mitigates the weak stationary effect, and ensures the error bars are approximately Gaussian [115].

7.2.2.1 Variability Spectra

Variability spectra provide a powerful discriminator of physical models (see e.g. [41, 94]). The energy dependence of the variability at a given timescale can be assessed by computing $F_{\rm var,E}$ from a series of energy bands $x(t, E)$, referred to as rms-spectra. These are typically determined in fractional units to allow comparison between different data sets, as is shown in Fig. 7.6.

The timescales over which the variability is investigated can be changed by changing the segment length $T = N\delta t$ as well as changing the binwidth δt. This can be translated to the Fourier domain, where the lower frequency-bound is given by $1/(N\delta t)$, and the upper frequency-bound is $1/(2\delta t)$. When estimated over a range of energies these are referred to as frequency-resolved spectra.

7.2.3 Time Evolution

For the statistics described in Sect. 7.2.2 the time ordering of the measured quantities is irrelevant, hence they provide no information about the time evolution of the process. For a univariate time series we can gain such information from the autocorrelation function (Fig. 7.7)

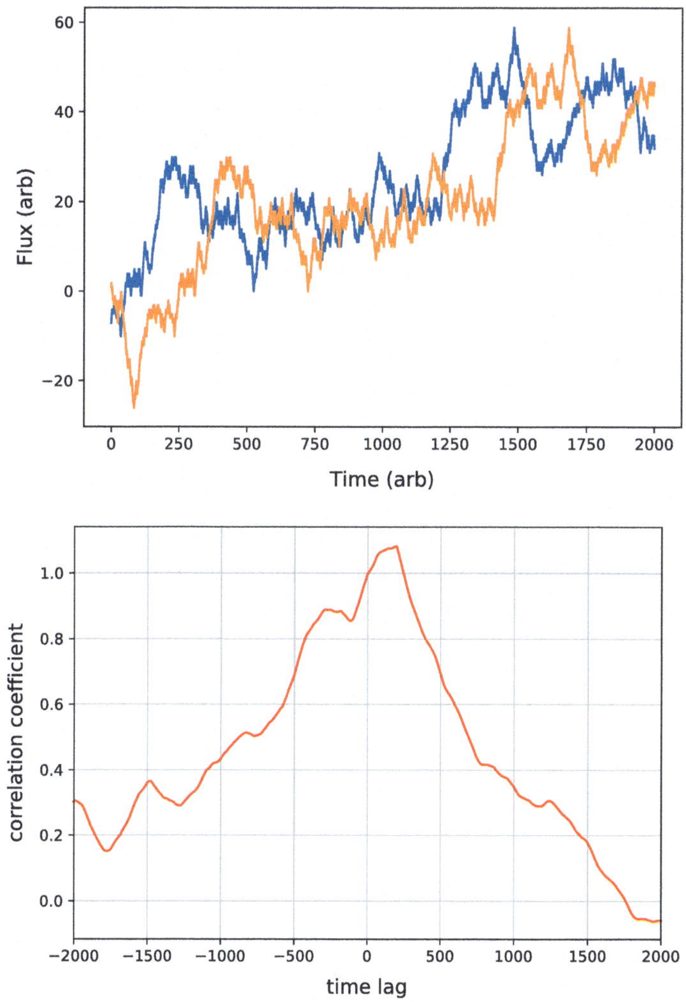

Fig. 7.7 Random walk simulation (blue), lagged by 200 time steps (orange). The correlation coefficient as a function of τ is shown below

$$\rho_{xx}(\tau) = \frac{E[(x_t - E[x_t])(x_{t-\tau} - E[x_t])]}{\sigma_x^2} \tag{7.7}$$

where E is the expectation value and $x_{t-\tau}$ is a lagged version of the time series, $x(t-\tau)$. The autocorrelation function is symmetric $\rho_{xx}(\tau) = \rho_{xx}(-\tau)$.

For multivariate data, where we have two simultaneously sampled time series x(t) and y(t)—such as observations of the same target in two different energy bands, we have the cross-correlation function:

$$\rho_{xy}(\tau) = \frac{E[(x_t - E[x_t])(y_{t-\tau} - E[y_t])]}{\sigma_x \sigma_y}. \tag{7.8}$$

The numerator in Eq. 7.8 is the cross-covariance function (CCF).

The irregular sampling nature of astronomical data often limits the use of the CCF and other statistical tools that require evenly sampled data. Interpolating between the gaps is commonly performed if the variability is slow (i.e. red noise), but is likely to introduce spurious correlations when the PSD has substantial amplitude at frequencies higher than the mean sampling rate. Edelson et al. [53] devised the discrete correlation function (DCF) which is defined for lags between each pair (a_i, b_j) of measured data points, and hence does not invent data.

$$DCF_{xy}(\tau) = \frac{1}{M} \sum \frac{(x_i - \bar{a})(b_j - \bar{b})}{\sqrt{(\sigma_a^2 - e_a^2)(\sigma_b^2 - e_b^2)}} \tag{7.9}$$

where the sum is over all pairs within the time interval $\Delta t_{ij} = t_j - t_i$, defined by time lag τ for $\tau - \Delta\tau/2 \leq \Delta t_{ij} \leq \tau + \Delta\tau/2$. Here, e are the measurement errors and M is the number of pairs within $\pm\Delta\tau$ of each time lag τ.

The Bartlett formula can be used to estimate the variance on the CCF estimates, and can therefore be used like an error bar [44].

$$\sigma^2\{CCF(\tau)\} = \frac{1}{N} \sum_i \mathrm{ACF}_x(\tau)\mathrm{ACF}_y(\tau) \tag{7.10}$$

where N is the number of data points contributing to each time lag τ. This formula can also be applied to the DCF if the ACF is formed from the DCF of the uneven time series with itself.

7.2.4 *Estimating the Power Spectrum*

When the stochastic process contains variations over a broad range of temporal timescale, it is often useful to study the process in the Fourier domain (see Fig. 7.8). The power spectrum (or power spectral density: PSD) is the Fourier counterpart of the ACF, and describes the amount of variability *'power'* (squared variability amplitude) as a function of Fourier (temporal) frequency (~1/timescale). There are a variety of ways to estimate the *spectral* content of a time series, but we focus here on the most commonly used for accreting sources.

The standard procedure for estimating the power spectrum from astronomical time series is to use the Bartlett method [44]. An evenly sampled and contiguous time series $x(t_i)$ with time bin size Δt and sampling length $N\Delta t$, can be broken into M non-overlapping segments of length N (where $i = 1, 2, \ldots, N$). The periodogram

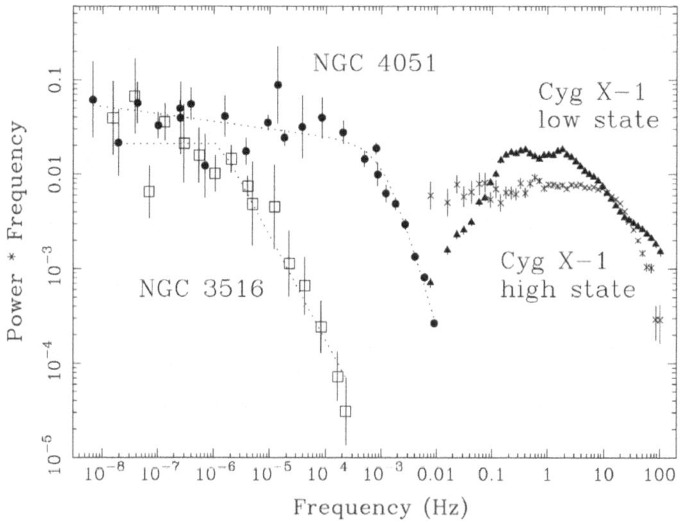

Fig. 7.8 Example PSD for two AGN and an XRB in a low and high state (reproduced from [84])

(e.g. [97]) of each segment is computed and the average of the M periodograms is taken to produce an estimate of the true PSD.

The periodogram $P(f)$ is the normalised modulus-squared of the discrete Fourier transform (DFT), which is a discrete estimator of the underlying continuous PSD, $\mathscr{P}(f)$. The DFT of $x(t_i)$ is defined as

$$X(f_j) = \mathscr{F}[x(t_i)](f) = \sum_{i=1}^{N} x(t_i)e^{-i2\pi f_j t_i} = \Re[X(f_j)] + \Im[X(f_j)] \tag{7.11}$$

with

$$\begin{aligned}\Re[X(f_j)] &= \sum_{i=1}^{N} x(t_i)\cos 2\pi f_j t_i \\ \Im[X(f_j)] &= \sum_{i=1}^{N} x(t_i)\sin 2\pi f_j t_i\end{aligned} \tag{7.12}$$

where $\Re(\cdot)$ and $\Im(\cdot)$ indicate the real and imaginary components of some variable. $X(f)$ is calculated at $N/2$ evenly spaced frequencies $f_j = j/N\Delta t$ (where $j = 1, 2, \ldots, N/2$). The highest frequency that can be sampled is the Nyquist frequency: $f_{N/2} = f_{Nyq}$. Because astronomical time series are real valued, the resulting complex Fourier transform is symmetric around $f_{j=0}$ (i.e. $X(f_{+j}) = X^*(f_{-j})$), meaning the Fourier transform of the negative frequencies provide no additional

Table 7.1 Commonly used periodogram normalisations for input light curves in cts^{-1} with Poisson errors. The Poisson noise level is different for light curves with Gaussian errors $\sigma_{err,i}$ (e.g. [115])

Normalisation	Ordinate	Integrated	Poisson noise
$A_{\text{abs}} = 1$	$[\text{ct s}^{-1}]^2\text{Hz}^{-1}$	$[\text{ct s}^{-1}]^2$	$P_{\text{noise}} = 2\Delta t\overline{\sigma_{\text{err}}^2}$
$A_{\text{Leahy}} = 1/\bar{x}$	$\text{ct s}^{-1}\text{Hz}^{-1}$	ct s^{-1}	$P_{\text{noise}} = 2$
$A_{\text{rms}}^2 = 1/\bar{x}^2$	$[\text{rms/mean}]^2\text{Hz}^{-1}$	$[\text{rms/mean}]^2$	$P_{\text{noise}} = \frac{2(\bar{x}+B)}{\bar{x}^2}\frac{\Delta t}{\Delta t_{\text{bin}}}$

information. The Fourier transform at $f_{j=0}$ is known as the direct component (DC) and represents the non-variable part of the signal (i.e. the mean level). The negative part of the Fourier transform $X(f_{j\leq 0})$ is generally neglected from variability studies. Additionally, the constant power from the DC component may leak into neighbouring frequencies. It is therefore standard practice to subtract the mean $\langle x_t \rangle$ from each segment M before computing the DFT.

The periodogram can then be calculated choosing an appropriate normalisation:

$$P(f) = A\frac{2\Delta t}{N}|X(f)|^2 = A\frac{2\Delta t}{N}X^*(f)X(f) \tag{7.13}$$

where $*$ denotes the complex conjugate and for brevity the subscript j has been dropped. The factor of 2 in Eq. 7.13 makes the periodogram 'one-sided', meaning that the original signal power (variance) is recovered when integrating over only positive frequencies. The factor $(\Delta t/N)$ ensures that the amplitudes at each f_j are independent of the sampling length of the time series.

Three commonly used normalisations in astronomy are: $A_{\text{abs}}{=}\bar{x}_t\, A_{\text{Leahy}}{=}\bar{x}_t^2\, A_{\text{rms}^2})$. These are shown in Table 7.1, along with the periodogram ordinate units and the result of integrating the periodogram between f_a and f_b (i.e. the sum $S_{b-a} = \sum_a^b P(f)\Delta f$).

The A_{abs} normalisation gives the periodogram in absolute units and the integrated periodogram gives the total variance in absolute units. Using the A_{Leahy} normalisation [81], the integrated periodogram gives the total rms amplitude. The A_{rms^2} normalisation [86, 124] is the most commonly used for AGN and XRB analysis due to the integrated periodogram giving the fractional variance of the data for a given range of frequencies.

7.2.4.1 Poisson Noise Level

The effect of Poisson noise is to add a continuous floor to the power spectrum. The power spectrum of Poisson noise is constant at all frequencies (white noise), so this typically dominates at higher frequencies.

7.2.4.2 Scatter and Bias in the PSD Estimate

The stochastic nature of underlying process means that each periodogram, $P(f)$, measured from a single realisation of the noise process, shows a large amount of scatter around the true underlying PSD, $\mathscr{P}(f)$, even when $\mathscr{P}(f)$ shows no time evolution:

$$P(f) = \frac{1}{2}\mathscr{P}(f)\chi_2^2 \tag{7.14}$$

where χ_2^2 is a random variable following a chi-square distribution with 2 degrees of freedom. The expectation value and variance of χ_ν^2 are ν and 2ν respectively, so the measured periodogram exhibits a large amount of scatter around the underlying PSD (e.g. [106, 115, 123]). This intrinsic scatter does not decrease by adding more points to the light curve as the standard deviation of the PSD is insensitive to the length of the observation.

Under the assumption of (weak) stationarity, averaging over M independent light curve segments or by binning over N adjacent frequencies produces a consistent estimate of the PSD with Gaussian errors for $n = M \times N \gtrsim 30$ [115]. The error in the estimate of the true PSD is now:

$$\delta\mathscr{P}(f) = \frac{\langle P(f)\rangle}{\sqrt{n}} \tag{7.15}$$

where $\langle\cdot\rangle$ indicates the average over M periodograms. The frequency binning can be done e.g. geometrically or logarithmically.

Another method for reducing the scatter in red noise light curves is to bin the logarithm of the periodogram (e.g. [88, 103]). Here, the errors converge to a Gaussian distribution with fewer ($n \geq 20$) points in each frequency bin.

Each periodogram measurement is biased away from the true PSD due to the finite sampling window. A finite time series can be written in terms of an infinite, continuous function $l(t)$ and a window or sampling function $w(t)$ that is non-zero only over a finite range and encodes the time sampling of the observed light curve $o(t)$:

$$o(t) = l(t) \cdot w(t) \Longleftrightarrow O(f) = L(f) \cdot W(f) \tag{7.16}$$

where as this is a multiplication in the time domain, it gives a convolution in the Fourier domain (by the convolution theorem). The periodogram (formed from $O(f)$) is distorted due to the convolution of the FT of the true light curve $L(f)$ with the spectral window function $W(f)$ (see e.g. [115, 123]). The two main effects of this distortion are *aliasing* and *leakage*.

If the light curve is not continuously and contiguously sampled then power from above the Nyquist frequency ($f_{N/2} = f_{Nyq}$) is reflected back (*aliased*) about f_{Nyq}. Fortunately, aliasing is negligible for continuous binned light curves, and in all but the steepest slope PSDs, the Poisson noise level dominates at f_{Nyq}. See [106] for a discussion on correcting for aliasing when it presents a problem.

Leakage or red-noise leak results from the finite duration of the sampling window, $w(t)$. Here, variability power is transferred (*leak*) from frequencies below the lowest sampled frequency through the lobes of the spectral window function (i.e. the *Fejer kernel*). Lower frequency power (i.e. below the bandpass) is apparent as quasi-linear (i.e. very slow/smooth) trends; these cause leakage because they add variance to the observed frequency bandpass, which adds a constant component to the variance in $P(f)$, biasing the slope of the PSD towards $\alpha = 2$ [115].

7.2.5 The Cross Spectrum

The properties of two simultaneous and time series can also be compared in the Fourier domain using multivariate methods such as the cross-spectrum (e.g. [46, 97]). From two evenly sampled time series $x(t)$, $y(t)$ we can compute the Fourier transforms $X(f) = |X(f)|e^{i\phi_x(f)}$ and $Y(f) = |Y(f)|e^{i\phi_y(f)}$, which have complex amplitude and phase. We can estimate the power spectrum for each time series using $|X(f)|^2$ and $|Y(f)|^2$, after subtracting the Poisson noise and applying some normalisation factor. The complex-valued cross-spectrum can be written as

$$C_{xy}(f) = X^*(f)Y(f) = |X(f)||Y(f)|e^{i(\phi_{y(f)} - \phi_{x(f)})} \tag{7.17}$$

following [87, 112]. Like the PSD, this is averaged over M non-overlapping segments, $M \geq 20$ [34].

7.2.6 Coherence

We can define the coherence or squared coherency as

$$\gamma^2_{xy}(f) = \frac{|\langle C_{xy}(f)\rangle|^2}{\langle|X(f)|^2\rangle\langle|Y(f)|^2\rangle} \tag{7.18}$$

The coherence is a measure of the linear correlation between two time series as a function of Fourier frequency [112]. It takes on values in the range [0,1], where 1 is perfect coherence and 0 is incoherent. In the absence of observational (Poisson) noise the coherence can be estimated using M non-overlapping segments with a standard deviation of $\delta\gamma^2(f) = \sqrt{2/M}(1 - \gamma^2(f))/|\gamma(f)|$ (e.g. [46]).

The linearity property of the FT means that Eq. 7.4 can be written for each energy band as

$$X(f) = S_x(f) + N_x(f) \tag{7.19}$$
$$Y(f) = S_y(f) + N_y(f). \tag{7.20}$$

Allowing for the noise contribution to the measured signal, the cross-spectrum in Eq. 7.17 can now be expressed as

$$|\langle C_{xy}\rangle|^2 = |\langle (S_x^* + N_x^*)(S_y + N_y)\rangle|^2 \tag{7.21}$$
$$= |\langle S_x^* \cdot S_y\rangle + \langle S_x^* \cdot N_y\rangle + \langle N_x^* \cdot S_y\rangle + \langle N_x^* \cdot N_y\rangle|^2 \tag{7.22}$$

where we have dropped the f subscript for brevity. The quantities $s_{xy} = \langle S_x^* \cdot S_y\rangle$ and $n_{xy} = \langle S_x^* \cdot N_y\rangle + \langle N_x^* \cdot S_y\rangle + \langle N_x^* \cdot N_y\rangle$ can be understood in terms of the intrinsic signal and the noise contribution of the measured cross-spectrum $\langle C_{xy}(f)\rangle$. As the signal in each band is uncorrelated with the noise component in either band, $\langle \Re(n_{xy}(f))\rangle = \langle \Im(n_{xy}(f))\rangle = 0$ for sufficiently large M [112].

For the case of high measured coherence, the squared amplitude of the cross-spectrum can be written as $|\langle C_{xy}(f)\rangle|^2 = |\langle S_x^*(f) \cdot S_y(f)\rangle|^2 + n_{xy}^2(f)$ [112]. The Poisson-noise corrected coherence then becomes:

$$\gamma^2_{cor,xy}(f) = \frac{|\langle C_{xy}(f)\rangle|^2 - n_{xy}^2(f)}{\langle P_x(f) - |N_x(f)|^2\rangle\langle P_y(f) - |N_y(f)|^2\rangle} \tag{7.23}$$

where $P_x = |X(f)|^2 = |S_x(f)|^2 + |N_x(f)|^2$ and $P_y = |Y(f)|^2 = |S_y(f)|^2 + |N_y(f)|^2$ are the noisey PSDs. The error on the Poisson-noise corrected coherence depends on the amplitude of $S_x(f)$ and $S_y(f)$ and the measured coherence, and are given in Eqs. 8 and 9 of [112].

There are several possible sources of bias on the coherence. The coherence is a squared quantity and so cannot take values below zero. This results in the coherence estimates being biased upwards slightly, although this effect is only significant when the intrinsic coherence is low (and so the true coherence values are close to zero). This is given by $B[\gamma^2] = 1/M(1 - \gamma^2)^2$ where γ^2 is the coherence and M is the number of segments (see e.g. [45], Sect. 9.2.3), which can be subtracted from the estimated value. The choice of segment length and binning over adjacent frequencies was also found to bias the estimates. This is due to red noise leak also distorting the cross spectrum (see [34, 56]).

The dominance of the Poisson noise and drop in variability power at high frequencies results in a critical frequency, $f_{\rm crit}$, above which the coherence and time lag (see below) error bars become unreliable. Epitropakis and Papadakis [56] give a formula to calculate this.

7.2.7 Fourier Time Lags

From the complex-valued cross-spectrum estimate (Eq. 7.17) we obtain a phase difference (or phase lag) between two signals $x(t)$, $y(t)$:

$$\Delta\phi(f) = \arg\langle C_{xy}(f)\rangle = \arctan\frac{\Im(\langle C_{xy}(f)\rangle)}{\Re(\langle C_{xy}(f)\rangle)} \tag{7.24}$$

where $\Delta\phi(f)$ is limited to $[-\pi, +\pi]$ (or $[0, 2\pi]$ depending on convention) so cannot distinguish $\pi/2$ from $-3\pi/2$. $\Delta\phi(f)$ can be transformed into a corresponding time lag

$$\Delta\tau(f) = \frac{\Delta\phi(f)}{2\pi f} \tag{7.25}$$

This recovers the (time averaged) frequency-dependent time-lags between any correlated variations in $x(t)$ and $y(t)$. The standard deviation of phase-lag estimates—when computed using M non-overlapping segments—are estimated following (e.g. [46, 87]):

$$\delta\phi(f) = \frac{(1-\gamma^2(f))^{1/2}}{\gamma\sqrt{2M}} \tag{7.26}$$

with the error on time lag at each frequency naturally given by

$$\delta\tau(f) = \frac{\delta\phi(f)}{2\pi f}. \tag{7.27}$$

Binning over N frequencies which are e.g. geometrically spaced and averaging over M segments gives $N \cdot M$ measurements. This improves the error $\delta\tau$ by a factor $1/\sqrt{N \cdot M}$.

7.2.7.1 Time Lags as a Function of Energy

The energy dependence of the time lags can be investigated using *lag-energy* spectra (e.g. [77, 108]). The time lag is computed between two bands $x(t)$ and $y(t)$, however $x(t)$ is a broad reference band, typically 0.3–10.0 keV, and $y(t)$ is the narrow comparison band of interest. If the reference band $x(t)$ contains $y(t)$ then the reference band is $x(t) - y(t)$, in order to avoid correlated errors [108].

Recently, [71] provided an updated way to determine uncorrelated error bars when calculating the time lag as a function of energy. This method also allows for non unity coherence to be used in the error bar calculation.

7.2.8 Transfer Functions

Two linearly correlated time series are related to each other through a response function. This is a filter responsible for any delay or smoothing between the input and output signals. This is potentially a powerful method for understanding any reprocessing occurring between two physically separated emitting regions, such as the disc and corona of an accreting black hole disc. In principle, this will provide valuable information about the spatial extent of the reprocessing medium and can encode relativistic effects (see below). In signal processing parlance, 'impulse response function', i.e. the response to an instantaneous impulse (usually a δ-function) is used for time domain, and 'transfer function' for the Fourier domain.

Two continuous time series $x(t)$ and $y(t)$ may be related via

$$y(t) = \int_{-\infty}^{+\infty} \psi(t-\tau)x(\tau)\mathrm{d}\tau = \psi(t) \otimes x(t) \tag{7.28}$$

where the function $\psi(t)$ relating the two signals is the 'response function'. Using the convolution theorem we can re-write this in the Fourier domain as

$$Y(\omega) = \Psi(\omega) \cdot X(\omega) \tag{7.29}$$

where $\omega = 2\pi f$, and

$$\Psi(\omega) = \int_{-\infty}^{+\infty} \psi(t)e^{-i\omega t}dt, \tag{7.30}$$

where $\Psi(\omega)$ is the 'transfer function' in the Fourier domain. If $\Psi(f)$ is linear and independent of the time series used to estimate it, then the two signals $X(f)$ and $Y(f)$ will have unity coherence (Eq. 7.18). The phase difference between $X(f)$ and $Y(f)$ (Eq. 7.24) is contained within $\Psi(f)$, therefore we can get the frequency dependent time delay from the Fourier transform of the response function:

$$\tau(\omega) = \frac{1}{\omega} \arg\left(\int_{-\infty}^{+\infty} \psi(t)e^{-i\omega t}dt\right) \tag{7.31}$$

7.2.8.1 A Simple Top-Hat Response Function

One of the simplest forms of response function is the 'top hat', which has width a and is centred on $t = b$, such that

$$\psi(t) = TH(t; a, b) = \begin{cases} 1/a & \text{if } b - a/2 \le t \le b + a/2 \\ 0 & \text{elsewhere} \end{cases} \tag{7.32}$$

which is normalised to unity. Substituting this into Eq. 7.30 we can find the Fourier transform of this function

$$\Psi(w) = e^{-i\omega b}\text{sinc}\left(\frac{a\omega}{2}\right) \tag{7.33}$$

To get the time delay we compute the argument of $\Psi(w)$

$$\tau(\omega) = \frac{1}{\omega}\arg\left[e^{-i\omega b}\text{sinc}\left(\frac{a\omega}{2}\right)\right] = \frac{1}{\omega}\arg\left[e^{-i\omega b}\right] = \frac{1}{\omega}(-\omega b) = -b \tag{7.34}$$

The sinc function is real so is dropped from the argument. This now produces a constant time delay $-b$ at all frequencies.

Since the Fourier transform is linear, we can compute the phase shift produced by a response function that is a combination of a δ-function and a top hat. The combined response has the form

$$\psi(t) = f \times TH(t; a, b) + \delta(t - c) \tag{7.35}$$

where f is a simple scaling factor that sets the fraction of the TH in the combined response. Taking a zero-centred δ-function (i.e. $c = 0$) the Fourier transform of the combined response is now

$$\Psi(w) = 1 + fe^{-i\omega b}\text{sinc}\left(\frac{a\omega}{2}\right) \tag{7.36}$$

Following Eq. 7.25 we can find the argument of the transfer function, yielding the time lag as a function of frequency (or lag-frequency spectrum):

$$\tau(w) = \frac{1}{\omega}\arctan\left(\frac{-f\sin(b\omega)\text{sinc}(a\omega/2)}{1 + f\cos(b\omega)\text{sinc}(a\omega/2)}\right). \tag{7.37}$$

Figure 7.9 shows the lag-frequency spectrum of a top hat plus δ-function response model for a variety of parameter values. When the top hat start time $t_0 = 0$ (i.e. $a = 2b$) no negative delay is observed and the first oscillation with zero delay occurs at $\sim 1/\tau(\text{max})$. In contrast, when $t_0 > 0$ (i.e. $a < 2b$) the time delay now goes negative at certain frequencies, with the first oscillation again corresponding to $\sim 1/\tau(\text{max})$. This is the 'ringing' effect, related to the Gibbs phenomenon, which arises due to the separation of the delta function and t_0 changing the frequency at which the phase wrapping occurs. The negative time delay oscillation covers a very narrow range of frequencies, with width $\sim 1/\tau(\text{max})$.

The measured time series in a given energy band is related to some underlying 'source' variability, $z(t)$, via

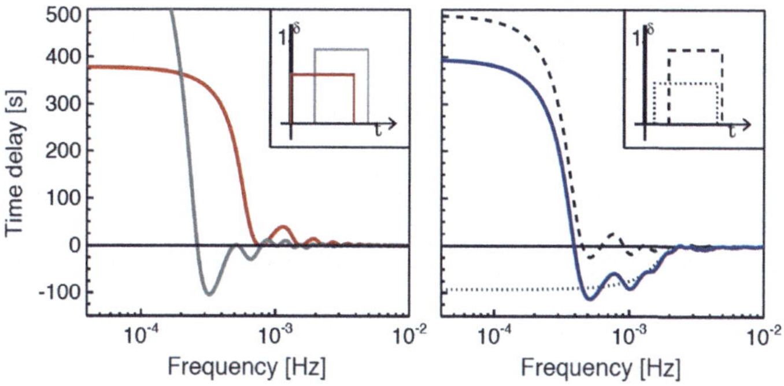

Fig. 7.9 Examples of transfer functions in the lag-frequency plot. **a** A single top hat plus delta function response function between the two time series, with the top hat starting at $t_0 = 0$ (red) and $t_0 > 0$ (grey). **b** A top hat plus delta function response function is present in each time series, with the overall transfer function (blue) resulting from the difference between the transfer function in each time series (see Sect. 7.2.8.1 for details). The insets show the time domain representation of the response functions

$$
\begin{aligned}
x(t) = \psi_x \otimes z(t) &\Longleftrightarrow X(f) = \Psi_x(f) \cdot Z(f) \\
y(t) = \psi_y \otimes z(t) &\Longleftrightarrow Y(f) = \Psi_y(f) \cdot Z(f)
\end{aligned} \tag{7.38}
$$

where it is assumed that $z(t)$ is some unseen driving process occurring before both $x(t)$ and $y(t)$. Each measured time series now has a response function relative to $z(t)$, which itself is not observed. The phases obtained from the cross-spectrum $C_{xy}(f)$ of the observed time series, which give the time lag-frequency spectrum, are (on average) the differences between the phases of the Fourier transforms of each response, i.e. $\phi(f) = \arg \Psi_y(f) - \arg \Psi_x(f)$. This is illustrated in Fig. 7.9b, where the overall transfer function results from the difference between the transfer function in each time series. Here, the negative time delay is much broader in frequency than in the single transfer function case, due to a genuine response function component present in each time series.

7.2.8.2 Modelling with Physical Transfer Functions

The top hat example is strictly only valid for describing the response for a thin shell, i.e. a spherical shell with $dr \gg r$, which is likely only applicable for distant scattering (e.g. [85]). A more realistic physical model is that of the illumination of an accretion disc by a hot corona, within $\sim 15R_g$ of the central compact object (e.g. [41, 47, 49, 52, 55, 73, 89, 118]).

An example of the impulse response from the disc and the corresponding transfer function in the frequency domain is shown in Fig. 7.10 (see also [41]). The response is formed using ray tracing of all the light paths between the source and reprocessor before being directed towards the observer at infinity. The delay in the onset of

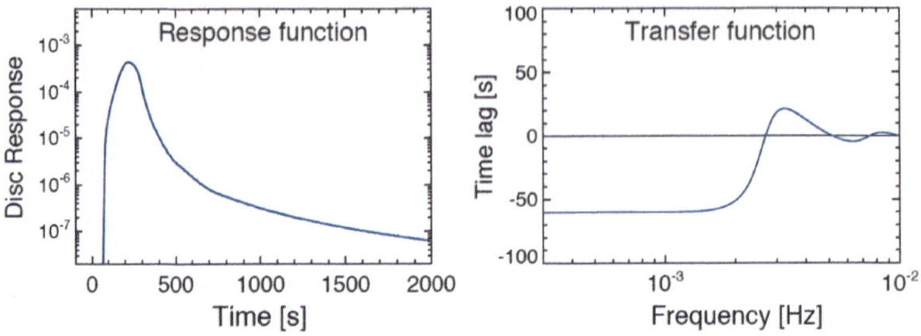

Fig. 7.10 An example response function (left) and corresponding transfer function (right) for a model of a point-like corona above a black hole illuminating an accretion disc (see e.g. [41, 52])

the response is caused by the separation between the source and the inner edge of the accretion disc. The response increases as the reprocessed emission arrives from the far side of the disc, then decays away as the emission is received from progressively larger radii. The models include all the special and general relativistic effects occurring near the black hole. Included in the model, is the 'dilution' factor, which is caused from the light curves $x(t)$ and $y(t)$ containing a mixture of continuum emission from both the direct and reprocessed components. This technique is also applicable for studying thermal disc reverberation and any time lags associated with absorbing clouds and outflows (e.g. [35, 61, 76, 122]).

7.2.9 Covariance Spectrum

We may wish to explore the energy dependence of the variability at a particular timescale or Fourier frequency. The rms-spectrum, or equivalently, frequency-resolved spectra are one way to do this (Sect. 7.2.2). Another approach is to use the *covariance* spectrum. The covariance spectrum [119] acts as a 'matched filter', picking out the correlated variability between a comparison energy band $x(t)$ and broad reference band $y(t)$. This method is similar to the rms-spectrum, except that any uncorrelated white noise is removed, providing a less noisy estimate of the variability spectrum, which is a particular issue at higher energies. Wilkinson and Uttley [119] presented a definition of the zero-time lag covariance spectrum in the time-domain. The covariance between $x(t)$ and $y(t)$ can be computed in the frequency domain for a given frequency range Δf_j (see [108] for details on this).

$$cov(\Delta f_j) = \sqrt{\gamma_{xy}^2(f)\langle|X(f)|^2\rangle\Delta f_j} \tag{7.39}$$

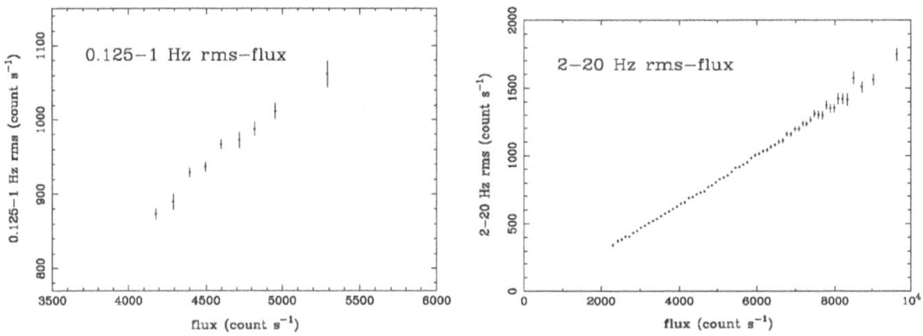

Fig. 7.11 The linear rms-flux relation for two variability timescales in the accreting X-ray binary source, Cygnus X-1, using *RXTE* data (reproduced with permission from [107])

which has an error

$$\delta cov(\Delta f_j) = \sqrt{\frac{n_{xy}^2 \Delta f_j}{2\langle |Y(f)|^2 \rangle}}. \tag{7.40}$$

7.2.10 The Linear Rms-Flux Relation

The linear rms-flux relation is a universal feature of accreting objects, where a linear relationship between the rms amplitude on short timescales and mean source flux on longer timescales is observed [105]. An example of the relation is shown in Fig. 7.11. Most importantly, this linear relation is observed over a broad range of timescales in a given source.

This property of the light curves is seen in active galactic nuclei (AGN, e.g. [116]), black hole (BH) and neutron star (NS) X-ray binaries (XRBs, e.g. [63, 65]) and ultraluminous X-ray sources [64, 66]. This phenomena is not limited to the X-ray band: it is also present in the fast optical variability from XRBs [59], blazars [54], accreting white dwarfs [51, 100, 109] and young stellar objects [101].

7.2.11 Non-stationary Time Series

A time series is considered stationary when its first order moments (i.e. mean and variance) do not evolve with time. If the PSD is not sampled well enough down to low frequencies—that is, the light curve is not long enough to measure the roll over to zero power—then the time series is considered only *weakly* stationary. This is typically the case for AGN data, due to the limited observation lengths. The effect of not sampling the PSD to low frequencies is illustrated in Fig. 7.12. When the

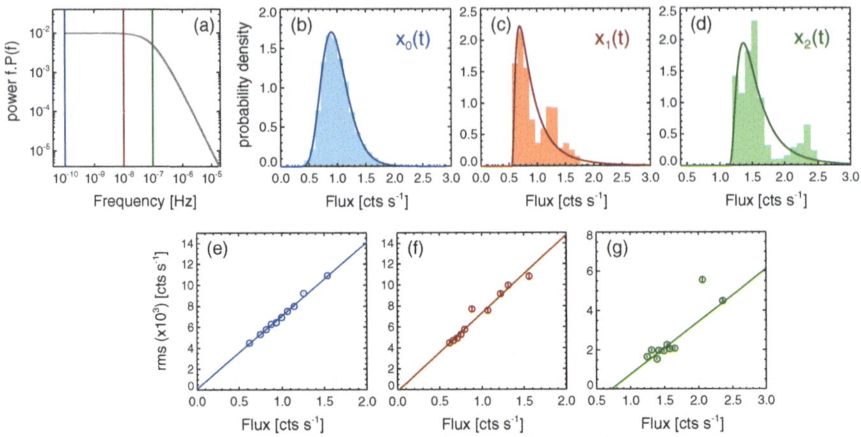

Fig. 7.12 Simulated PSD illustrating the effect of weak non-stationarity (see Sect. 7.2.13 for details). The recovered lognormal flux distribution and rms-flux relation are shown for the full simulated PSD in blue. As a progressively smaller part of the PSD is sampled, the observed quantities deviate from their expected values, even though the simulation includes them (reproduced from [40])

PSD extends down to very low frequencies, the lognormal distribution of counts and linear rms-flux relation are recovered. As a progressively smaller part of the light curve (higher frequency of the lower bound) is sampled the resulting distribution and rms-flux relation deviate from the true expectation values. The red and green selections are typical for AGN due to the long timescales required to fully sample the PSD.

The rms-flux relation is source of strong non-stationarity in the light curve. This can be factored out by investigating the PSD using fractional ([rms/mean]^2) units. For stationary data (even weakly stationary) the resulting PSDs should show no significant changes in the PSD shape and normalisation. If there are significant change sin the PSD shape and normalisation between two or more observations, then this indicates another source of strong non-stationarity, in addition to the rms-flux relation. This was clearly demonstrated for the first time in the AGN, IRAS 13224-3809 using a 2 Ms *XMM-Newton* observation [39, 40]. This change is occurring on ∼days timescales, which for a black hole mass $\sim 10^6\,M_\odot$ corresponds to ∼10 s in a $10\,M_\odot$ XRB. This shows the non-stationarity is different to the time dependent PSD changes observed in XRBs on longer timescales, which are associated with source state evolution.

7.2.12 *Variability Models*

The origin of the underlying variability process and origin of X-ray emission mechanisms are not necessarily the same thing: the variability can originate in a different

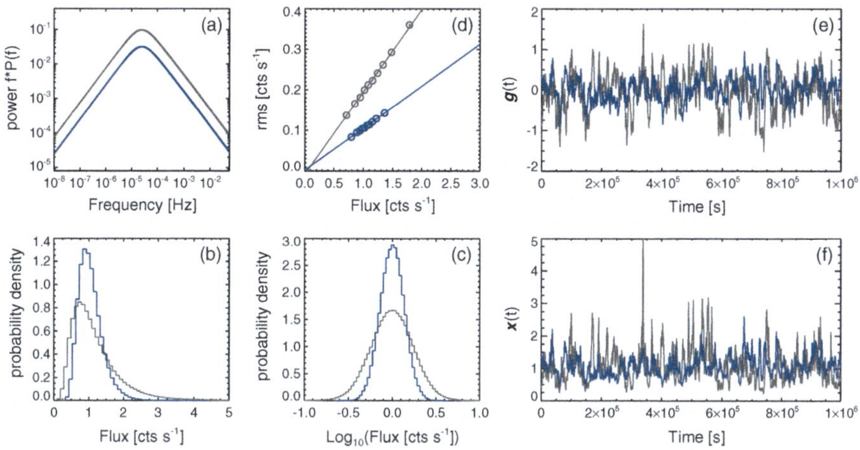

Fig. 7.13 Simulated light curves for a broad noise component, where $x(t) = \exp[g(t)]$ to imprint the rms-flux relation and naturally produce a lognormal distribution of flux. The affect of increasing the variance (integral of the PSD) can be seen (reproduced from [40])

location to where it is emitted. It is therefore important to understand the properties of the variability process if we want to (i) understand the underlying accretion physics, and (ii) use the variability to correctly perform analysis such as modelling the time lags with transfer functions as discussed in Sect. 7.2.8.2.

The PSD of many different types of variability processes can be identical, meaning this alone is not an adequate discriminator between models. The non-linear behaviour, lognormal flux distribution and rms-flux relation are a more fundamental feature of the underlying process in accreting objects. These properties naturally rule out many types of models, including additive shot-noise, self-organised criticality (SOC) and any additive model. The variability model must be multiplicative in order to reproduce the observed variability characteristics.

Uttley et al. [107] proposed the "exponentiation" model to describe the key observed light curve properties. Here, an underlying (and unseen) Gaussian distributed light curve, $g(t)$ is exponentiated to produce a non-linear light curve with the same properties as the data: a lognormal distribution in fluxes as well as a linear rms-flux relation. Importantly, the model produces the desired linear rms-flux relation over all timescales.

The leading physical model to describe these properties of the variability, in addition to the low-frequency hard band lags, is the propagation of accretion rate fluctuations (e.g. [42, 69, 70, 78–80, 82, 99, 121]). In this model, long timescale fluctuations in the local mass accretion rate at large radii propagate through the accretion flow and modulate shorter timescale fluctuations originating at smaller radii. Numerical simulations (e.g. [50]) and those including magnetohydrodyamics (MHD; [67]) are able to produce the rms-flux relation (Fig. 7.13).

7.2.13 *Simulating Time Series Data*

Simulating time series with the statistical properties of observed data is a powerful tool to have at our disposal. It allows us to perform parametric tests of the analysis methods, place confidence intervals on observed measurements, and make robust simulated estimates for expected data quality for telescope proposals and future missions.

The method of simulating time series depends on the desired PSD properties, which are typically approximated with a $\sim 1/f^{\beta}$ spectrum. A time series with constant power at each frequency (white noise; $\beta = 0$) can be simulated by randomly sampling numbers from a Gaussian distribution. Red noise ($\beta = 2$) can be simulated using a random walk, where each time series point is drawn from a Gaussian distribution and added to some fraction of the previous time series value.

For a more complicated PSD shape (e.g. a bending power-law) the algorithm of [104] is commonly used. For a desired power spectral shape we sample random complex phase and amplitudes from a Normal distribution. These are then inverse Fourier-transformed back into the time domain to produce a simulated time series with the properties of the defined power spectrum (see e.g. [33] for more details).

The effects of leakage can be accounted for using Monte Carlo simulations of light curves with duration much longer (e.g. $50\times$) than the observed light curve (e.g. [106]). Observational noise can be added to each simulated light curve by drawing a Poisson random deviate with mean equal to the mean count-per-bin in the real light curves.

To produce the observed log-normal distribution of fluxes on all timescales (as observed in the rms-flux relation) we expatiate the Gaussian distributed time series (i.e. $x(t) = \exp[g(t)]$) at each point [107, 113]. This has the effect of adding additional power to $x(t)$ than was expected from the input PSD. One way to mitigate this is to rescale the variances to the expected value (see [40, 107] for more details).

7.2.13.1 Periodicity Searches

Periodic or quasi-periodic variability from astrophysical objects can provide important information about the underlying physics occurring in the system. In accreting systems, these are known as quasi-periodic oscillations (QPOs; e.g. [36–38, 62, 98]). The characteristic variability timescale of an object scales as $1/M$, meaning that a typical observation will span many more characteristic timescales in a smaller mass object. This improvement in the number of segments, and therefore $n = N \times M$ frequency estimates, means narrower and weaker features are easier to detect in the PSD. These are typically fit with a narrow coherent Lorentzian component in addition to the broadband noise.

Periodicity searches are more complicated in AGN owing to the lower number of periods that can potentially be observed in a given observation. This is complicated further by the red noise nature of the underlying broadband noise: the increased power at low frequencies means that those timescales dominate and produces light curves with variability pattern close to a sin wave of the dominating frequency.

There are several methods for testing the significance of any periodic-like features. A robust Maximum Likelihood Estimation (MLE) method for detecting coherent narrow peaks in power was presented in ([110], with application in [36, 37]).

Irrespective of the method, it is important to model the underlying broadband noise properly, use a well calibrated test statistic and correctly calculate the false detection probability. This includes accounting for the number of global frequency trials. This is also the case if the data are sliced further by epoch or energy band. The dangers of period searches in large datasets are described in detail in [117]. The high probability for red noise PSDs to produce spurious periodic looking light curves can be verified by the reader using the simulation methods discussed in Sect. 7.2.13. We caution the user here that the popular Lomb-Scargle periodogram should not be used for periodicity searches in the presence of red noise (e.g. [117]).

7.2.14 Timing Analysis Summary

I have outlined the important aspects of some of the common timing analysis tools used in astronomy. Other important tools which are routinely applied to the data include; Principle Components Analysis (PCA; e.g. [90]), Auto-Regressive methods such as ARMA and CARMA (e.g. [39, 78, 117]), line searches for coherent pulsations (e.g. [102]), phase-resolved spectroscopy (e.g. [72]) and higher-order moments such as the bispectrum (e.g. [83]). Other methods, such as wavelet transforms, are often applicable, however the statistical properties and reliability of such methods are far less understood. With well understood methods and well calibrated statistical tests, timing analysis offers powerful and complementary techniques for understanding the source behaviour and their underlying physics.

7.3 Spectral Analysis

This section will take the form of a walk-through analysis of one specific AGN, although examples from other sources will be used to supplement this. The AGN in question, IRAS 13349+2438, is a good example of many of the phenomena seen in AGN, and will allow us to discuss how the different spectral components interact and must be accounted for.

7.3.1 Selection of the Source and Background Regions

The selection of appropriate source and background extraction regions is extremely important for X-ray spectroscopy. The source region must be sufficiently large to capture the bulk of the photons from the source, but the larger the region the more background contamination becomes a problem. Similarly, background regions should be large enough that a good average spectrum can be obtained, but not so large that they include other point sources, detector chip-gaps, or additional sources of background.

The ideal sizes and shapes of the regions will typically depend on the type of source, the brightness, the environment, and the instrumentation (including the operating mode of the instrumentation). For example, gratings only have one spatial dimension, so the circular extraction regions commonly used with CCD images are not appropriate.

In Fig. 7.14 we show two images from the XMM-Newton EPIC-pn camera, of two AGN taken in different window modes. In the first, from IRAS 13349+2438, the observation was taken in small window mode. The source extraction region is a $40''$ radius circular region centred on the source, and the larger background region is positioned as far from the source as possible, to minimise the number of source photons in the background spectrum.

In the second image, of IRAS 13224-3809, the detector was operated in large window mode, so much more space is available. However, the background region is still positioned close to the source, to avoid contaminating the background with instrumental background emission, which is stronger in the outer parts of the detector.

7.3.2 Binning the Data

Choosing an appropriate level of binning for spectral data is also an important issue, which is often neglected. Again, there is a balance to be found: too much binning removes information from the spectrum, but too little will disguise a poor fit and potentially miss genuine features of the data.

A good rule of thumb for X-ray CCD data is to bin to oversample the instrumental resolution by a constant factor, and then to a minimum signal to noise ratio. This will generally insure a good compromise. Signal to noise alone (or any binning scheme relying on photon counts) will fail to bin data at all at high count rates, while oversampling along results in extremely noisy spectra at low count rates. Given the range of spectral shapes seen in X-ray spectra, and the strong dependence on energy of telescope effective areas.

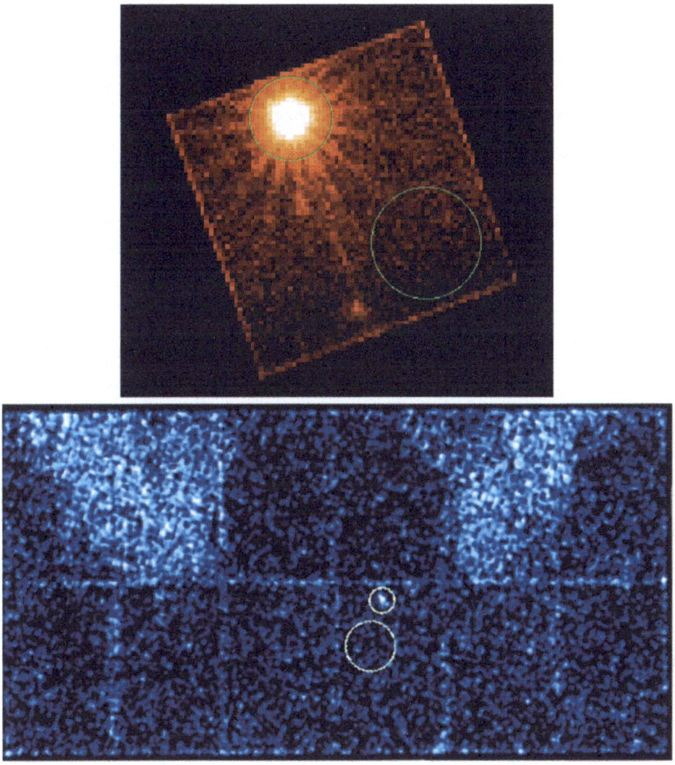

Fig. 7.14 XMM-Newton images of two AGN, with source and background regions marked. Top: IRAS 13349+2438, taken in small window mode. The source PSF fills much of the field of view, so the background must be taken from the furthest corner of the image. Bottom: narrow-band (8–8.5 keV) image of IRAS 13224-3809, from [91], taken in full-window mode. This energy band contains significant contamination from the Cu Kα line from copper in the detector, which is visible in the bright regions at the top of the image

7.3.3 Choice of Fitting Package

There are several dedicated fitting packages for use with X-ray data. The most common are XSPEC [43], ISIS [68], and SPEX [75].

Of these, SPEX is the most specialised. It was explicitly designed to work with high resolution X-ray spectra, primarily from gratings, and as such contains a range of sophisticated and powerful tools for modelling high resolution spectra of ionised absorption and emission. The main downside of SPEX is that it not compatible with the wider range of models available for XSPEC and ISIS, so may not be suitable for fitting broad band spectra.

XSPEC is the oldest fitting package, although it has been continuously updated (at the time of writing, the latest verion is 12.10.1), and as such it is also the most commonly used. It lacks some of the more advanced features of SPEX and ISIS (the ability to rebin spectra on the fly, for example), but has access to a huge library of models, and can be easily integrated with PYTHON scripts via the PYXSPEC interface.

Finally, ISIS provides the most powerful scripting tools, built around S- LANG, and extremely powerful parallelisation methods. ISIS is compatible with all XSPEC models, and enables more advanced functionality in some more recent models developed for both packages.

In practise, a users choice of fitting package likely depends on the package used by their supervisor, but users should bear in mind that other options are available and may be more suitable for specific applications.

The remainder of this section will refer to examples in XSPEC, but much of this is applicable to all three programs with only minor differences.

7.3.4 Model Fitting

Here, we will discuss one approximate procedure for fitting the X-ray spectrum of an AGN in XSPEC.

7.3.4.1 Preliminary Examination

Before investing a large amount of time and effort in fitting complex models, it is a good idea to get a general feel for the data by fitting simple phenomenological models to the spectra.

In Fig. 7.15 we show two such fits to the time-averaged EPIC-pn data of IRAS 13349. In the first, the data are fit with a power-law, modified by Galactic absorption.[13] Looking at this plot, we can see an excess at low energies, a deficit at intermediate energies, a peak around 7 keV, and an absorption-line like feature at 8 keV. This immediately suggests the presence of a soft excess, warm absorption, relativistic reflection, and an ultra-fast outflow (UFO).

However, if we add a black body to phenomenological fit, the residuals change greatly. Most obviously, the deficit from 0.8–2 keV disappears. Clearly, the strength of warm absorption in the EPIC-pn spectrum is model dependent, and cannot be trivially determined without additional information.

7.3.4.2 Fitting Warm Absorption

Accurate modelling of warm absorption in AGN spectra requires high resolution data, otherwise the absorbing spectrum cannot be uniquely determined—there are too many free parameters (velocity, ionization, density, abundances etc.) to be confident

[13]The Galactic column density for a given source can be estimated using the HEASARC N_H tool: https://heasarc.gsfc.nasa.gov/cgi-bin/Tools/w3nh/w3nh.pl/.

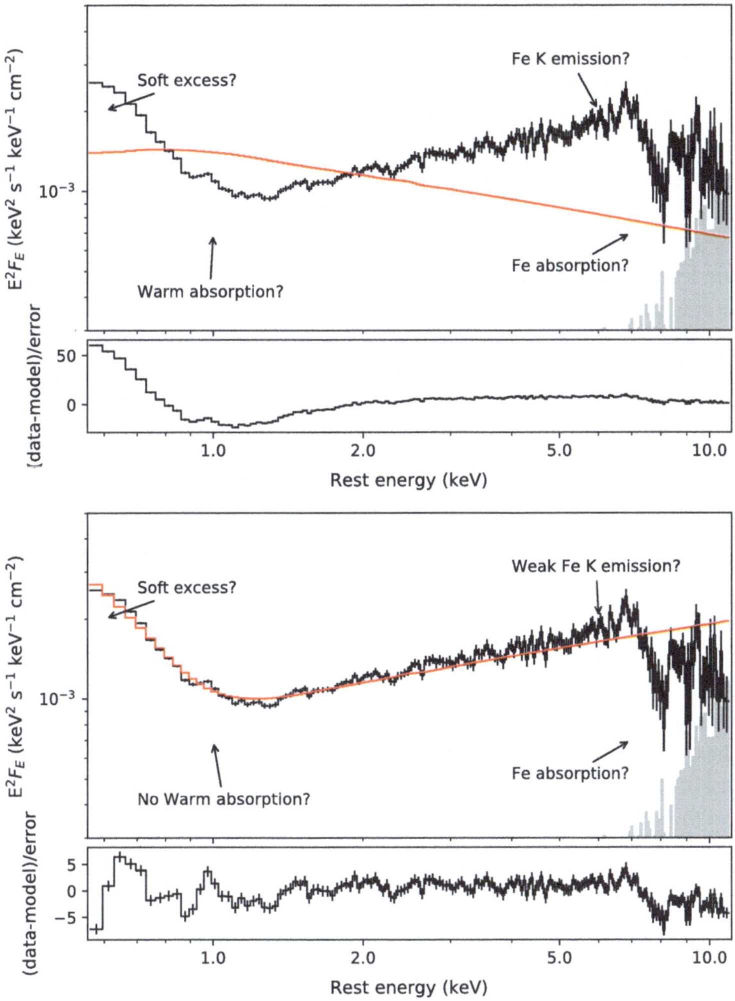

Fig. 7.15 Preliminary phenomenological fits to the EPIC-pn spectrum of IRAS 13349+2438. Data are corrected for the effective area of the instrument, using the SETPLOT AREA command in XSPEC, but not unfolded. Top: fit with a powerlaw, absorbed by Galactic absorption. Bottom: as top, but with a black body to model the soft excess. In each case, the residuals are shown in units of standard deviations from the model

of a fit without directly fitting absorption lines. With current instrumentation this means using grating spectra, which can easily resolve complex absorption spectra.

As a rule, when fitting grating spectra the χ^2 statistic may not be valid. χ^2 statistics are only valid when the data is well approximated by Gaussian uncertainties. At low count-rates this is not the case, so it is recommended to use Cash statistics [48], which are valid for Poisson distributed data.

An example fit is shown in Fig. 7.16, building a description of the absorption in the RGS spectrum of IRAS 13349+2438 by gradually increasing the complexity of the model. In the top panel, the data are fit with a power-law, and only galactic absorption is included, modelled with TBNEW [120] (note that TBNEW is both more accurate and faster than its predecessor TBABS, so there is no reason not to use it).

Clearly this fit is not good, but it helps to highlight the absorption lines present in the spectrum. The obvious next step is to add some absorption to the model. For this, we use the SPEX XABS photoionised absorption model, converted to an XSPEC table model.[14] This conversion allows the model to be used in conjunction with the more complex models available in XSPEC, at the cost of some flexibility and accuracy.

The fit with a single layer of absorption is shown in the middle panel. This provides a much better description of the data, but still leaves some obvious residuals, most strikingly around the Fe unresolved transition array (UTA) at ~18 Å. Adding a second layer improves this fit again (bottom panel), reducing the residuals to an acceptable level across the bandpass.

Obviously, this fit is still not perfect: there are still residuals around the UTA and Fe L edge, and the model underpredicts the strength of the Fe absorption lines below 15 Å. These kinds of effects are to be expected: it is very difficult to obtain a perfect fit to spectra of this complexity with physical models, as there are so many factors that can affect the spectrum (non-equilibrium gas, non-solar abundances, complex atomic physics etc.). For our purposes, this fit is sufficient to give a working description of the absorption, which we can now apply to the broad-band spectrum.

7.3.4.3 Fitting Relativistic Reflection

The next logical step is to reintroduce the EPIC data and add a relativistic reflection component to our spectral model (replacing the power-law), to fit the broad iron line. We can also keep the RGS spectrum in our fitted dataset, so that the warm absorption is well constrained.

There are several different models for fitting X-ray reflection, the most commonly used of which is RELXILL [60]. There are two main flavours of RELXILL, based on the assumed emissivity profile across the disk: the broken power-law and lamp-post models. From a purely spectral fitting perspective, it is usually safer (particularly for new users) to use the lamp-post variant, as this model is unlikely to get stuck in unphysical regions of parameter space. However, this model is less flexible, and makes very specific assumptions about the source geometry.

[14] http://bit.ly/xabs_tables.

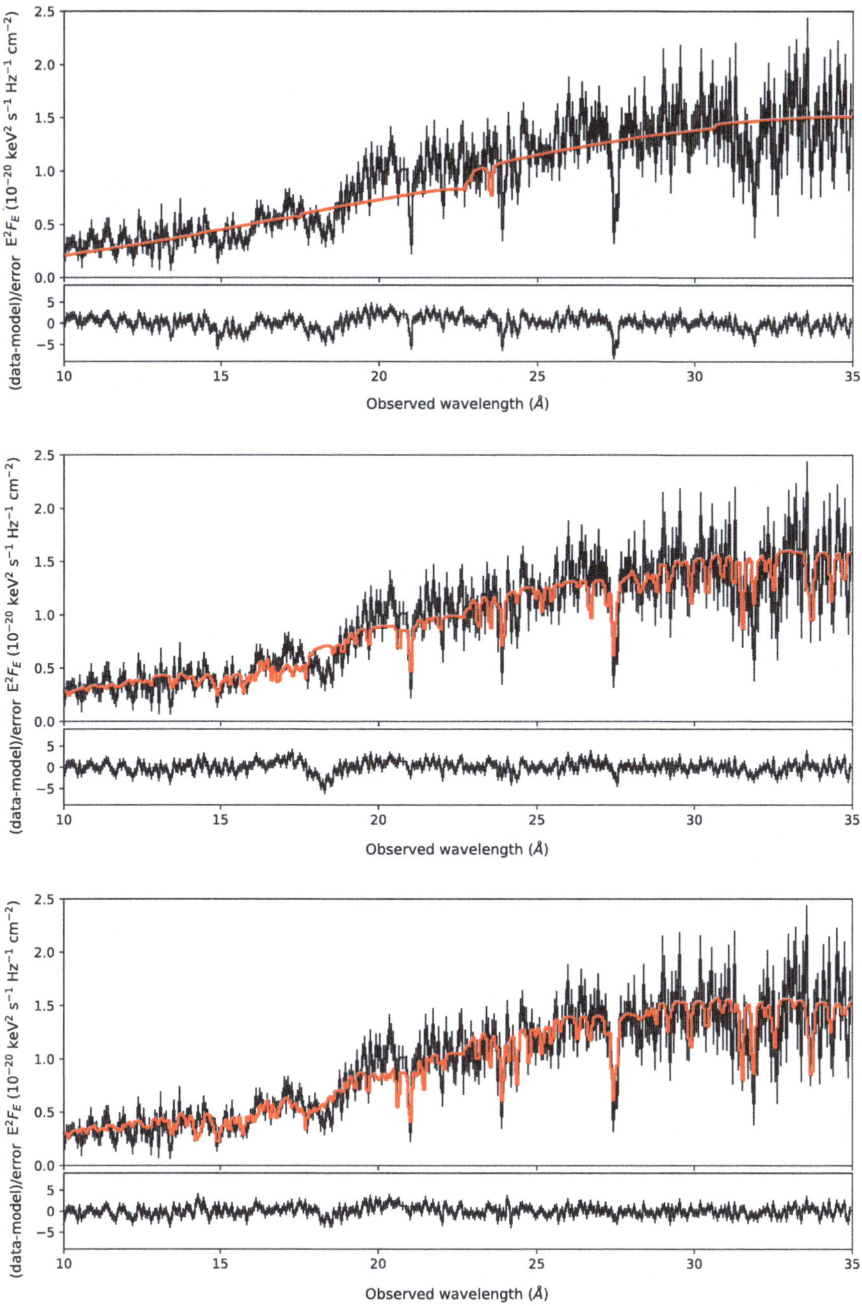

Fig. 7.16 XMM-Newton RGS spectrum of IRAS 13349+2438, fit with absorption models of increasing complexity. In the top panel, the model is a power-law with Galactic absorption applied. The middle panel adds a single layer of warm absorption, and the bottom panel adds two. In each case, the residuals are shown in units of standard deviations from the model

In this case, we will use the broken power-law version, fixing the outer emissivity index to the classical solution of 3 and the break radius to 6 R_G. This keeps these parameters from wandering to unphysical regions of parameter space.

We also keep the phenomenological soft excess black-body component in the model. How much of the soft excess is due to reflection is a subject of debate still, as the soft excess is generally smoother than expected from reflection spectra. Because of the shape of AGN spectra, and the effective area curves of X-ray telescopes, the soft excess usually contains much more signal than the Fe K band. Therefore, to avoid possibly biasing important parameters by forcing the reflection model to fit the soft excess, it can be useful to keep a phenomenological soft excess component in the model.

A good rule of thumb for fitting the soft excess with reflection is to see whether the blurring parameters (spin, emissivity, inclination) change when the soft excess is included in the fit [92]. If very different parameters are returned, then the soft excess is likely biasing the fit, potentially away from the correct solution.

When a normal reflection model is not fitting the soft excess well, it may be worth investigating high-density reflection models, which have shown a great deal of promise in fitting the soft excess in AGN [74].

In this case, we have enough signal in the Fe K band to get good constraints on the reflection parameters without the soft excess, so we will take the phenomenological approach here.

The fit adding in RELXILL is shown in the top panel of Fig. 7.17. Note that plotting the residuals in units of standard deviations is extremely useful for joint RGS/EPIC fits, where the two instruments have extremely different signal to noise ratios. By plotting in this way, both sets of residuals can be easily seen and compared.

Clearly, this model gives a far superior fit to the simple phenomenological models we used earlier. However, there remain some residual features in the EPIC-pn spectrum that are not accounted for in the model, and cannot be attributed to warm absorption.

7.3.4.4 Fitting Relativistic Outflows

The final thing to account for in our model is this strongly blueshifted absorption. It is worth bearing in mind that detections of these ultra-fast outflows (UFOs) must be undertaken with great care: it is very easy to make a false detection of an Fe absorption line at high energies, where the statistics are generally low and calibration and background contamination become less predictable (this is potentially the explanation for the noisy feature around 9 keV in the EPIC-pn spectrum). A good explanation of the likely large number of false detections of UFOs in AGN can be found in [114].

The easiest way to confirm the presence of a UFO is to compare the spectra from multiple instruments. With XMM-Newton, the EPIC-MOS can be used to confirm the presence of a UFO absorption line (as in the case of IRAS 13349+2438, see [93]). Additionally, if multiple lines are present with consistent velocities, this can

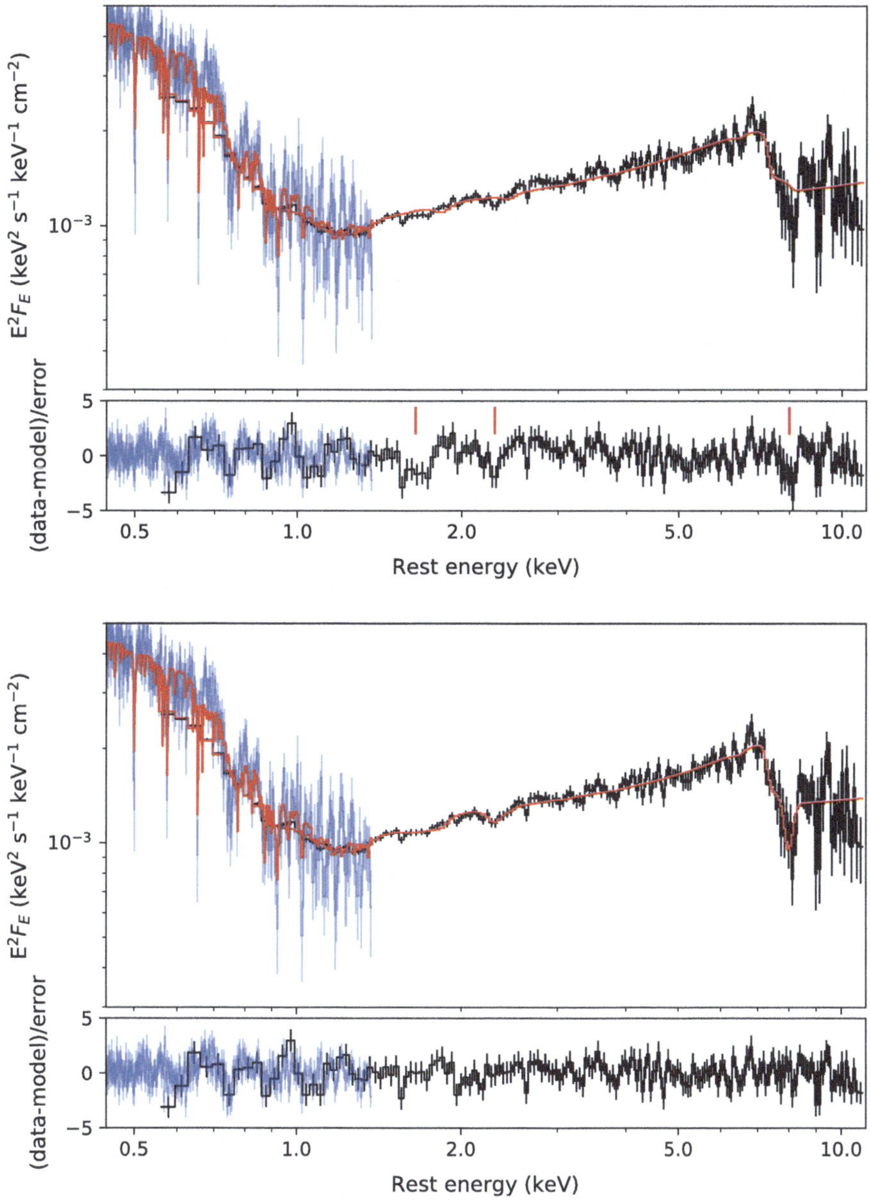

Fig. 7.17 Top: Joint RGS/EPIC-pn fit, including relativistic reflection. This model generally gives a good description of the spectra, but leaves a few absorption lines (marked with red lines in the residuals). Bottom: the same model, with four Gaussian absorption lines added, corresponding to Fe XXV, FeXXVI, Si XIV and Mg XII. All four lines have the same blueshift

reliably confirm the presence of an outflow [96]. However, UFOs are usually very highly ionised, so often only high ionisation states of Fe can be seen. This means that the absence of lower energy lines cannot be used to disprove the presence of a UFO. When such lines do appear, the most likely features are the Hydrogen-like lines of O, Ne, Mg, Si, S, Ar and Ca. Several of these lines are found at energies that are usually unaffected by warm absorption, in the EPIC band, so can be cleanly detected if present.

Because UFOs are characterised by a small number of absorption lines, it is trivial to fit them with a phenomenological approach. The primary aim of this approach is to make a detection, and to constrain the velocity. This can then be followed up with more detailed physical modelling to constrain the column density and ionization. For demonstration purposes, we will show a phenomenological approach.

We add four Gaussian absorption lines to the model, with their energies a fixed ratio to each other: 1.47:2.01:6.70:6.97, corresponding to Mg XII, Si XIV, Fe XXV and Fe XXVI. By fitting for one energy and keeping the others at a fixed ratio to that, we can ensure that all lines have the same blueshift.

The results of this fit are shown in the lower panel of Fig. 7.17. The three features are well fit by the phenemenological absorption model, so the only remaining residuals are at low energies where the warm absorption model under- or over-predicts some features, and the calibration/background feature at 9 keV. Our model is now giving a good description of the complex X-ray spectrum, accounting for multiple different physical phenomena.

7.3.5 Parameter Uncertainties

Having obtained a good model fit, describing all the main features of the spectrum, we now need to extract the physical parameters of interest from the fit, and their uncertainties. There are three main approaches, which we will approach in order of complexity.

7.3.5.1 Error Commands

All three fitting packages discussed above have a variant of a simple error calculation. This works by taking a parameter of interest, then slowly moving it away from the best fit value, re-fitting at each step, until a threshold fit statistic is exceeded.

This will quickly return positive and negative errors, and is fairly reliable. It has the added advantage that it will often stumble on a better fit, as part of the process of moving the parameters around. It is therefore a good idea to always run this kind of analysis after finding an initial best fit.

For simple models or parameter spaces, there is often no need for a more sophisticated analysis tool. For example, if we consider the outflow velocity of the UFO absorption from the previous section, it is obvious that there will only be one possible

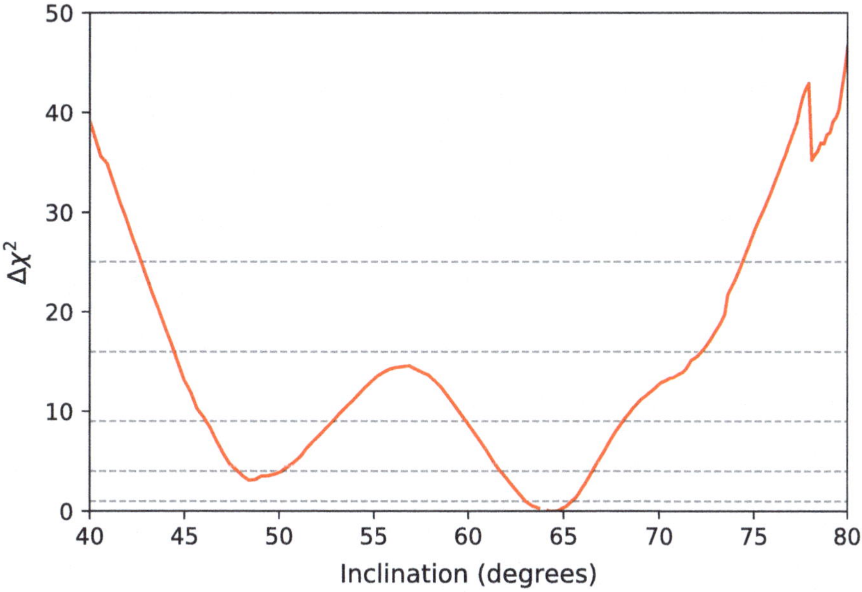

Fig. 7.18 Results of stepping the inclination parameter away from the best fit value using the XSPEC `steppar` command. Dashed horizontal lines indicate the 1–5σ confidence levels for a single parameter

solution, and the parameter space is not complex, so it is safe to rely on simple error calculations.

7.3.5.2 Stepping Parameters

A simple extension of the error routine is to step a parameter across a user-defined range. This is useful for parameters that have a more complex parameter space, and which may have multiple minima in the fit statistic. Standard error commands may not be able to escape from a minimum, so stepping a parameter may allow a fuller exploration of the parameter space.

We show an example of this in Fig. 7.18, for the disk inclination parameter from the joint EPIC-pn/RGS fit. For this plot we step the inclination parameter twice, moving away from the best-fit value in either direction. This ensures that the best fit solution is not missed by the stepping algorithm, which can happen when stepping from an extreme value.

The most obvious result from this is that there are two possible solutions for the disk inclination, one at 65° and one at 50°, with only a small statistical difference between them. Based on this plot alone, there is no way to say which of these is the correct solution.

We can also see a discontinuity in the contour at 77°. This is a common artifact of stepping algorithms, and occurs when the fit gets stuck in a false minimum for some part of the parameter space, before suddenly discovering a better solution and jumping to it. One workaround for this is to use a custom stepping routine that runs the error calculations at each step, which is much more likely to find the best fit solution. However, this will make the routine many times slower.

As a general rule, the more complex the model and corresponding parameter space, the more likely stepping routines are to get stuck in false minima. However, for simple models this should be easily sufficient and there is no real reason to use more complex techniques.

7.3.5.3 Markov Chain Monte Carlo

The most reliable way of determining the uncertainties on model parameters is to use MCMC to explore the parameter space. Unfortunately, this is also by far the most computationally expensive, although it can be parallelised very effectively.

XSPEC and ISIS both include native MCMC routines, and various alternatives have been developed over the years. For our example, we will use the XSPEC_EMCEE program written by Jeremy Sanders,[15] based on the EMCEE code [58].

We run the MCMC with 100 walkers and 10000 steps, after an initial 1000 step burn-in. The resulting contours for a few parameters are shown in Fig. 7.19. MCMC chains are very effective at identifying parameters with degeneracies: for example, the photon index Γ is strongly degenerate with the reflection fraction R. Once the MCMC is complete, it is trivial to check through parameters to find such degeneracies.

This plot also illustrates a weakness of the MCMC approach, at least in this implementation. Because the walkers are initialised around the best-fit value, they struggle to find alternative solutions that are cleanly separated. For example, two solutions for inclination are found by the stepping algorithm, but the MCMC chains do not manage to make the jump from one to the other, because of the low likelihood region in between. This can be avoided by using more walkers, running the chains for longer, or starting the walkers more widely distributed over the parameter space, but all of these will significantly increase the already long run-time for this kind of analysis.

A pragmatic solution to this is to identify minima with a stepping routine, then use MCMC to explore those which give a plausible best fit. Usually, one combination of parameters will give a unique best fit and MCMC can be used without worrying about this issue, but it is worth checking this with simpler, faster techniques.

[15]https://github.com/jeremysanders/xspec_emcee.

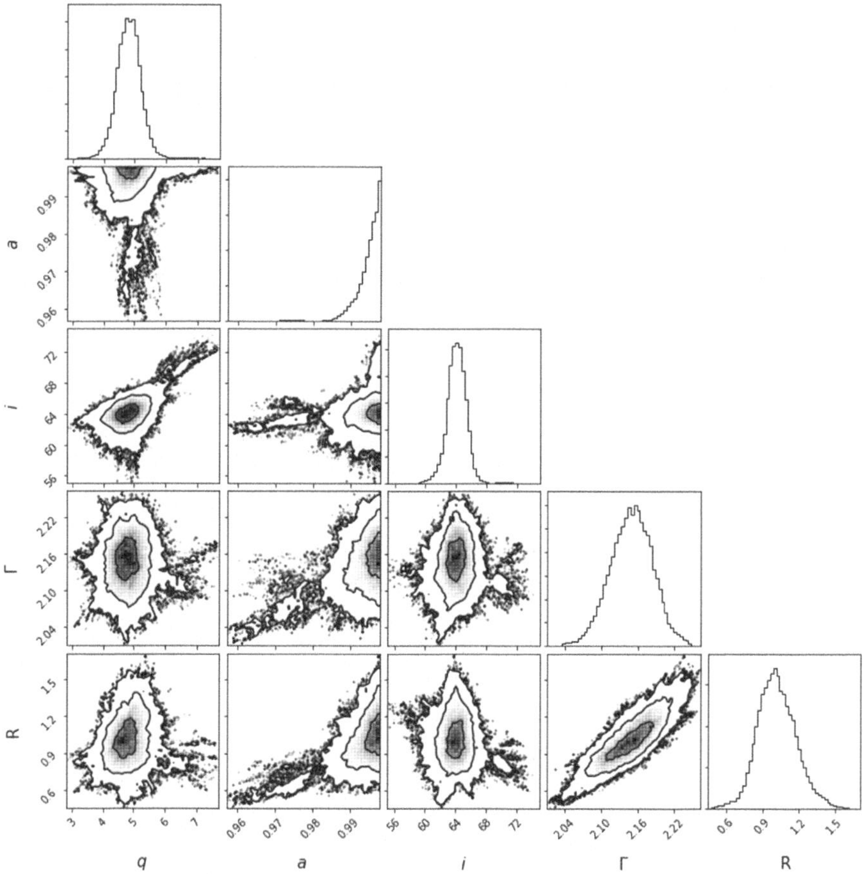

Fig. 7.19 Example MCMC contours for some of the reflection parameters from the broadband joint EPIC-pn/RGS fit. Contours correspond to 1–3σ

7.4 Likelihood Analysis for Gamma-Ray Data

7.4.1 Likelihood Analysis

A low detection rate characterises γ-ray satellites with an extent of the point spread function (PSF) of some degrees at 100 MeV. For this reason, statistical techniques are required to analyse γ-ray data. Likelihood analysis is the predominant analysis method of point-sources and extended sources. The *likelihood* term was introduced by Fisher [11] to quantify the relevant extent to which the data support a statistical hypothesis, i.e. the likelihood is the joint probability of the observed data given the hypothesis. It is possible to use a parametric model of the hypothesis that must be estimated considering the data. The estimated value of the parameters that maximise the likelihood are considered the best fit for the data. This method is called *maximum likelihood estimator (MLE)* or *method of maximum likelihood*. MLE methods find the value of model parameters that maximise the likelihood of producing the data.

The *likelihood ratio test* [21], that is the ratio of the likelihood of two hypotheses (that can be described with two maximized parametric models) for the same data, also called the null and the alternative hypothesis, is used for hypothesis testing. Let L_0 be the likelihood for the null hypothesis and L_1 be the likelihood for the alternative hypothesis.

The likelihood ratio is then simply the ratio of these two maximum likelihoods with the test statistic T_s defined as

$$T_s = -2ln\frac{L_0}{L_1} \tag{7.41}$$

where L_0 and L_1 are the maximum value of the likelihood function for the null hypothesis and the alternative hypothesis, respectively.

From Wilks's theorem [31], the T_s distribution is expected to follow asymptotically χ^2_{n-m} in the null hypothesis (assuming the null hypothesis is true), where $n - m$ is the number of additional parameters that are optimized in the alternative hypothesis. In the most simple case, $n - m = 1$ (e.g., in the case of the determination of the flux of a single source); this means that from Wilks's theorem, T_s is expected to be asymptotically distributed as χ^2_1 in the null hypothesis.

The application of likelihood to photon-counting experiments is described in [5] and has been used in many observatories: COS-B γ-ray data [22, 23] and Einstein Observatory [24], EGRET data [19], COMPTEL [27], AGILE [4], and Fermi [1].

For γ-ray analysis, two alternative candidate likelihood analyses are possible: binned and unbinned. Binned likelihood analysis is the preferred method for most types of γ-ray analysis. However, when analyzing data over short periods (with few events), it is better to use the unbinned analysis. In this book, we consider the binned likelihood analysis because it covers the majority of scientific cases. For a description of the unbinned likelihood analysis see [28].

7.4.2 Data Description

The likelihood analysis depends on the isotropic and Galactic diffuse emission, γ-ray photon statistics, the instrument response functions (IRFs) as functions of energy and off-axis angle, and on the background filtering for event selection.

The data acquired from the satellite, after background event filtering, is a photon list. To perform a binned likelihood analysis the data are binned into FITS count maps, while γ-ray exposure maps and Galactic diffuse emission maps are used to calculate the models.

In the γ-ray data, it is possible to find events classified as photons coming from

1. γ-ray point and extended sources;
2. the Galactic diffuse emission (that is a background component with respect to the celestial point sources);
3. The isotropic diffuse emission (that is a background component for the celestial point sources);
4. other sources of background.

Besides the other sources of background, we can find the Earth limb. At the ~550 km altitude of AGILE and Fermi, the limb of the Earth is an intense source of γ-rays from CR collisions with the upper atmosphere, and during the observations the FoV generally subtends part of the Earth limb. Even if an effective on-board background rejection filtering is present, a residual component of limb emission remains in the data; to further reduce γ-ray Earth-albedo contamination the data selection and exposure calculations limit the FoV excluding photons from the Earth albedo. Other data selection can be applied on the original photon list depending on specific characteristics of the satellite.

7.4.3 Model Selection

7.4.3.1 Instrument Response Functions

A critical component of γ-ray data analysis is the instrument response function (IRF), that is a parameterised representation of instrument performance that can be factored into three parts given E the true energy and θ the true direction of the γ-ray in the satellite reference frame:

1. Effective area, $Aeff(E, \theta, s)$, the product of the cross-sectional geometrical collection area, γ-ray conversion probability, and the efficiency of a given event selection s for a γ-ray with energy E.
2. Point-spread function (PSF), $PSF(\widehat{\theta}, E, \theta, s)$, the probability density to reconstruct an incident direction $\widehat{\theta}$ for a γ-ray with (E, θ) for event selection s.
3. Energy dispersion, $E_{edp}(\widehat{E}, E, \theta, s)$, the probability density to measure an event energy $\widehat{E}$ for a γ-ray with (E, θ) for event selection s.

7.4.3.2 Background Components

The *Galactic diffuse γ-ray emission* is the primary component of the background for the analysis of point and extended γ-ray sources. It is assumed to be produced by the interaction of CRs with the Galactic interstellar medium, the cosmic microwave background (CMB), and the interstellar radiation field (ISRF) through three physical processes: hadron-hadron collision, Bremsstrahlung, and inverse Compton emission. The *isotropic emission* includes both a contribution from the cosmic extragalactic diffuse emission and a component of noise due to residual CR induced backgrounds at the detector level. More details are provided in Chap. 4.

Each background model is a linear combination of isotropic and Galactic diffuse components of the γ-ray emission. Among the parameters that may be varied to find the maximum likelihood are the coefficients of the diffuse and isotropic components. The result of the model fitting is the value of these parameters if they are kept free.

7.4.3.3 Point-Like Sources

To describe a single point source many parameters, in general, are used: the predicted source counts s_c, two parameters corresponding to the position of the source (e.g. $glon$, $glat$, in Galactic coordinates), and some parameters of the spectral model. It is possible to keep each parameter either free or fixed; a free parameter is allowed to vary to find the maximum likelihood. The result of the model fitting is the value of these parameters if they are kept free, in particular the predicted source counts, the values of the spectral shape parameters, and the position of the source in Galactic coordinates.

7.4.3.4 Hypothesis Formulation

We report here some examples of hypothesis formulation.

In the context of the γ-ray transient analysis, the null hypothesis is defined as an analysis region containing only steady sources with no flaring events. We can translate this into the ensemble of models by keeping the flux of a possible flaring source fixed to zero, and the fluxes of the remaining sources fixed to their known fluxes; background parameters can be kept free or fixed. In the alternative hypothesis that a flaring source is present, the flux of this source is allowed to be free.

In general, the null hypothesis corresponds to the absence of the point source, while the alternative hypothesis corresponds to its presence. The null hypothesis is a subset of the alternative hypothesis, corresponding, e.g. to a source with zero flux. The Galactic diffuse and isotropic coefficients, as well as the parameters of other known point sources in the field of view, must be kept either fixed or free in the same manner when evaluating both the null and alternative hypotheses.

7.4.4 Model Fitting

The parameters kept free are estimated by the MLE. It is possible to keep each parameter either free or fixed; a free parameter is allowed to vary to find the maximum likelihood.

The values of the parameters that maximize the likelihood are those describing the model in the ensemble that are most likely to reproduce the data.

To limit the effect of systematic errors far from the position of the candidate source, the analysis is limited to an analysis region of some degrees centered on the source position of the source under evaluation.

7.4.5 Source Detection and Localization

The likelihood ratio test is used to compare two ensembles of models, one of which is a subset of the other, each of which can be characterized by a set of parameters. In the most common case, one of the ensembles of models is the null hypothesis, while the other, of which the null hypothesis is a subset, is the alternative hypothesis, corresponding to, for example, the hypothesis of the existence of a source.

Each model is a linear combination of parameters for point-like and extended sources, isotropic, and Galactic diffuse γ-ray background components of the γ-ray emission.

The expected departure of the T_s distribution from χ_1^2 is of order $(N)^{-1/2}$, where N is the number of samples. In our context, the number of samples is the number of photons that carry information about all the parameters; these are all the photons in the analysis region. This is true regardless of the number that is eventually estimated to come from the point source.

If the result of the MLE constrains the flux of a source, and therefore the source counts s_c, to be either greater than or equal to zero, the consequence is that the ensemble of models considered is half of the theoretically possible number, and the shape of the T_s distribution differs from that of Wilks's theorem by being asymptotically distributed as $\frac{1}{2}\chi_1^2$ instead of χ_1^2 [19].

The probability that the result of a trial in an empty field has $T_s \geq h$ (that is the complement of the cumulative distribution function) is

$$P(T_s \geq h) = \int_h^{+\infty} \varphi(x)dx \tag{7.42}$$

which is also called the p-value $p = P(T_s \geq h)$, where $\varphi(x)$ is the T_s distribution. This is the pre-set (pre-trial) type-1 error (a false positive, rejecting the null hypothesis when in fact it is true). Given a statistical distribution, a "p-value" assigned to a given value of a random variable is defined as the probability of obtaining that value or larger when the null hypothesis is true. This value may be interpreted as an "occurrence-rate", that is, how many trials occur on average before obtaining a false detection at a level equal or greater to h.

7.4.5.1 Localisation

The position of each source was determined by maximising the likelihood with respect to its position, keeping the other parameters of the point-like source free. For each source, an elliptical or circular confidence regions can be evaluated (e.g. at the 95% confidence level).

7.4.5.2 Post-trial Probability

Since the probability of not making a false positive error in a single trial is $1 - p$, the probability of not making any false positive error is $(1 - p)^K$ (type-I error), hence the probability of making at least one false positive error is $\pi = 1 - (1 - p)^K$. This is defined as the post-trial probability, which is also referred to as the *experiment-wide error rate*, while p is denoted as the pre-trial probability, or comparison-wise error rate. For an experiment-wide false positive rate of π, we constrain the comparison-wise error rate with the Dune-Ŝidák correction $p \leq 1 - (1 - \pi)^{1/K}$.

The final evaluation of the significance of detection must be done considering the post-trial probability.

7.4.6 *Science Tools*

7.4.6.1 AGILE-GRID Science Tools

The AGILE Team and the SSDC work together to develop, maintain, and publicly distribute a suite of instrument-specific science analysis tools (hereafter AGILE-GRID Science Tools[16]) that can be used to perform standard astronomical analyses.

The AGILE Science Tools perform the positional and spectral shape optimisation at the same time. AGILE Science Tools supports only binned likelihood analysis.

A Python version of the Science Tools can be downloaded.[17]

7.4.6.2 Fermi-LAT Science Tools

The LAT team and the FSSC work together to develop, maintain, and publicly distribute a suite of instrument-specific science analysis tools (hereafter Fermi-LAT Science Tools[18]) that can be used to perform standard astronomical analyses.

---

[16]https://agile.ssdc.asi.it/.

[17]https://agilepy.readthedocs.io/en/latest/.

[18]http://fermi.gsfc.nasa.gov/ssc/data/analysis/software.

References

1. M. Ackermann et al., ApJS **203**, 4 (2012)
2. D.S. Aguado, R. Ahumada, A. Almeida, ApJS **240**, 23 (2019)
3. H. An, K.K. Madsen, N.J. Westergaard, SPIE **9144**, 1QA (2014)
4. A. Bulgarelli et al., A&A **540**, A79 (2012)
5. W. Cash, ApJ **228**, 939 (1979)
6. C. Carter, M. Karovska, D. Jerius, ASPC **295**, 477 (2003)
7. J.E. Davis, M.W. Bautz, D. Dewey, SPIE **8443E**, 1AD (2012)
8. I.N. Evans, F.A. Primini, K.J. Glotfelty, ApJS **189**, 37 (2010)
9. H. Ebeling, G. Wiedenmann, PhysRev **47**, 704 (1993)
10. G. Fabbiano, M. Elvis, A. Paggi, ApJL **842**, L4 (2017)
11. R.A. Fisher, *Statistical Method for Research Workers* (1925)
12. P.E. Freeman, V. Kashyap, R. Rosner, ApJS **138**, 185 (2002)
13. B.A. Grefenstette, F.W. Harrison, S.E. Boggs, Nature **506**, 339 (2014)
14. R.P. Kraft, D.N. Burrows, J.A. Nousek, ApJ **374**, 344 (1991)
15. S.M. Lamassa, T. Yaqoob, N.A. Levenson, ApJ **835**, 91 (2017)
16. G.B. Lansbury, D.M. Alexander, A. Del Moro, ApJ **785**, 17 (2014)
17. T.-P. Li, Y.-Q. Ma, ApJ **272**, 317 (1983)
18. K.K. Madsen, F.A. Harrison, C.B. Markwadt, ApJS **220**, 8 (2015)
19. J.R. Mattox et al., ApJ **461**, 396–407 (1996)
20. A. Moretti, S. Campana, T. Mineo, SPIE **5898**, 360 (2005)
21. J. Neyman, E.S. Pearson, Biometrika **20A**, 175 (1928)
22. A.M.T. Pollock et al., A&A **94**, 116 (1981)
23. A.M.T. Pollock et al., A&A **146**, 352 (1985)
24. A.M.T. Pollock et al., ApJ **320**, 283 (1987)
25. A.M. Read, S.R. Rosen, R.D. Saxton, A&A **534A**, 34 (2011)
26. S.R. Rosen, N.A. Webb, M.G. Watson, A&A **590**, A1 (2016)
27. V. Schonfelder et al., ApJS **86**, 657–692 (1993)
28. W. Tompkins, https://arxiv.org/pdf/astro-ph/0202141.pdf (2002)
29. G. Vianello, ApJS **236**, 17 (2018)
30. M.C. Weisskopf, W. Kinwah, V. Trimble, ApJ **657**, 1026 (2007)
31. S.S. Wilks, Ann. Math. Stat. **9**, 60 (1938)
32. E.L. Wright, P.R.M. Eisenhardt, A.K. Mainzer, ApJ **140**, 1868 (2010)
33. W.N. Alston, S. Vaughan, P. Uttley, MNRAS **429**, 75 (2013)
34. W.N. Alston, S. Vaughan, P. Uttley, MNRAS **435**, 1511 (2013b)
35. W.N. Alston, C. Done, S. Vaughan, MNRAS **439**, 1548 (2014a)
36. W.N. Alston, J. Markevičiūtė, E. Kara, A.C. Fabian, M. Middleton, MNRAS **445**, L16 (2014)
37. W.N. Alston, M.L. Parker, J. Markevičiūtė, A.C. Fabian, M. Middleton, A. Lohfink, E. Kara, C. Pinto, MNRAS **449**, 467 (2015)
38. W. Alston, A. Fabian, J. Markevičiutė, M. Parker, M. Middleton, E. Kara, [Astronomische Nachrichten] **337**, 417 (2016)
39. W.N. Alston et al., MNRAS **482**, 2088 (2019)
40. W.N. Alston, MNRAS **485**, 260 (2019)
41. W.N. Alston et al., Nat. Astron. (2020). https://doi.org/10.1038/s41550-019-1002-x
42. P. Arévalo, P. Uttley, MNRAS **367**, 801 (2006)
43. K.A. Arnaud, ASPC **17**, ASPC..101 (1996)
44. M. Bartlett, Nature **161**, 686 (1948)
45. J. Bendat, A. Piersol, *Random Data: Analysis and Measurement Procedures* (A Wiley-Interscience Publication, Wiley, 1986)
46. J.S. Bendat, A.G. Piersol, *Random Data: Analysis and Measurement Procedures* (Wiley, 2010)
47. E.M. Cackett, A. Zoghbi, C. Reynolds, A.C. Fabian, E. Kara, P. Uttley, D.R. Wilkins, MNRAS **438**, 2980 (2014)
48. W. Cash, ApJ **228**, 939 (1979)

49. P. Chainakun, A.J. Young, E. Kara, MNRAS **460**, 3076 (2016)
50. P.S. Cowperthwaite, C.S. Reynolds, ApJ **791**, 126 (2014)
51. A. Dobrotka, J.-U. Ness, MNRAS **451**, 2851 (2015)
52. M. Dovciak, B. De Marco, E. Kara, G. Matt, V. Karas, G. Miniutti, W. Alston, in *The X-Ray Universe* (2014), p. 244
53. R.A. Edelson, J.H. Krolik, G.F. Pike, ApJ **359**, 86 (1990)
54. R. Edelson, R. Mushotzky, S. Vaughan, J. Scargle, P. Gandhi, M. Malkan, W. Baumgartner, ApJ **766**, 16 (2013)
55. A. Epitropakis, I.E. Papadakis, M. Dovčiak, T. Pecháček, D. Emmanoulopoulos, V. Karas, I.M. McHardy, A&A **594**, A71 (2016)
56. A. Epitropakis, I.E. Papadakis, MNRAS **468**, 3568 (2017)
57. E.D. Feigelson, *Time Series Problems in Astronomy: An Introduction* (1997), p. 161
58. D. Foreman-Mackey, D.W. Hogg, D. Lang, J. Goodman, PASP **125**, 306 (2013)
59. P. Gandhi, ApJL **697**, L167 (2009)
60. J. García et al., ApJ **782**, 76 (2014)
61. E. Gardner, C. Done, MNRAS **448**, 2245 (2015)
62. M. Gierliński, M. Middleton, M. Ward, C. Done, Natur **455**, 369 (2008)
63. T. Gleissner, J. Wilms, K. Pottschmidt, P. Uttley, M.A. Nowak, R. Staubert, A&Ap **414**, 1091 (2004)
64. L.M. Heil, S. Vaughan, MNRAS **405**, L86 (2010)
65. L.M. Heil, S. Vaughan, P. Uttley, MNRAS **422**, 2620 (2012)
66. L. Hernández-García, S. Vaughan, T.P. Roberts, M. Middleton, MNRAS **453**, 2877 (2015)
67. J.D. Hogg, C.S. Reynolds, ApJ **826**, 40 (2016)
68. J.C. Houck, L.A. Denicola, *Astronomical Data Analysis Software and Systems IX*, vol. 216 (2000), p. 591
69. A. Ingram, C. Done, MNRAS **405**, 2447 (2010)
70. A. Ingram, M. van der Klis, MNRAS **434**, 1476 (2013)
71. A. Ingram, MNRAS **489**, 3927 (2019)
72. A. Ingram, M. van der Klis, M. Middleton, D. Altamirano, P. Uttley, MNRAS **464**, 2979 (2017)
73. A. Ingram, G. Mastroserio, T. Dauser, P. Hovenkamp, M. van der Klis, J.A. García, MNRAS **488**, 324 (2019)
74. J. Jiang, et al., arXiv:1908.07272 (2019)
75. J.S. Kaastra, R. Mewe, H. Nieuwenhuijzen, *UV and X-ray Spectroscopy of Astrophysical and Laboratory Plasmas*, vol. 411 (1996)
76. E. Kara et al., MNRAS **446**, 737 (2015)
77. E. Kara, W.N. Alston, A.C. Fabian, E.M. Cackett, P. Uttley, C.S. Reynolds, A. Zoghbi, MNRAS **462**, 511 (2016)
78. B.C. Kelly, M. Sobolewska, A. Siemiginowska, ApJ **730**, 52 (2011)
79. A.R. King, J.E. Pringle, R.G. West, M. Livio, MNRAS **348**, 111 (2004)
80. O. Kotov, E. Churazov, M. Gilfanov, MNRAS **327**, 799 (2001)
81. D.A. Leahy, W. Darbro, R.F. Elsner, M.C. Weisskopf, S. Kahn, P.G. Sutherland, J.E. Grindlay, ApJ **266**, 160 (1983)
82. Y.E. Lyubarskii, MNRAS **292**, 679 (1997)
83. T.J. Maccarone, P.S. Coppi, MNRAS **336**, 817 (2002)
84. I.M. McHardy, I.E. Papadakis, P. Uttley, M.J. Page, K.O. Mason, MNRAS **348**, 783 (2004)
85. L. Miller, T.J. Turner, J.N. Reeves, A. Lobban, S.B. Kraemer, D.M. Crenshaw, MNRAS **403**, 196 (2010)
86. S. Miyamoto, K. Kimura, S. Kitamoto, T. Dotani, K. Ebisawa, ApJ **383**, 784 (1991)
87. M.A. Nowak, B.A. Vaughan, J. Wilms, J.B. Dove, M.C. Begelman, ApJ **510**, 874 (1999)
88. I.E. Papadakis, A. Lawrence, MNRAS **261**, 612 (1993)
89. I. Papadakis, T. Pecháček, M. Dovčiak, A. Epitropakis, D. Emmanoulopoulos, V. Karas, A&A **588**, A13 (2016)
90. M.L. Parker et al., MNRAS **447**, 72 (2015)
91. M.L. Parker et al., Natur **543**, 83 (2017)

92. M.L. Parker, J.M. Miller, A.C. Fabian, MNRAS **474**, 1538 (2018)
93. M.L. Parker, G.A. Matzeu, M. Guainazzi, E. Kalfountzou, G. Miniutti, M. Santos-Lleó, N. Schartel, MNRAS **480**, 2365 (2018)
94. M.L. Parker, W.N. Alston, Z. Igo, A.C. Fabian, MNRAS **492**, 1363 (2020)
95. D. Percival, A. Walden, *Spectral Analysis for Physical Applications* (Cambridge University Press, 1993)
96. K.A. Pounds, J.N. Reeves, A.R. King, K.L. Page, P.T. O'Brien, M.J.L. Turner, MNRAS **345**, 705 (2003)
97. M. Priestley, *Spectral Analysis and Time Series* (Academic Press, London, 1981)
98. R.A. Remillard, J.E. McClintock, Ann. Rev. A&A **44**, 49 (2006)
99. S. Scaringi, MNRAS **438**, 1233 (2014)
100. S. Scaringi, E. Körding, P. Uttley, C. Knigge, P.J. Groot, M. Still, MNRAS **421**, 2854 (2012)
101. S. Scaringi, et al., Sci. Adv. **1**, e1500686 (2015)
102. R. Sathyaprakash et al., MNRAS **488**, L35 (2019)
103. L. Stella, E. Arlandi, G. Tagliaferri, G.L. Israel, Proceedings of the "International Conference on Applications of Time Series Analysis in Astronomy and Meteorology" (TSAM93), https://ui.adsabs.harvard.edu/abs/1994astro.ph.11050S/abstract (1994)
104. J. Timmer, M. König, A&A **300**, 707 (1995)
105. P. Uttley, I.M. McHardy, MNRAS **323**, L26 (2001)
106. P. Uttley, I.M. McHardy, I.E. Papadakis, MNRAS **332**, 231 (2002)
107. P. Uttley, I.M. McHardy, S. Vaughan, MNRAS **359**, 345 (2005)
108. P. Uttley, E.M. Cackett, A.C. Fabian, E. Kara, D.R. Wilkins, A&Apr **22**, 72 (2014)
109. M. Van de Sande, S. Scaringi, C. Knigge, MNRAS **448**, 2430 (2015)
110. S. Vaughan, MNRAS **402**, 307 (2010)
111. S. Vaughan, eprint arXiv:1309.6435 (2013)
112. B.A. Vaughan, M.A. Nowak, ApJL **474**, L43 (1997)
113. S. Vaughan, P. Uttley, eprint arXiv:0802.0391 (2008)
114. S. Vaughan, P. Uttley, MNRAS **390**, 421 (2008)
115. S. Vaughan, R. Edelson, R.S. Warwick, P. Uttley, MNRAS **345**, 1271 (2003)
116. S. Vaughan, P. Uttley, K.A. Pounds, K. Nandra, T.E. Strohmayer, MNRAS **413**, 2489 (2011)
117. S. Vaughan, P. Uttley, A.G. Markowitz, D. Huppenkothen, M.J. Middleton, W.N. Alston, J.D. Scargle, W.M. Farr, MNRAS **461**, 3145 (2016)
118. D.R. Wilkins, E.M. Cackett, A.C. Fabian, C.S. Reynolds, MNRAS **458**, 200 (2016)
119. T. Wilkinson, P. Uttley, MNRAS **397**, 666 (2009)
120. J. Wilms, A. Allen, R. McCray, ApJ **542**, 914 (2000)
121. A.A. Zdziarski, MNRAS **360**, 816 (2005)
122. A. Zoghbi et al., ApJ **836**, 2 (2017)
123. M. van der Klis, in *Timing Neutron Stars*, ed. by H. Ögelman, E.P.J. van den Heuvel (1989), pp. 27–+
124. M. van der Klis, in *Statistical Challenges in Modern Astronomy II*, ed. by G.J. Babu, E.D. Feigelson (1997), p. 321

The manufacturer's authorised representative in the EU is Springer Nature Customer Service Centre GmbH, Europaplatz 3, 69115 Heidelberg, Germany. If you have any concerns regarding our products, please contact ProductSafety@springernature.com

Printed and bound by CPI Group (UK) Ltd, Croydon, CR0 4YY
30/09/2026
02225559-0011